THE PHYSICIST AND THE PHILOSOPHER

THE PHYSICIST & THE PHILOSOPHER

EINSTEIN, BERGSON, AND THE DEBATE THAT CHANGED OUR UNDERSTANDING OF TIME

JIMENA CANALES

PRINCETON UNIVERSITY PRESS

PRINCETON AND OXFORD

Published by Princeton University Press, 41 William Street,
Princeton, New Jersey 08540
In the United Kingdom: Princeton University Press, 6 Oxford Street,
Woodstock, Oxfordshire OX20 1TW

press.princeton.edu

Jacket photographs: Albert Einstein, detail from photograph of Albert Einstein and Others, 1931. Courtesy of the Dibner Library of the History of Science and Technology, Smithsonian Institution Libraries. Henri Bergson, from *The Outline of Science: A Plain Story Simply Told*, by J. Arthur Thomson, 1922.

Library of Congress Cataloging-in-Publication Data
Canales, Jimena, author.
The physicist and the philosopher : Einstein, Bergson, and the debate that changed our understanding of time / Jimena Canales.
pages cm
Includes bibliographical references and index.
ISBN 978-0-691-16534-9 (hardcover : alk. paper) 1. Time—Philosophy.
2. Relativity (Physics) 3. Einstein, Albert, 1879–1955. 4. Bergson, Henri, 1859–1941.
5. Physicists—United States—Biography. 6. Philosophers—France—Biography. I. Title.
BD638.C326 2015
115—dc23
2014047686

British Library Cataloging-in-Publication Data is available

This book has been composed in Sabon Next LT Pro

Printed on acid-free paper. ∞

Printed in the United States of America

10 9 8 7 6 5 4 3 2 1

CONTENTS

PREFACE

"I cannot get it into my head," wrote Einstein, that the last thirty years "make up almost 10^9 seconds." What makes a moment meaningful, haunting our past and our future? April 6, 1922 was a significant date for Einstein; it was the day he met Henri Bergson, one of the most respected philosophers of his era.

In a widely publicized meeting in Paris, the philosopher congratulated the physicist for having discovered a stunning theory but chastised him for having lost aspects of time that were intuitively important for us. Appalled to see a theory ignore what attracted our attention toward certain events and not to others, Einstein's critic sketched out the principles of an alternative cosmology that would neither fall prey to the arid precision of the sciences nor wallow in poetic rhetoric. Applauded for his "full-blooded" notion of time, his objections would inspire generations to come.

During the face-to-face encounter "between the greatest philosopher and the greatest physicist of the twentieth century," his audience learned how to become "more Einsteinian than Einstein." Bergson did not contest any experimental results; he accused the physicist of "grafting upon science" a dangerous "metaphysics." The physicist responded swiftly, enlisting allies against the man who refused to grant to science—and physics—the power to reveal the time of the universe.

"The time of the universe" discovered by Einstein and "the time of our lives" associated with Bergson spiraled down dangerously conflicting paths, splitting the century into two cultures and pitting scientists against humanists, expert knowledge against lay wisdom. With repercussions for American pragmatism, logical positivism, phenomenology, and quantum mechanics, a series of intrigues and alliances explain why longstanding rivalries between science and philosophy, physics and metaphysics, objectivity and subjectivity are still so passionately fought. By the end of their lives, Bergson reconsidered Einstein and Einstein reconsidered Bergson, but their views remained irreconcilable.

The Physicist and the Philosopher is divided into four main parts. The first opens with three chapters that take us directly to the meeting between Einstein and Bergson. Part 2 then focuses on the *men*. It details the various contexts where Einstein's contributions were considered in direct relation to Bergson's critique. We follow the debate as it reverberated from France to England, Germany, and America. In each of these places, we meet some of the major players involved in the conflict, such as the Catholic Church, and see how it affected various scientific and philosophical movements, such as American pragmatism, logical positivism, and quantum mechanics. Some of these chapters focus on key moments before and after April 6, 1922, when similar arguments to those delivered that day were advanced.

Part 3 centers on the *things*. It investigates why Einstein and Bergson remained so divided by zooming into particular examples that came up again and again—explicitly and repeatedly—in their discussions and those of their interlocutors. Certain things, such as the telegraph, telephone, radio, film, and automatic registering devices, played salient roles. Microscopic particles, tiny microbes, immense observers, superfast beings, animals and ghosts entered their discussions as well.

Part 4 concludes with *words*—the last comments they made about each other. At that time, Bergson was nearly eighty, witnessing the rise of Nazism in Germany, the occupation of Paris, and a new era of conflict and unrest. Einstein was well into his seventies. He had retired from the Institute of Advanced Studies in Princeton and was reminiscing about Bergson a few months before the Americans detonated the world's first hydrogen bomb. In the end, we encounter a story of the rise of science in a divided century, of misunderstanding and mistrust, and of the everyday things that tear us apart.

PART 1

THE DEBATE

CHAPTER 1

Untimely

On April 6, 1922, Einstein met a man he would never forget. He was one of the most celebrated philosophers of the century, widely known for espousing a theory of time that explained what clocks did not: memories, premonitions, expectations and anticipations. Thanks to him, we now know that to act on the future one needs to start by changing the past.

Why does one thing did not always lead to the next? The meeting had been planned as a cordial and scholarly event. It was anything but that. The physicist and the philosopher clashed, each defending opposing, even irreconcilable, ways of understanding time. At the Société française de philosophie—one of the most venerable institutions in France—they confronted each other under the eyes of a select group of intellectuals. The "dialogue between the greatest philosopher and the greatest physicist of the 20th century" was dutifully written down. It was a script fit for the theater.[1] The meeting, and the words they uttered, would be discussed for the rest of the century.

The philosopher's name was Henri Bergson. In the early decades of the century, his fame, prestige, and influence surpassed that of the physicist—who, in contrast, is so well known today. Bergson's reputation was at risk after he confronted the younger man. But so was Einstein's. The criticisms leveled against the physicist were immediately damaging. When the Nobel Prize was awarded to Einstein a few months later, it was not given for the theory that had made the physicist famous: relativity. Instead, it was given "for his discovery of the law of the photoelectric effect"—an area of science that hardly jolted the public's imagination to the degree that relativity did. The reasons behind the decision to focus on work *other* than relativity were directly traced to what Bergson said that day in Paris.

The president of the Nobel Committee explained that although "most discussion centers on his theory of relativity," it did not merit the prize. Why not? The reasons were surely varied and complex, but the culprit mentioned that evening was clear: "It will be no secret that the famous philosopher Bergson in Paris has challenged this theory." Bergson had shown that relativity "pertains to epistemology" rather than to physics—and so it "has therefore been the subject of lively debate in philosophical circles."[2]

The explanation that day surely reminded Einstein of the previous spring's events in Paris. Clearly, he had provoked a controversy. These were the consequences. He had been unable to convince many thinkers of the value of his definition of time, especially when his theory was compared against that of the eminent philosopher. In his acceptance speech, Einstein remained stubborn. He delivered a lecture that was not about the photoelectric effect, for which he had been officially granted the prize, but about relativity—the work that had made him a star worldwide but which was now in question.

The invocation of Bergson's name by the presenter of the Nobel Prize was a spectacular triumph for the philosopher who had lived his life and made an illustrious career by showing how time should *not* be understood exclusively through the lens of science. It had to be understood, he persistently and consistently insisted, *philosophically*. But what exactly did he mean by that? As it turns out, Bergson's philosophy was as controversial as Einstein's physics.

What led these two brilliant individuals to adopt opposite positions on nearly all the pertinent issues of their era? What caused a century to end as divided as the twentieth did? Why did two of the greatest minds of modern times disagree so starkly, dividing intellectual communities for years to come?

THAT EVENING

On that "truly historic" day when the two met, Bergson was unwillingly dragged into a discussion he had explicitly intended to avoid.[3] The philosopher was by then much more senior than Einstein. He spoke for about half an hour. He had been prodded by an impertinent colleague, who had been in turn pressured to speak by the event organizer. "We are more Einsteinian than you, Monsieur Einstein," he said.[4] His objections would be heard far and wide. "Bergson was supposed by all of us to be dead," explained the writer and artist Wyndham Lewis, "but Relativity, oddly enough at first sight, has resuscitated him."[5]

The physicist responded in less than a minute—including in his answer one damning and frequently cited sentence: *"Il n'y a donc pas un temps des philosophes."*[6] Einstein's reply—stating that the time of the philosophers did not exist—was incendiary.

Einstein had traveled to the City of Lights from Berlin. When his train arrived at the Gare du Nord, "photographers, reporters, filmmakers, officials and diplomats awaited him in imposing numbers." The scientific celebrity decided to descend by the other side of the tracks, escaping surreptitiously, like a robber. He made his way through dangerous cables and warning signs before arriving at a tiny door that led to the boulevard de la Chapelle, which, in the afternoon, was as empty as the Sahara Desert. Safe from the cameras and the crowds, Einstein laughed like a child.[7]

The physicist's visit was "a sensation that the intellectual snobbery of the capital could not pass up."[8] Intellectuals were not the only ones excited by his presence. It literally set off "crowds in a craze," quickly enthralling unsuspecting Parisians.[9] An observer described an "unfettered frenzy by the public at large around certain of Einstein's commentators."[10] Einstein's trip "reanimated and brought to the stage of a paroxysm the curiosity of the public for the scientist and his work."[11]

What Einstein said next that evening was even more controversial: "There remains only a psychological time that differs from the physicist's." At that very moment, Einstein laid down the gauntlet by considering as valid only two ways of understanding time: physical and psychological. These two ways of examining time, although scandalous in the particular context that Einstein uttered them, had a long history. With Einstein, they would have an even longer one—becoming two dominant prisms inflecting most investigations into the nature of time during the twentieth century.

The simple, dualistic perspective on time advocated by Einstein appalled Bergson. The philosopher responded by writing a whole book dedicated to confronting Einstein. His theory is "a metaphysics grafted upon science, it is not science," he wrote.[12]

Einstein fought back with all his energy, strength, and resources. In the years that followed, Bergson was largely perceived to have lost the debate against the younger physicist. The scientist's views on time came to dominate most learned discussions on the topic, keeping in abeyance not only Bergson's but many other artistic and literary approaches, by relegating them to a position of secondary, auxiliary importance. For many, Bergson's defeat represented a victory of "rationality" against "intuition."[13] It marked a moment when intellectuals were no longer able to keep up with revolutions in science due to its increasing complexity. For that reason, they should stay out of it. Science and its consequences should be left to those who arguably knew something about it—the scientists themselves.[14] Thus began "the story of the setback, after a period of unprecedented success, of Bergson's philosophy of absolute time—unquestionably under the impact of relativity."[15] Most important, *then* began the period when the relevance of philosophy declined in the face of the rising influence of science.

Biographers who write about Einstein's life and work rarely mention Bergson. One exception, a book written by a colleague, paints a picture of eventual rapprochement between the two men.[16] But other evidence shows just how divisive their encounter was. A few years before their deaths, Bergson wrote about Einstein (1937), and Einstein mentioned Bergson (1953) one last time. They underlined—once again—just how wrong the perspective of the other remained. While the debate was for the most part removed from Einstein's legacy, it was periodically brought up by many of Bergson's followers.[17] The simple act of reviving the discussion that took place that day in April 1922 was not a matter that could be taken lightly. Not only is the incident itself divisive—its relevance *for history* is still contested.

The two men dominated most discussions about time during the first half of the twentieth century. Thanks to Einstein, time had been finally "deposed from its high seat," brought down from the lofty peak of philosophy to the practical down-to-earth territory of physics. He had shown that "our belief in the objective meaning of simultaneity" as well as that of absolute time had to be forever "discarded" after he had successfully "banished this dogma from our minds."[18] The physicist had shown that "space by itself, and time by itself" were two concepts "doomed to fade away into mere shadows."[19]

Bergson, in contrast, claimed that there was more to Time than scientists had ever wagered—and he meant scientists of all stripes, ranging from Darwinian evolutionists to astronomers and physicists. To explain those aspects of Time that were most important and that scientists constantly disregarded, Bergson would frequently capitalize the term. He associated it with *élan vital*, a concept translated worldwide as "vital impulse." This impulse, he argued, was interwoven throughout the universe giving life an unstoppable impulse and surge, ever productive of new unexpected creations, and imperfectly grasped by science. Although science could only deal with it imperfectly, it was the backbone of artistic and creative work. Bergson's influence on literature was seen as spreading to Gertrude Stein, T. S. Elliot, Virginia Woolf, William Faulkner, and numerous others who introduced breaks, twists, and turns in narratives where the future appeared before the past and the past after the future.[20]

Einstein's and Bergson's contributions appeared to their contemporaries forcefully at odds, representing two competing strands of modern times. Vitalism was contrasted against mechanization, creation against ratiocination, and personality against uniformity. During these years, Bergson's philosophy was often placed next to the first in these pairs of terms; Einstein's work frequently appeared alongside the second.[21] Bergson was associated with metaphysics, antirationalism, and vitalism, the idea that life permeates everything. Einstein with their opposites: with physics, rationality, and the idea that the universe (and our knowledge of it) could stand just as well without us. Each man represented one side of salient, irreconcilable dichotomies that characterized modernity.

This period consolidated a world largely split into *science* and *the rest*. What is unique about the appearance of these divisions and subsequent incarnations is that after the Einstein and Bergson encounter, science frequently appeared firmly on one side of the dichotomy. Other areas of culture appeared on the other side—including philosophy, politics, and art.

The stature of both men was envied by many of their contemporaries. Sigmund Freud, the founder of psychoanalysis, once described himself as having "little claim to be named beside Bergson and Einstein as one of the intellectual sovereigns" of his era.[22] The confrontation between them was "a controversy that presently separates the two most renowned men of our times."[23] Although Einstein's brain was paraded in formaldehyde as the perfect embodiment of the organ of genius, the locks of Bergson's hair kept at his barbershop were "treated as holy relics."[24]

"Early in this century, two very prominent, and originally independent, lines of thought collided," explained a physicist and historian who put his career on the line by siding with Bergson. "On the one hand . . . was the system of Bergson. . . . On the other hand, the physical theory of relativity, which . . . dominated scientific thought," he continued. "It was inevitable that one or the other of these views should give way," he concluded.[25] More recently, the debate between them continues to be widely perceived as *inevitable*. "Bergson's confrontation with Einstein was inevitable," wrote the famous philosopher Gilles Deleuze, more than half a century after their meeting.[26] And thus we find these two men playing key roles in the salient divisions of modern times. Can we move beyond them?

Bergson's defeat was a decisive turning point for him personally, when the fame, wisdom, and caution of the elder was tested by the impetuous braggadocio of the younger, but it was also a key moment marking the rise of the authority of science vis-à-vis other forms of knowledge. In the years that followed their meeting, the philosopher and physicist became engaged in numerous other disputes that would touch on just about everything. Some of their differences were highly abstract—about the nature of time, the role of philosophy, and the reach and power of science. Others were more concrete, such as the role of the government, the place of religion in modern societies, and the fate of the League of Nations. But almost anywhere that we look—from vegetarianism to war, from race to faith—we find that the two men took pretty much opposite stands on almost all pertinent issues of their time.

There are many reasons why we know much about Einstein and little about Bergson. Most of them have to do with how the debate intensified after their first meeting; the debate took off like wild fire.[27] The tension between the two men escalated after Bergson published a no-holds-barred book devoted to relativity theory. The controversial tome, designed to be carefully followed with pencil or pen in hand, appeared later that year. *Duration and Simultaneity* inspired hundreds of responses by prominent thinkers centrally engaged with the disagreement between the physicist and the philosopher. The book was as contentious as it was successful. Nearly a decade after its publication, a writer and eager reader of the work of both men still asked: "Would the book by the most brilliant of the contemporary philosophers clarify the ideas of the most brilliant of the scientists?"[28] In 1936, less than a decade and a half after it first appeared, a successful biologist warned prospective buyers that they "might have difficulty in finding" a copy of *Duration and Simultaneity* "as the last edition is exhausted."[29]

Einstein is well known and respected today; Bergson is much less. Yet at the time of their meeting the situation was quite the opposite. Bergson was an established figure as a public intellectual and philosopher, hobnobbing in the mornings with heads of state, filling lecture rooms in the afternoon, and providing bedtime reading for many at night; Einstein had only recently become a rising star in the eyes of the public and was still finding his voice outside of scientific spheres.

Bergson and Einstein met a few more times and exchanged a couple of letters. Einstein sent a friendly postcard from Rio de Janeiro to Bergson after their problematic encounter in Paris.[30] They never debated publically again. Instead, they propagated their respective positions in publications and letters to others. Some of these letters eventually reached the public; others remained in private hands until they found their way to archives. Through them, we can trace clear instances of highly effective backbiting. A number of prominent disciples took it upon themselves to end the debate in favor of the man they supported. The debate grew to engulf the public at large. Few remained neutral.

After their first encounter, Einstein insisted that the philosopher simply did not understand the *physics* of relativity—an accusation with which most of Einstein's defenders agreed and which Bergson forcefully resisted. In light of these accusations, Bergson revised his argument in three separate appendices to *Duration and Simultaneity* that he included in a second edition and in a separate paper published in a specialized journal. Bergson's response has frequently been ignored. By taking it in consideration, we can see that their dispute hinged on a lot more than mere technical disagreements pertaining to factual details of relativity theory. Bergson never acknowledged defeat. According to him, it was Einstein and his interlocutors who did not understand *him*.

In one sense this book is about one day, but in another it is much broader. Before the two men actually met, it seemed nearly impossible to foresee such a strong potential for conflict between them, their science, and their philosophies. We find some evidence of animosity on Einstein's part in 1914, when in a letter to a friend he described Bergson's philosophy as "flaccid" and not even worth reading for the purpose of improving his command of the French language.[31] For Bergson, evidence reveals the contrary: an initial fascination with Einstein and his theory. A friend of his recalled how, upon hearing about it, the philosopher plunged himself into a careful study of its mathematics. At that time, Bergson thought he would publish only a "note" on it, with an overall positive assessment. It would "show the agreement between relativity and my views on space and spatial time," he confided to a friend. But these conciliatory intentions soon waned. It became clear that Bergson's concept of *duration*—a label used by the philosopher to describe aspects of time that could never be grasped quantitatively—had to be "set apart."[32]

At the 1921 Oxford Congress of Philosophy, papers on Bergsonian philosophy and Einsteinian physics were delivered side-by-side with no apparent problems. What happened on that sixth of April that changed the status quo?

This book is about two men and one day. But it is also about what these two men have come to represent. Most important, it is about how these men and their respective advocates *came to be who they were*. Specific events and interactions shaped them as much as they, in turn, shaped the world around them. After arguing for nearly a century in terms of *for* or *against*, we can now search for a third route: to understand *both* of their positions, their emergence, and their context.

A REVOLUTION AGAINST BERGSON

We know Einstein by reputation—a man frequently compared to Newton and to Columbus. By publishing "what is arguably the most famous scientific paper in history," he created a revolution comparable to that of Copernicus.[33] In 1919 an eclipse expedition brought international fame to the controversial scientist. Partly because of his vocal pacifist and antinationalist stance, Einstein was one German-born scientist supported by many members of war-torn countries and admired by those who shunned the dangerous rising tide of German nationalism. As one scientist of the period put it, when talking about time, one needed to talk about Einstein. Otherwise it would be like "not talking about the sun when discussing daylight."[34] Since then, Einstein was crowned as the man whose work took "sensorial perception and analytical principles as sources of knowledge," nothing more and nothing less.[35] The theory of relativity broke with classical physics in three main respects: first, it redefined concepts of time and space by claiming that they were no longer universal; second, it showed that time and space were completely related; and third, the theory did away with the concept of the ether, a substance that allegedly filled empty space and that scientists hoped would provide a stable background to both the universe and their theories of classical mechanics.

In combination, these three insights were related to a startlingly new effect, *time dilation*, which profoundly shocked scientists and the general public. In colloquial terms, scientists often described it by saying that time slowed down at fast velocities and, even more dramatically, that it completely stopped at infinite ones. If two clocks were set at the same time with respect to each other, and if one of them separated from the other traveling at a constant speed, they would mark different times, depending on their respective velocities. Although observers traveling with the clocks would be unable to notice any changes in their own system, one of them was slow *in comparison* to the other. Researchers calculated a striking difference between "time$_1$," as measured by a stationary clock when compared to "time$_2$" as measured by a clock in motion. Which of these referred to time? According to Einstein, *both*—that is, all frames of references should be treated as equal. Both quantities referred equally to time. Had Einstein found a way to stop time?

Bergson was not convinced. Claiming that the sensational conclusions of the physicist's theory were not so unlike the fantastical searches for the fountain of youth, he concluded: "We shall have to find another way of not aging."[36]

Relativity scientists argued that our common conception of "simultaneity" needed to be upgraded: two events that seemed to occur simultaneously according to one observer were not necessarily simultaneous for another one. This effect was connected to other aspects of the theory: that the speed of light (in vacuo and in the absence of a gravitational field) was constant.[37] The velocity of most physical objects could successively be increased by piggy-backing on other fast-moving objects. For example, a train traveling at a certain speed could be made to travel faster if placed on top of another fast train. While the first train could be traveling at, say, 50 mph, the one on top would go at 100 mph, the next one at 150 mph, and so on. Not so with light waves. The speed of light, in Einstein's account of special relativity, was not only constant; it was an unsurpassable velocity. This simple fact led scientists not only to abandon the concept of absolute simultaneity, it also led them to a host of additional paradoxical effects, including *time dilation*.

As with Einstein, we also know Bergson mostly by reputation.[38] Bergson was compared to Socrates, Copernicus, Kant, Simon Bolívar, and even Don Juan.[39] The philosopher John Dewey, known as one of the main representatives of American pragmatism, forcefully claimed that "no philosophic problem will ever exhibit just the same face and aspect that it presented before Professor Bergson."[40] William James, the Harvard professor and famed psychologist, described Bergson's *Creative Evolution* (1907) as "a true miracle," marking the "beginning of a new era."[41] For James, *Matter and Memory* (1896) created "a sort of Copernican revolution as much as Berkeley's 'Principles' or Kant's Critique did."[42] The philosopher Jean Wahl once said that "if one had to name the four great philosophers one could say: Socrates, Plato—taking them together—Descartes, Kant and Bergson."[43] The philosopher and historian of philosophy Étienne Gilson categorically claimed that the first third of the twentieth century was "the age of Bergson."[44] He was simultaneously considered "the greatest thinker in the world" and "the most dangerous man in the world."[45] Students described him as "an enchanter" credited with "saving France and the liberty of Europe."[46] Many of his followers embarked on "mystical pilgrimages" to his summer home in Saint-Cergue, Switzerland.[47] Lord Balfour followed his work carefully, and "when a past prime minister of England engages in a controversy with the principal philosophical thinker of the era, everyone should listen."[48] Theodore Roosevelt, the president of the United States, was one of the many who listened carefully to what Bergson had to say, writing an article directly addressing Bergson's philosophy.[49] Yet others considered his work as marking the passing of winter and the coming of a new spring for Western civilization.[50]

Bergson was widely viewed as the main man leading the "insurgence against reason" that many diagnosed as a contemporary disease of the interwar period. As a result, he was accused of denigrating the "physical sciences" to "at best a merely practical device for manipulating dead things."[51] The historian and theorist Isaiah Berlin associated him with the "abandonment of rigorous critical standards and the substitution in their place of casual emotional responses."[52] The mathematician and philosopher Bertrand Russell charged him with anti-intellectualism, a dangerous disease affecting "ants, bees and Bergson" in which intuition ruled over reason.[53] Bergson's *Introduction to Metaphysics* was "the *Discourse on Method* for modern anti-rationalism."[54] He was reputed to be spiritualist, antiscience, and the leading representative of the "modern occult revival," the "revolt against mechanism," and the "new spiritualism."[55] Believed to have been influenced by religious beliefs, and frequently associated with the Catholic Church, Bergson was Jewish. Rumors circulated that he had converted to Catholicism. Were they true? Yet his work was also placed on the Catholic Church's Index of Prohibited Books, forbidding believers from reading it and disseminating it.[56]

At the Lycée Condorcet, Bergson obtained prizes in English, Latin, Greek, and philosophy. He was acclaimed for his mathematical work, receiving a national prize and publishing in the *Annales de mathématiques*. He published two theses, one a highly specialized dissertation on Aristotelian philosophy and another, titled *Time and Free Will*, which would go through countless editions. In 1898 he became a professor at the École Normale; in 1900, he moved to the prestigious Collège de France.

His fifth book, *Creative Evolution* (1907), brought him universal fame. His lectures were so crowded with *tout Paris*, that his students could not find seats. It was rumored that socialites sent their servants ahead of time to reserve them, and "in illustrations of the time, we see people climbing windows to get a glimpse of the celebrated philosopher."[57] During his reception at the Académie française he received so many flowers and applauses that underneath the clamor he was heard protesting "But I am not a ballerina!" Even the Paris Opera, it was evident, was not spacious enough for him.[58] Two thousand students turned up for a lecture at New York's City College (1913).[59]

This universal fame followed him until 1922, when he published *Duration and Simultaneity*, a book that he described as a "confrontation" against Einstein's theory. It unabashedly intended to out-Einstein Einstein by interpreting all known scientific facts associated with relativity theory in a new way. It was in press during their meeting and appeared later that year. It did not produce the author's hoped-for effect.

"The Jew is told: 'You're not at the level of the Arab because at least you are white, and you have Bergson and Einstein,'" explained Frantz Fanon, who fought for decolonization and for Algeria's independence from France. For him, the two men exemplified the racial tensions of the post–World War II era.[60] The French allegedly used them to foster the "so-called dependency complex of the colonized" to prove the superiority of whites against blacks, and to play Jews and Arabs against each other. Bergson and Einstein were frequently cited together as icons of modernity and of cultural and literary modernism. Their fame reached across the world.[61]

The confrontation between the two intellects was particularly shocking because those involved believed that agreement in scholarly matters, especially in scientific ones, should be reached. We were all accustomed to "endless discussion without resolution over the best structure to give a government, or over the most perfect form of art, or over a certain problem of metaphysics or ethics," but this should *not* happen in a case "dealing only with logical deductions based on *facts* that none of the adversaries even dream of contesting."[62] This was a "disconcerting thing, and perhaps, without precedent."[63] There needed to be an end to something that could only be explained as "a colossal misunderstanding" or a "monstrous mistake." Something urgent had to happen in order to have "everyone agree."[64] The arguments advanced had the disconcerting flavor of a "double monologue" that seemed to resemble those of "the tower of Babel" filled with "contradictory discussions where the affirmations are as categorical on one side as they are on the other."[65] "Bergson and the relativists might both be wrong but cannot be right," explained a physicist who dedicated most of his adult life to figuring out who should be the winner.[66] By the end of the twentieth century, the debate was still a "head-on clash of rival conceptions."[67]

To this day one can safely refer to it as a "*locus classicus*" and conclude that "The historical debate between Bergson and Einstein on the theory of relativity is . . . a classic."[68] In the words of the poet Paul Valéry, their confrontation was the singular "*grande affaire*" of the twentieth century.[69] Did their debate end a "golden age before the divorce between the two cultures?"[70] It opened up a veritable "can of worms" that lasted for the next hundred years.[71]

Einstein, on that day, had good reasons to be worried about how the philosopher's attack would affect him. He had promised to give the money from the Nobel Prize, which he was expecting to get, as alimony to his ex-wife. But before the prize was awarded that same year, some wondered if Bergson's critique had thrown "the whole relativity doctrine into the lap of metaphysics, from which . . . Einstein was determined to rescue it."[72] Others started to consider Einstein's theory as simply irrelevant for everyday human concerns. Alain, a widely read author who would become an important antifascist writer, claimed that, "from an algebraic point of view all [of Einstein's work] is correct; from a human point of view all is puerile."[73]

The years that followed their encounter in Paris can be compared to those of the religious wars—with one major difference: instead of debating about how to read the Bible, thinkers across a wide variety of disciplines debated about how to read the complex *unfolding of nature through time*.

CHAPTER 2

"More Einsteinian than Einstein"

"When Albert Einstein left for Paris in March 1922, he knew that he would be skating on thin ice," wrote one biographer.[1] Einstein's visit was highly symbolic for the two countries.[2] This was a period of extreme tension between France and Germany, which were still recoiling from the Great War (1914–1918) and under the spell of lingering resentments and violent accusations. A German ultranationalist opponent of the physicist who commented on the visit complained how this was simply "not the proper time" for Einstein to go:

> Since the end of the war the French have suppressed the German people in the most brutal manner. They have torn away piece after piece of their body, have engaged in one act of extortions after another, they have placed colored troops to watch over the Rhineland, and they have made insufferable demands on the German people through the reparation commission. And just at this very time Mr. Einstein travels to Paris to deliver lectures.[3]

The scientist Max Planck described Einstein's decision to travel there as "heroic" yet likely to cause even more problems. "Despite the advantages that it presents," it will bring to you "a thousand enmities written and not written," he explained to Einstein.[4] Others had exactly the opposite view, believing that Einstein's visit could help relations among nations, heralding "the victory of the Archangel over the Demon of the Abyss."[5]

Einstein had protested the Great War; Bergson had patriotically defended the actions of his country. Einstein had turned forty-three the previous month; Bergson was sixty-two.

After intense coverage of Einstein's work in newspapers and learned circles, here was the first opportunity to discuss relativity "in the presence of the monster himself."[6] Many hoped that in an intimate question-and-answer forum Einstein would reveal "more than through his written work, his intimate principles and true driving ideas."[7] They hoped they could obtain "clarifications from the mouth of the author himself" on the most controversial aspects of his theory.[8] That Einstein would meet Bergson only made his visit even more exciting, leading to "a debate that, in its eternal interest, infinitely surpasses the mediocre political imbecility [*politico-nigologiques*] and the lowly pecuniary controversies of the common fare in which we are accustomed to partake."[9]

After receiving three invitations, Einstein declined all of them.[10] He had, however, second thoughts about the last one, coming from a friend of his at the Collège de France. These doubts intensified after a conversation with the foreign minister, Walther Rathenau, who worked to improve relations between these two countries before he was brutally murdered. Rathenau urged him to attend. Shortly thereafter Einstein withdrew his previous declination, notified the Prussian Academy of Sciences, and started preparing his trip.[11]

Einstein was invited to France with the express purpose that his visit would "serve to restore relations between German and French scholars." In his travel notification to the Prussian Academy of Sciences, he quoted the letter of invitation from Paul Langevin: "The interests of science demand that relations between German scientists and us be reestablished." Langevin, future host, close colleague, and old friend, firmly believed that Einstein "will contribute to this better than anyone else."[12]

Einstein had become a veritable celebrity a few years before the encounter in Paris, when he was catapulted to fame in 1919, at the end of the war.[13] His name appeared on the cover of numerous newspapers around the world that charged him with revolutionizing—not only physics—but everyday notions of time and space. The headline of the *Times* on November 7 read "Revolution in Science/New Theory of the Universe/Newtonian Ideas Overthrown," and three days later the *New York Times* announced "The Lights of the Heavens Askew."[14] Newspapers recounted how observations of an eclipse expedition had proved that traditional concepts of time and space needed to be completely overhauled. A recent historian argued that the "modern world began on 29 May 1919 when photographs of a solar eclipse confirmed the truth of a new theory of the universe."[15] By the fall of 1920, Einstein saw how "presently every coachman and waiter is debating whether the theory of relativity is correct."[16] In the first six years after the eclipse more than six hundred books and articles on relativity were published.[17]

Before becoming a worldwide star, Einstein worked hard to expand the relevance of his theory of relativity *beyond* the community of physicists. In 1917 he published a *gemeinverständlich* version of both the special and general theory. His newfound fame soon dwarfed his own popularization attempts. Popular and specialized expositions of relativity proliferated almost automatically after this date. His *Über die spezielle und die allgemeine Relativitätstheorie (gemeinverständlich)* was translated into English, French, Spanish, and Italian. Then came Einstein's famous *Four Lectures on Relativity*, presented at Princeton University in 1921.

But, at least in philosophical circles, Bergson had good reasons to feel stronger than his rival. During their meeting, the scientist was grilled about almost everything, from the mathematical details of his theory to its broader philosophical implications.[18] The forum was challenging for him linguistically, since he was competent in French, but not fluent. "The language will certainly cause me some trouble," he explained to Langevin, who had generously invited him.[19] Before it all started, Einstein strategized to minimize the "disturbing" effects that could arise from "the deficiencies in expressing" himself in French. "I have to speak in Paris at the Collège de France in—I shudder to say—*French*," he confessed.[20] "If only my beak were better polished in French," he lamented.[21] That language, after all, had always been Einstein's least favorite subject in school. He consistently obtained his worst grades in it, leading many since then to believe that he had been a bad student.[22] A listener during the meeting remarked how Einstein pronounced "relativity" with two accents and mispronounced "equations." It actually sounded as if he said "rélativité" and "ékations."[23] Bergson, in contrast, was a noted and experienced orator who knew how to speak impeccable French and English.

The event organizer at the Société française de philosophie, Xavier Léon, introduced the scientist as the "genial author" of relativity theory, remarking how "the date of April 6 will make history in the annals of our society."[24] Some of the most important French intellectuals of the time were in the room. Langevin spoke first after the introduction.

Langevin had been one of Einstein's first supporters in France. He presented the scientist and his theory in a way that was already familiar to many, including Einstein. But less familiar to him were some of the philosophers in attendance, such as Léon Brunschvicg, who asked a difficult question about the relation of Einstein's theory to a "Kantian conception of science." While Brunschvicg wanted clarification on highly technical aspects of Kant's philosophy in relation to relativity, the physicist's answer was nothing but dismissive. Each philosopher had "his own Kant," he told Brunschvicg, so he "could not respond" because he "did not know how *you* interpret Kant."[25]

Others who initially did not want to speak were prodded by the organizer who wanted—and was expecting—a lively meeting. Édouard Le Roy, a student of Bergson, made this clear: "Our friend Xavier Léon *really* [*à toute force*] wants me to speak. Faced with his polite insistence, I cannot refuse. But, deep down, I have nothing to say." Nonetheless in the "two words" uttered by Le Roy, Bergson was dragged into the discussion.

Le Roy believed that "the point of view of philosophers and physicists were both equally legitimate" but were—in the end—different: "I believe in particular that the problem of time is not the same for Einstein and Bergson." Le Roy concluded his commentary by saying that since "Bergson was among us" it would be more appropriate for "Bergson himself to take the floor."[26]

After having sat silently during Einstein's Collège de France lecture the previous day, Bergson now responded begrudgingly, insisting that he "had come here to listen." When he first spoke, he lavished praise on the foreign physicist. The *last* thing he intended to do was to engage Einstein in a debate. With regard to Einstein's theory Bergson had no objections: "I do not raise any objections against your theory of simultaneity, any more than I do so against the theory of relativity in general."[27] What Bergson wanted to say was that "all did not end" with relativity. He was clear: "All that I want to establish is simply this: once we admit the theory of relativity as a physical theory, all is not finished."[28] Philosophy, he modestly argued, still had a place.

Einstein disagreed with Bergson and replied with a provocative phrase: "The time of the philosophers does not exist." Einstein was facing an audience mainly composed of philosophers and *hosted* by philosophers. By and large, philosophers had even shown themselves to be one of the most open and inviting communities in France toward the German-born physicist. Was Einstein unappreciative of their good will?

What did Einstein intend by uttering that phrase? Einstein fought against giving philosophy (and by inference Bergson) a predominant role in matters of time. His objections were based on his views about the role of philosophy and philosophers in society—views that differed from Bergson's.

ONLY "OBJECTIVE EVENTS"

During his meeting with Bergson, Einstein defended *his* definition of time as having a clear "objective meaning" in contrast to other definitions. "There are objective events that are independent of individuals," he insisted that day, implying that his notion of time was one of them.[29] His theory was *not* just a fruitful hypothesis or one convenient explanation that could be chosen out of many. "One can always choose the representation one wants if one believes that it is more comfortable to do so than another one for the task at hand, *but that does not have any objective sense*," he insisted.[30] The astronomer Charles Nordmann, who followed Einstein's visit closely, explained the physicist's intentions. "If there is in fact an opinion against which Einstein fought vigorously and notably, one can remember, right after the debates at the Collège de France, it is one that gave his theory only a formal or mathematical importance," he recounted.[31]

"All went brilliantly well," wrote Einstein to his wife that evening. Einstein eagerly prepared his trip back home, holding a "not-empty leather bag" filled with money given to him by the Baron de Rothschild. Back in Germany, inflation was out of control. After "the last discussion" ended, he felt good about his performance and proud to have served his country's interest. "If the Germans only knew what services I performed for them here by this visit," they would clearly thank him, he told her. "But they are too small-minded to grasp it," he concluded.[32]

The debate between the two men escalated quickly. After their first meeting, Bergson and Einstein were scheduled to meet again in a few months, this time, for an entirely different purpose. Bergson was president of the International Committee on Intellectual Cooperation, one of the most prestigious branches of the League of Nations. Einstein was one of its members. While the participation of Bergson and Einstein augmented the prestige of the League of Nations, their heated encounter in Paris only intensified doubts about the viability of international collaborations—even those founded for the express purpose of reducing conflict in Europe. Would conflict be reduced, or would it escalate? In 1922, it was hard to foresee what would happen.

By the beginning of the fall, *Duration and Simultaneity*, the contentious book that had been in press during the Paris meeting, appeared in print. Bergson expressed the "duty" to defend philosophy from the encroachment of science. These were strong words: "The idea that science and philosophy are different disciplines meant to complement each other . . . arouses the desire and also *imposes on us the duty* to proceed to a confrontation."[33] Bergson chastised relativity theory for "stopping to be a physics to become a philosophy" and a deeply flawed one at that.[34]

Although Einstein's simple statement that day—"the time of the philosopher does not exist"—served as a detonator, many additional factors intensified the conflict between the two men and the views they represented. Bergson and Einstein belonged to different communities with different cultural and intellectual heritages.

Einstein obsessively searched for unity in the universe, believing that science could reveal its immutable laws and describe them in the simplest possible way.[35] Bergson, in contrast, claimed that the ultimate mark of the universe was just the opposite: never-ending change. Philosophies that did not stress the fluctuating, contingent, and unpredictable nature of the universe—as well as the essential place of human consciousness in it and its central role in our knowledge of it—were, according to him, retrograde and unlearned. While Einstein searched for consistency and simplicity, Bergson focused on inconsistencies and complexities.

The German scientist was deeply steeped in an elite *Kultur* tradition, considering himself a member of "a supratemporal community of exceptional minds that existed in a universe parallel to that of the philistine masses."[36] Bergson also belonged to a cultural elite, but a very different one from Einstein's. He saw himself as the continuator of a school of French, post-Cartesian philosophy. Bergson studied and continued the work of his teacher Émile Boutroux, of Boutroux's teacher Jules Lachelier, and of the man who inspired all of them equally, Félix Ravaisson.[37] Einstein focused on an entirely different tradition that revolved largely around the German classics: Lessing, Kant, Schiller, and Goethe. While Einstein's sources were widely read within and outside of Germany, Bergson's, in contrast, were studied by a much smaller circle of philosophy specialists.

Einstein's leftist politics and his pacifism during the Great War contrasted starkly with Bergson's vocal nationalism during that same period. Einstein's personal understanding of himself as a marginalized Jewish outsider clashed with Bergson's comfort as an assimilated French Jew. Bergson was not only a famous professor in one of France's most elite institutions, he also belonged to a small inner circle of well-placed intellectuals and politicians. Even during the virulently anti-Semitic Vichy period (1940–1944), the philosopher was well looked after.[38]

"Remarkably, throughout my long existence, I have not collaborated with anyone but Jews," explained Einstein to his close friend Besso in 1937, already exiled in Princeton and decades after he had completed his most brilliant work.[39] Einstein's personal identity was defined against that of the dominantly Christian European bourgeoisie. After the horrors of the Great War, a conflict that he forcefully opposed, he came to think "that our kinfolk really are more sympathetic (less brutal) than these horrid Europeans."[40] In contrast, when Bergson prohibited the publication after his death of his correspondence and notes (and even their availability to the public through libraries) his reasons were clear: he had to protect himself against his "mortal enemies (among which there are all types of Jews, my coreligionists)."[41] While a veritable Einstein industry continues to glorify the physicist, promoting and controlling his image through well-funded institutions, Bergson's followers are few and far between.

The physicist's bohemian lifestyle contrasted with Bergson's monastic asceticism. Einstein's rural south German Swabian origins on the margins of mainstream bourgeois culture and the precarious boom-or-bust business initiatives of his father fostered in the young scientist a contradictory disdain for financial comfort as well as a profound longing for it. His social status clashed with Bergson's, who had an influential Polish banker as a paternal grandfather and a doctor from Yorkshire as a maternal one. The physicist's messy, peripatetic personal life remained at odds with Bergson's privileged stability.

Bergson was born in Paris. As a child he lived with his family for a few years in London and in Geneva, before returning to France. When his family moved back to England, he stayed behind in a boardinghouse so that he could continue his studies. From that moment on, he remained in France, visiting his parents abroad during the summers. During his twenties, he spent a few years teaching in the provinces. After that, he lived in Paris for the rest of his life.

The physicist, in turn, lived and spent time in many different places in Germany, Switzerland, Italy, and Eastern Europe as an adult and as a child. At age sixteen, he lived by himself when his parents moved to Italy for a business opportunity that proved disastrous. Distinctly handsome, he broke hearts as a teenager, had a daughter out of wedlock (who was most probably given away), was accused of adultery by his first wife, went through a prolonged legal battle over divorce and alimony, and collected more than a few amorous peccadilloes along the way. Einstein came close to bankruptcy a number of times, finding it difficult to come up with money to support two families. During his Parisian sojourn he was able to save "one piece of fine soap and a tube of toothpaste" generously giving it to his wife back home.[42] Shocked to discover that the postage for a single letter "costs 17 marks," he was forced to use the post sparingly. "In consideration of this," he explained to his wife, "I'm not going to write very often."[43]

Bergson led an exemplary private life, doting on a daughter who was born deaf and who became a gifted artist. Close friends described his marriage as one of "uninterrupted happiness."[44] He was well-to-do, leading the quiet life of a comfortable university professor. He was, as one student put it, the "picture of sobriety," who when confronted with a "number of dishes that is called a banquet," preferred to eat "a bun and a glass of milk."[45] "With a happy consistency in habitat and theory which few philosophers attain, he resides in *Rue Vital*," explained a contemporary.[46] While Bergson extolled the virtues of vegetarianism, Einstein longed for the delicious goose cracklings that his second wife sent him through the mail. For a number of years, Bergson lived in a bourgeois house on a beautiful boulevard (at Villa Montmorency, 18 avenue des Tilleuls, from 1902 to 1915), cultivating roses and taking care of two cats.[47] In contrast, the actor Charlie Chaplin, who visited Einstein's home in Berlin, thought that "one could find the same apartment in the Bronx."[48] Follow them inside their respective homes: Bergson's apartment was decorated with "some drawings of his deaf-mute daughter, who had talent and had taken lessons with Rodin." George Oprescu, an art historian who knew both of them, compared their different decorating styles: "Einstein, in his modest quarters in Berlin, also did not have any art, but I remember the appreciative look he gave me when I offered him some lithographs by Daumier, which one could buy in Paris for a few francs."[49]

As with infinitely nesting matryoshka dolls, differences between them were readily apparent from the most intimate to the most public of perspectives. Were the differences between Einstein and Bergson mainly cultural, personal, political, and ideological?[50] Psychological, intellectual, social, institutional, political, and national differences proved a most fertile ground for an expanding conflict. But the shape of this expanding conflict nonetheless remained surprisingly fixed when considered across one single axis: Einstein and Bergson differed in their views about the nature of time and the power of science to reveal it. Discussions on and references to time appeared from the most private to the most public, from the most scientific to the most political and philosophical, from the most profound to the most informal, of places. Every time, Einstein and Bergson disagreed.

TIME, WITH A CAPITAL *T*

Bergson capitalized "Time" in the foreword to the second edition of *Duration and Simultaneity*. By capitalizing the term, he signaled to his readers that he was including something larger in the concept than if he had referred to mere, lowercase "time."[51] The rest of the book made it clear that he was *not* referring to the same category used by physicists. For years, he and his students had been trying to separate *their* concept of Time from that of others. "Time," for them, included aspects of the universe that could never be entirely captured by instruments (such as clocks or recording devices) or by mathematical formulas. Confusing clock time with time-in-general, and judging one by the standards of the other, could not be more abhorrent for Bergson. But these differences were subtle, and many readers intent on evaluating Bergson's argument ignored these distinctions.

Most interpretations of the Einstein-Bergson debate insist that Bergson made a mistake in *Duration and Simultaneity* because he did not entirely understand the physics of relativity. One claim in the book is frequently cited as the main culprit: that time is not altered according to the velocity of a system. In the foreword to his second edition, Bergson explained that the book's central message was to "explicitly prove that there is no difference, in what concerns Time, between a system in motion and a system in uniform translation."[52] Elsewhere in the book, he categorically stated that if a clock traveling close to the speed of light is later compared to a stationary clock, it "does not present a delay when it finds the real [stationary] clock, upon its return."[53] This claim, taken at first blush and in isolation, was completely at odds with the concept of *time dilation* in the theory of relativity. Because of this particular claim, many readers insisted that Bergson "was not sufficiently conversant with the outlook and problems of mathematics and physics."[54] Later commentators cited Bergson's remark that "once reentering [Earth], it [one clock] marks the same time as the other" as proof of his profound misunderstanding of relativity.[55] This single statement about clock delay has been enough to discredit him in the eyes of most scientists and some philosophers.[56]

What exactly did Bergson not accept about relativity? In a footnote to the main text, he explained that he fully accepted "the invariance of the electromagnetic equations."[57] Elsewhere in the book he insisted that his qualms were not *at all* with any technical results or conclusions. None of them were meant to bear on physics: "The theory was studied with the aim of responding to a question posed by a philosopher, and no longer by a physicist." "Physics," he added, "was not responsible for answering that question."[58] His statement that there was "no difference" in the Time of a system in motion and a stationary one should not be taken literally. "It is just a manner of explaining oneself," so that he could get to "the depth of the matter," he explained to the scientist Hendrik Lorentz.[59]

Bergson's protests notwithstanding, most of the educated public concluded that he had made a glaring mistake during his debate with

Einstein and in his subsequent publications on relativity theory: "These attempts [Bergson's] . . . have totally failed: science, on this issue, has passed purely and simply to become the order of the day."[60] By the 1960s Bergson's fate as somebody who simply did not understand science was sealed: "The best explanation for Bergson's impressive failure as a scientific theoretician is the same as that for his failure to succeed as a metaphysician: he was not sufficiently conversant with the outlook and problems of mathematical physics."[61] Even a writer in the *Annales Bergsoniens*—an ongoing series solely dedicated to his philosophy—stated that "Bergson could not understand him [Einstein]."[62]

Yet in his face-to-face debate and in the book that followed, Bergson repeatedly stressed that he was *not* contesting any of Einstein's *scientific* claims. He repeatedly explained that he was referring to "Time"—something different from the physicists' "time." He often chose a completely different word—*duration*—to underline the aspects of time that concerned him. Why, then, has the debate been so often understood in terms of Bergson's error? There are many reasons why and how people came to believe that Bergson made a mistake. Essential clues are hidden in archives, such as in Einstein's correspondence. There we learn that it was Einstein himself who first and forcefully propagated this view, and that deep down (as we can read in his journal and later correspondence), he knew better.

DURATION AND ÉLAN VITAL

What drew Bergson's students to call him an enchanter? What motivated socialites to send servants ahead of time to reserve seats for his lectures? Why was he meticulously read by presidents and prime ministers? Why did his enemies want to murder him? Why did others contemplate suicide before they were saved by reading Bergson? Why were his books placed on the index of forbidden texts? Why did the most important French philosophers and polemicists pen full monographs about him? How did his philosophy become more of a movement, often called *le Bergsonisme*, that sometimes escaped the intentions of the philosopher himself? Why did his work affect so many fields beside philosophy, from musicology to film theory? Why was his work relevant across the entire political spectrum, pleasing, equally, anarchists, syndicalists, and fascists? How did some of his key phrases end up in the Nuremberg trials, in advertisements, and in contemporary novels?

During the trial of major German war criminals after World War II, the chief prosecutor for the French Republic cited an excerpt from one of Bergson's last books:

> "Humanity," says our great Bergson, "groans, half crushed by the weight of the progress it has made. . . . The ever-growing body awaits the addition of a soul, and the machine requires a mystic faith."

What did Bergson mean by "mystic faith," and how did it relate to his philosophy of time? Both emerged from a vital impulse that pushed forcefully forward. The prosecutor explained it clearly: this "mystic faith" was allegedly the same force that drew the ancients to create civilization and the moderns to defend human rights and democracy:

> We know what it is, this mystic faith of which Bergson was thinking. It was there at the zenith of the Greco-Roman civilization, when Cato the Elder, the wisest of the wise, wrote in his treatise on political economy. . . .
>
> It is this mystic faith which, in the realm of politics, has inspired all the written or traditional constitutions of all civilized nations ever since Great Britain, the mother of democracies, guaranteed to every free man, by virtue of Magna Charta and the Act of Habeas Corpus, that he should be "neither arrested nor imprisoned, except by the judgment of his peers delivered by the due process of the law."
>
> It is this faith which inspired the American Declaration of 1776:
>
> "We hold these truths to be self-evident, that all men have been endowed by their Creator with certain inalienable rights."
>
> It is that which inspired the French Declaration of 1791:
>
> "The representatives of the French people . . . have resolved to set forth in a solemn declaration the natural, inalienable and sacred rights of man. Consequently, the National Assembly recognizes and declares, in the presence and under the protection of the Supreme Being, the following rights of the man and of the citizen."[63]

The same quote, which in the mid-1940s carried lessons as important as those legated by the Magna Carta, the Act of Habeas Corpus, the constitution of Great Britain, and the American and French Declarations, was expanded in the late 1950s for a commercial advertisement for the Rand Corporation that was dedicated to U.S. national security:

> Humanity is groaning, half-crushed under the weight of the progress it has made. Men do not sufficiently realize that their future depends on themselves. They must first decide whether they wish to continue to live. They must then ask whether they want merely to live, or to make the further effort necessary to fulfill, even on our unmanageable planet, the essential function of the universe, which is a machine for making gods.

A different, earlier Bergson was quoted in Haruki Murakami's famous novel *Kafka on the Shore* (2002) at the height of an explicit sexual scene: "The pure present is an ungraspable advance of the past devouring the future. In truth, all sensation is already memory."[64]

How could Bergson be quoted verbatim in criminal courts as well as in bedrooms during moments of erotic titillation? Why was he the topic of so many conversations from central Europe to the world beyond? Why did his work fascinate some of the main figures fighting for French decolonization? Why did writers in China and Japan cite him as exemplifying European modernity? Why did he enthrall young Latin American intellectuals, who upon arriving in Paris rushed to attend his lectures?[65] Why were reading clubs across the globe formed with the explicit purpose of discussing him?[66] Even in the imaginary café described in *The Institute for Clock Synchronization*, a 1954 novel written by the Turkish author Ahmet Tanpinar, conversations centered on "history, Bergson's philosophy, the logic of Aristotle, Greek poetry."[67]

For years to follow, the debate between the two men was carried out mainly through others. Einstein was not pleased. He and his followers did not permit the philosopher's comments to pass unchallenged. Bergson would eventually learn that Einstein had "intemperate disciples, who brought his views closer to philosophy, and who built on top of it an extravagant doctrine."[68] A bevy of "ballet dancers" around him were set loose to "fill the ears of the public" and promote his theories.[69] But Bergson was hardly alone.

Allies and enemies of one or the other sat on professional committees determining who would get coveted academic positions, opening or closing doors to professionals who supported one side or the other. They published books, articles, manifestos, and popular broadsheets defending or attacking one of the two men. Whole philosophical systems were developed as ways out of the impasse of the debate.

The repercussions of the confrontation between them spread across the globe. Einstein's trip to Paris was only one of many other trips on which he embarked during those years: Chicago, Washington, London, Rio de Janeiro, Japan, Spain, and Jerusalem were among his destinations.[70] Bergson was also a world traveler, but for very different reasons. Reputations traveled even faster. Books, articles, lectures, news, and correspondence about them moved quicker and farther than the men themselves. Telephone, telegraph, radio, and film (and later even television in Einstein's case) transmitted texts, images, and voices promoting their work and, sometimes, referring to the debate.

Three years after their meeting, a scientist and popular science writer from Barcelona could expect his readers to be "aware of Bergson's objections" to Einstein and of the particular occasion when the philosopher had shown "all his anger."[71] In Spain, the philosopher José Ortega y Gasset and the writer and politician Ramiro Ledesma Ramos wrote about both men. In Latin America, Alfonso Reyes, a young intellectual from Mexico charged with nationalizing his country's oil industry, explained in his notes on Einstein that "one day the time of physics, the time of psychology, and Bergson's concept of 'real duration,' must be reconciled, since now it has been provisionally put aside."[72] In Japan, during the summer of 1947, the physicist Satosi Watanabe delivered a talk exploring the connections between Bergson's work and quantum mechanics at the Maison Franco-Japonaise.[73] Often by direct reference to Bergson and Einstein, thinkers from Central Europe to Northern Africa and the Middle East adopted one of the "main themes" of the century: that the "lived time that they experience is different from measured clock time."[74]

The debate between Einstein and Bergson soon became entangled with larger discussions about the rise of fascism in Europe and about the proper role of philosophy and science in technology-driven societies. Thinkers returned again and again to the debate in high-voltage discussions between intellectuals working under the new National Socialist regime and those who were forced to emigrate. In all of these contexts, interpretations changed as starkly as the world itself in the decades spanning from the Belle Époque to the Cold War.

Bergson's philosophy appealed to the heart and not only to the mind. As such, it aspired to be more comprehensive than scientific knowledge. Beyond hearts and minds, it involved hands, eyes, and ears, inspiring numerous artists. As soon as technologically feasible, his texts were recorded on LPs and CDs. The reach of his philosophy was broad, dealing with ethics *and* aesthetics. It curbed the excesses of a cold, dry rationalism that was associated with the mechanistic universe of René Descartes and the stiff hierarchies of knowledge described by Auguste Comte. It was an antidote to a mathematical and static understanding of the universe, whose rigidity was widely despised, associated as it was with empty rationalism and the violent excesses of the French Revolution. It corrected the naive optimism of some Enlightenment figures, such as that of the Marquis de Condorcet (who after penning a treatise on progress ironically committed suicide) and of Jean-Jacques Rousseau, whose famous descriptions of "noble savages" simply did not match the realities of imperial encounters at the dawn of the twentieth century. It was as profound as religion, yet free from the control of a frequently reactionary, antimodern, and increasingly out-of-touch Church. It represented a new spirituality based on a new undogmatic ethical foundations. Instead of offering a philosophy in which God did not exist, he provided one in which Christ was rarely mentioned (and when he was, it was in the company of other religious figures). Although ostensibly nondenominational, it was largely adopted by citizens who were still baptized, confirmed, married, and buried by priests; by citizens who went to church on Sundays and who fasted during Lent, but who preferred to read Bergson instead of the Bible.

Among his offerings they found not only insights into the nature of time, but full treatises dedicated entirely to the pressing concerns of the flesh-and-blood inhabitants of the emerging century—topics that escaped the cold logic of science and the arid academic philosophy of the universities. Bergson was the paradigmatic philosopher of memories, dreams, and laughter.[75]

RISE AND FALL

Why, comparative to this great fame, is Bergson much less known today? The entry of "Time" in the *Stanford Encyclopedia of Philosophy* (2010) does not even include him. Scholars interested in the topic frequently turn to less controversial authors, such as John McTaggart, who were much less important during those years.[76]

How was it possible to write off from history a figure who was once so prominent? By the time Bergson died on January 3, 1941, the event was particularly shocking because the world had already fallen into the habit of thinking of him as dead.[77] His debate with Einstein precipitated this vertiginous downward fall. The philosopher's fame had peaked when he was almost fifty. But this fame left him as starkly as it had reached him. Einstein, in contrast, remained mostly unknown to the public at large until he entered his forties, yet he successfully maintained beyond his death the status of an icon.

The controversy affected how each man was perceived and remembered. Einstein was generally portrayed as a stalwart figure who stood against intuition and who held very strong views about the power of science as the privileged means for obtaining truth about the world. Partly because of the development and outcome of his debate with Bergson, we commonly believe that relativity theory received incontrovertible experimental proof: "Experiment has decided in favor of Einstein's conception," explained Hans Reichenbach, who was one of Einstein's strongest defenders but who, we should also remember, was a resolute enemy of Bergson.[78] In light of Reichenbach's personal relationship with Einstein and his negative conception of Bergson, we can see how tendentious his claim was.

When Einstein's theory was first developed in 1905, it was often criticized for lacking experimental evidence. Einstein continued to work hard on it, developing a more substantial theory, known as the general theory.[79] Some of his earlier claims, such as those on "the constancy of the speed of light" (approximately 300,000 km/s or 186,000 miles/s) that was also an "infinitely great" velocity (two claims that could seem contradictory) were explained in clearer ways. He later reclassified his early work as belonging to the "special" theory, which remained—much to his own surprise—completely valid under the new framework. The more "general" theory increased its range, applications, and number of experimental verifications.

A rapidly growing number of experiments have unambiguously proved the validity of Einstein's theory. A stunning confirmation took place in 1972, when scientists tested it by transporting an atomic clock eastward around the world and comparing it with one transported westward. The eastbound traveler lost 59 nanoseconds, while the one transported westward gained 273 nanoseconds.[80] Other experiments with cosmic-ray muons (particles that enter into the Earth's atmosphere from outer space) showed that their lifespan before decaying was noticeably increased. Scientists interpreted the particles' prolonged life as due to time dilation effects arising from traveling at speeds close to that of light.

Can we simply conclude, in light of this substantial experimental proof, that Bergson was simply wrong to object? For Bergson, the important questions at stake were *not at all* about the experimental validity of Einstein's theory—they were about the relation of science to metaphysics and about the relation of science to experiment more generally. How are abstract scientific concepts, such as the variable t for time in the relativity equations, related to concrete experimental facts? Can other theories explain those same facts? What is the connection of theoretical science (with its universal claims) to experimental work, concrete things, and local contexts? These questions, he argued, should not be left aside.

By following the debate and its unraveling we can understand why Einstein emerged as the man who set science apart from metaphysics and why he was considered to be doing secular science, although he was profoundly spiritual. Einstein often said he believed in God, albeit this was a deity who did not meddle directly in human affairs. The physicist would often interject references to religion in scientific discussions, as when he delivered his famous exclamation against quantum mechanics that "God does not play dice with the universe." A friend of his once recalled: "Many a time, when a new theory appeared to him or forced, [Einstein] remarked 'God doesn't do anything like that.'"[81] When the poet Paul Valéry, who at one point served as a liaison between Einstein and Bergson, asked him what proof he could evidence in favor of the unity of nature, Einstein answered that he took this unity as "an act of faith."[82]

Although we now remember Einstein as a revolutionary who overthrew previous theories, many observers at the time saw relativity theory as having more similarities than dissimilarities with classical physics such as Newtonian physics or Cartesian philosophy.[83] Bergson himself propagated the view of Einstein as *conservative* rather than revolutionary. Einstein was often accused of having introduced the question of "relativism" into science and knowledge more generally. Yet he directly opposed any blanket descriptions of his theory as consonant with other forms of artistic or cultural "relativism." Bergson, instead, gladly extolled the virtues of thinking about the world in terms of shifting relations and criticized Einstein for producing a theory based on absolute concepts.

While we also largely remember Einstein as a pacifist, we often forget how he did not support the League of Nations and that he renounced his pacifism after the Nazis gained power. Einstein's famous August 1939 letter to President Roosevelt urging the support of nuclear weapons research was written just one month after the physicist resigned from the League of Nations' International Committee on Intellectual Cooperation—an institution that had been headed by Bergson and that advocated a completely different approach to armament control. Bergson, who was accused of war mongering during the Great War, pushed for arms control and diplomacy for the rest of his life.

Perhaps the greatest misconception we have of Einstein is the common association of him with the development of the atomic bomb—an association that became engraved in the public's imagination after he was portrayed on the July 1946 cover of *Time* magazine with a mushroom cloud in the background inscribing the famous $E = mc^2$ equation. But Einstein never worked on the bomb project. To fully understand Einstein's position on weapons research and the abandonment of his early pacifism, his debate and animosity toward Bergson (and toward the institute within the League of Nations that Bergson headed) is essential.

Einstein's work has been widely considered as having had a decisive impact on modern art. The multidimensional views of reality described by relativity theory are considered to have laid the ground for important transformations in the visual arts, affecting important artists from Matisse to Picasso and entire movements, such as cubism.[84] But now it seems that stronger influences can be more directly traced to Bergson's philosophy.[85] In literature, the case for Bergson's pertinence is even clearer than in the visual arts. His influence on Marcel Proust's *À la recherche du temps perdu* is considered not only in terms of its thematic focus and organization, but also in terms of actual relations: Bergson was married to Proust's cousin. Proust followed Bergson's work closely, describing how Catholic priests were prohibited from reading his books and attending his seminars.[86] "Most of Proust's work is an exposition of Bergson's philosophy," expressed a renowned literary critic.[87]

Bergson is frequently remembered as a reactionary. But those most responsible for accusing him of being reactionary held extreme political views: the radical French writer Julien Benda and the fascist author Charles Maurras. These extremes figures were completely opposed to the other but united in their hatred of Bergson. Maurras belonged to the extreme Catholic right; Benda was a Jewish intellectual belonging to the anticlerical left. Both hated Bergson passionately. Bergson was "what brought Benda and Maurras together, two otherwise diametrically opposed people" backing two diametrically opposed political visions.[88]

We know that Einstein was often attacked just because he was Jewish, and some authors have speculated how Bergson's own attack fit within these broad fascist attacks. But Bergson was also attacked for this same reason.[89] The *Action française*, a right-wing racist forum, opposed his candidacy to the French Academy simply because of his background. The anti-Semite Maurras labeled Bergson the master of "Jewish France"—a "foreigner [*métèque*]" who had to be combated for that simple reason.[90] When Bergson's candidacy for the Académie française came up in 1914, Maurras blamed the outcome on a "Jewish intrigue designed to place Bergson in the Académie."[91] Another writer similarly claimed that "all of Israel has its eyes fixed on this election, considered by this enemy race of ours, as an important episode of the Franco-Jewish War."[92]

The attacks on Bergson from the Catholic right were followed by those from the Jewish left. Benda accused Bergson of not belonging to the "right" Jewish constituency. He thought that they were broadly two types of Jews: the Hebrews who worshipped Yahweh (and whose modern leader was Spinoza), and the "sensation-loving" Carthaginians, who worshiped Belphegor and whose modern leader was Bergson.[93] Benda's animosity was such that he once supposedly claimed that "he would have happily killed Bergson if this was the only way to destroy his influence."[94] Léon Daudet, a journalist and lifelong friend of the anti-Semite Maurras, disparagingly referred to Bergson as an "ornamented little Jew."[95]

Not only was Bergson attacked for being Jewish or for not being the "right" kind of Jew, he was also accused of being a propagandist for Catholicism. At one point he claimed that "my reflections have led me nearer and nearer to Catholicism, in which I find the absolute completion of Judaism."[96] He indeed asked to receive a Catholic burial, and at one point considered baptism. Yet the common association of Bergson with Catholicism also needs to be rethought. Bergson was mostly attracted to Christian mysticism, especially to the one represented by St. Theresa of Avila and St. John of the Cross. Mysticism was often seen with suspicion by the Catholic Church—it was a movement that was at times part of Catholicism, yet much broader and at times even a serious threat to the Church itself. Bergson was particularly attracted to mystics who were widely considered to be the most radical and whom the Catholic Church had condemned, such as Jeanne Guyon. Pre-Christian mysticism caught his attention as well.[97] He was officially added to the Index of Prohibited Books on June 1914.

Bergson was frequently described and remembered as holding science in disdain and for having an irrational abhorrence for scientific facts—a view that Einstein himself actively tried to forge. The opening sentence of *Duration and Simultaneity* carefully stated that he would *not* go against any facts of observation: "we take the formulas . . . term by term, and we find out to which concrete reality, to what thing perceived or perceptible, each term corresponds."[98] Bergson wanted *more* not *less* weight placed on experiment and mathematics. He wanted to return to the results of the Michelson-Morley experiment—an experiment that was central in discussions of relativity theory.[99] He combated the charge of anti-intellectualism and considered his philosophical project—and this is the word he used—to be "supra-intellectual."

The philosopher took pains to stress that he held no grudge against Einstein as an individual and had no qualms with the physical nature of Einstein's theory. He thus differentiated his position from the racist and nationalist attacks that Einstein encountered in Germany. He objected only to certain *philosophical* extensions of relativity. "Physics could do a service to philosophy by abandoning certain ways of speaking that induce philosophy into error, and that risk confusing physicists themselves about the metaphysical significance of their views."[100] These confused ways of speaking, he claimed, came from people, such as Einstein, who wanted to "transform this physics, *telle quelle*, into philosophy."[101] Bergson led the fight against a perceived encroachment of physics on philosophy—a fight that easily attracted young recruits.

Determining time, Bergson insisted, was a complex operation. "To know what time it is" was not simply about reading a number (the time) given by an instrument (the clock). It was an assessment of the overall *meaning* of that moment. The broader significance of certain events explained why clocks "work," why they are "fabricated," and why they are "bought." Yet these questions did not interest Einstein during those years, who believed that time was either what clocks measured or it was nothing at all. His mind had no room to explore the reasons why clocks may have been invented in the first place. The contrary was true for Bergson, who wanted to know what led us to live a clockwork-driven existence and to figure out how to break out of it: "Time is for me that which is most real and necessary; it is the necessary condition of action: What am I saying? It is action itself."[102]

Early on in his career Einstein was very modest about his knowledge of philosophy. He declined an offer to become chief coeditor of the *Annalen der Philosophie* because of his limited knowledge of the field. He was also reticent about publishing in philosophy journals even when requested. In 1919 he described himself as being "too little versed in philosophy" and merely "passively receptive" to it.[103] But when he confronted Bergson, his modesty was nowhere to be found.

Although frequently pigeonholed to fit into opposing categories (such as mechanism versus vitalism and objectivity versus subjectivity), none of these dichotomies do justice to the complexities of either men's work. Bergson never wanted to take *sides* in discussion organized around major divisions, understanding his philosophy instead as one that explained connections.[104] He especially never wanted to take the side that he was often accused of taking: that of spiritualism. His interest lay firmly in exposing "the point of contact between consciousness and things, between the body and spirit."[105] Einstein, as well, frequently made comments that make it impossible to confine him to these specific positions. As support for his work started to mount, the physicist's own views about his previous work and about science in general changed. He became more circumspect about how it related to other forms of knowledge. By the end of his life, the physicist placed the word "truth" in quotation marks, finding it to be bounded and constrained by much broader "conceptual systems" (*Begriffsystem*). "Scientific 'truth,'" he explained, was different from "empty fantasy" only in degree and not in kind. They simply differed in how much certainty they could attain.[106] A careful engagement with the debate between Einstein and Bergson reveals difficulties in separating these terms and—in particular—in placing Einstein or Bergson exclusively in one or the other. The task at hand is to problematize these divisions—that is, to overthrow a simplistic understanding of these men's positions—and delineate alternatives.

Bergson had the last word during their meeting at the Société française de philosophie. "It was," he told his audience, "getting late that day" when he offered his closing remarks. To prove that time was not *entirely* owned by physicists, he explained the reasons why he would have to cut his commentary short: "Because we cannot talk about time without noting the hour, and that it is late, I will limit myself to summary comments on one or two points."[107] Perhaps a trick by a great orator, but at least for some in the audience, it seemed to work.

CHAPTER 3

Science or Philosophy?

Time lies at the center of our modern hierarchies. For the powerful, time is most valuable. For the unemployed, it is a curse. The powerful make the powerless wait; the powerless wait on the powerful. How can debates about time help us understand these hierarchies? Benjamin Franklin warned young men that "time is money" urging them to use it wisely.[1] Writing almost a century and a half later, a less thrifty philosopher budgeted time differently. Although time was money, "everyone has this money in his purse" to keep on spending as they wish.[2] Since then, many others continue to offer advice for mastering time. Who can speak authoritatively about it?

"What, then, is time?" inquired Saint Augustine in his *Confessions*, famously pointing out a paradox in our conception of time. Nothing was more intuitive yet more complicated than time itself: "If no one asks of me, I know; if I wish to explain to him who asks, I know not." These words, written in northern Africa sometime between the year 397 and 398, seemed extremely prescient in Europe fifteen hundred years later. Satosi Watanabe, one of Bergson's staunchest followers, stated simply: "This view [Augustine's], after fifteen centuries of progress in the human sciences, must even now represent the view of today's philosophers."[3] It may be no exaggeration to describe modernity as an inability to cope with time as much as with money; it is a series of unsuccessful attempts to master both.

Can a new understanding of the debate between Einstein and Bergson shed light on the nature of time itself? Instead of continuing to debate the nature of time, to see it as entirely demystified by modern science or as eternally ungraspable, close attention to their debate reveals how some of the most portentous pronouncements on time appeared in much more mundane settings that funneled arguments into two opposite directions. To understand time we can return to April 6, 1922, and unpack this day as a moment when these discussions emerged most forcefully.

In one sense, the debate between Einstein and Bergson appears to be the opposite of another famed encounter in the history of science and philosophy, that between Thomas Hobbes and Robert Boyle at the end of the seventeenth century. At the Royal Society of London, Boyle and Hobbes debated about the existence of the vacuum. Boyle believed it could be created using an air pump; Hobbes did not. But neither of them disputed the matters of fact at hand. "Mr. *Hobbes*," claimed Boyle, "does not deny the truth of any of the matters of fact I have delivered."[4] The confrontation between Hobbes and Boyle resulted in an entente cordiale between experimental science (represented by Boyle) and political philosophy (represented by Hobbes) that lasted up to the first decades of the twentieth; the one between Einstein and Bergson led to a different outcome. Despite eventually reaching agreement on the facts at hand, both men and their defenders were still unable to agree on the right boundary between science and philosophy.

For better or for worse, the debate between Einstein and Bergson has still not ended, and probably never will. We cannot hope for a return to the days of the Royal Society, when scientists produced uncontested facts in laboratories, distributing them for general consumption. And we cannot hope to a return to the glorious period of the Scientific Revolution, with its triumphal belief in science as the panacea for all ills. Publications, theses, and essays continue to appear arguing passionately for one side or the other. Einstein or Bergson? Scholars continue to disagree about what the crux of the debate itself is. While some argue that experiments have decisively proved the physicist correct and the philosopher wrong, others insist that these issues are simply *not* up to experimental verification. This difficulty is in part understandable because arguments from both sides have shifted and evolved throughout a long period. New questions have been posed and new answers have been put forth. From 1922 to the time he died, Bergson presented and represented his case in myriad different forms. Relativity theory itself also changed radically from its first formulation in 1905 as experimental work increased, as it was expanded by Einstein into the broader general theory, and as it was slowly assimilated by a new generation of scientists.

Was the theory of relativity science, philosophy, or both? At the time of the debate, science and philosophy occupied a completely different place in society than they do now.[5] Science was not always a nice word. During the period of the French Revolution, the term *scientifiques* was used disparagingly by the revolutionary Jean-Paul Marat. He used it to sideline those involved in what he considered to be a wasteful, self-serving project to measure a portion of the Earth's circumference and use this measurement to define the meter.[6]

The term "scientist" started to be used more frequently—and started acquiring positive connotations—in the 1830s, when it was invoked to replace the previous designation of "natural philosophers." The poet Samuel Coleridge, at a meeting of the recently formed British Association for the Advancement of Science, complained that the term "philosopher" was "too wide and too lofty" for contemporary students of natural knowledge and sought a word that would deny them that "lofty" title. The mathematician William Whewell responded by proposing the term "scientist." With this neologism, "natural philosophy" would be more clearly distinguished from "philosophy," which would increasingly only mean moral, political, and metaphysical philosophy. Whewell gave further currency to the term "scientist" in his *Philosophy of the Inductive Sciences* (1840), where he also coined the term "physicist" to describe studies of "force, matter, and the properties of matter."[7]

It is also only recently that science became seen as more directly attached to truth than to philosophy. In the thirteenth century, Thomas Aquinas believed that theology was the highest of all sciences.[8] In 1750 the famous Enlightenment writer and author of the *Encyclopédie*, Denis Diderot, still claimed that the words "science" and "philosophy" were "synonyms."[9] In the aftermath of the debate between Einstein and Bergson, those words appeared more distinct than ever, close to being antonyms. Bergson's philosophy was considered by many to simply be antiscience.

Bergson was used to accusations that he was against science; they had started years before his debate with Einstein. He defended himself against the charge, saying that his work had "no other object than to bring about a rapprochement between metaphysics and science and to consolidate each by means of the other without sacrificing anything in either, after having in the first instance clearly distinguished them from one another." The accusation that he was against science, he insisted, was completely off: "Where, when in what terms have I ever said anything of that sort? Can anyone show me, in all that I have written, one line, one word, which can be interpreted in this way?"[10]

So how did Bergson differ with Einstein? Was he wrong because he did not cover or understand the general theory of relativity? If we examine all of Bergson's publications on the topic and not only those that are usually selected, we can see how the philosopher expanded his views and eventually explained how they would apply in a case that considered the general, and not only special, theory of relativity. He also showed how they fit perfectly well with all experimental facts that had been observed at the time. He eventually clearly stated that he wholly accepted the relativistic effects of *time dilation*—under certain conditions.

The philosopher André Lalande, one of the founders of the Société française de philosophie, was unusual in that he considered Bergson's arguments as a whole and across all of his publications. In contrast to most others who followed the debate, he did not limit himself to citing the usual out-of-context quotes used to prove Bergson wrong. He summarized the disagreement between the two men on the question of time as follows: "The chief question here, of course, is to know what sort of *reality* should be accorded to the various opposed observers who disagree in their measurement of time."[11] Another follower of the debate drew a similar conclusion: "Bergson admits all of the results of relativity, but only refuses to accord them the same real value."[12] Many commentators accepted Einstein's theory and its consequences but were loathe to ascribe an equal reality to the discordant or dilated times it described. Bergson was concerned with the questions of *how*, *why*, and *under what circumstances* should the clock-delays described by relativity theory be unambiguously considered as real temporal changes. Instead of taking in relativity theory lock, stock, and smoking barrel, his project was "therefore a question of allotting shares to the real and the conventional."[13]

The question of *how reality was bestowed* on certain effects and not on others was indeed at the crux of Bergson's philosophy. For him, the line between the real and the unreal could shift across time and history. For Einstein, it should not.[14] For many of the thinkers involved, the debate over relativity theory was *not* merely technical. It was about what importance should be accorded to the different times that appeared in the theory's equations and about their relation to our everyday common notion of time. Bergson protested against the theory's "authority for equating every system [as real] and declaring all times of equal worth."[15] Bergson refused to grant Einstein the authority to do this.

BERGSON'S TIME: OUTSIDE THE CLOCK

In his first important book, which was published before he had even reached the age of thirty, Bergson developed a philosophy of time that explicitly dealt with it in a way that differed from how it was usually considered. "The time that the astronomer uses in his formulas," he explained, "the time that clocks divide in equal parts, that time, one can say, is something else."[16] Could he be right? Could this young man have discovered an essential aspect of time that differed from how it was measured by clocks and used by scientists? As his work continued to grow, Bergson expanded his initial speculations to become an undisputed authority on the topic. Time, he argued, was not something out there, separate from those who perceived it. It did not exist independently from us. It involved us at every level.

Bergson found Einstein's definition of time in terms of clocks completely aberrant. The philosopher did not understand why one would opt to describe the timing of a significant event, such as the arrival of a train, in terms of how that event matched against a watch. He did not understand why Einstein tried to establish this particular procedure as a privileged way to determine simultaneity. Bergson searched for a more basic definition of simultaneity, one that would not stop at the watch but that would explain why clocks were used in the first place. If this, much more basic, conception of simultaneity did not exist, then "clocks would not serve any purpose." "Nobody would fabricate them, or at least nobody would buy them," he argued. Yes, clocks were bought "to know what time it is," admitted Bergson. But "knowing what time it is" presupposed that the correspondence between the clock and an "event that is happening" was *meaningful* for the person involved so that it commanded their *attention*. That certain correspondences between events could be significant for us, while most others were not, explained our basic sense of simultaneity and the widespread use of clocks. Clocks, by themselves, could not explain either simultaneity or time, he argued.

If a sense of simultaneity more basic than that revealed by matching an event against a clock hand did not exist, clocks would serve no meaningful purpose:

> They would be bits of machinery with which we would amuse ourselves by comparing them with one another; they would not be employed in classifying events; in short, they would exist for their own sake and not serve us. They would lose their raison d'être for the theoretician of relativity as for everybody else, for he too calls them in only to designate the time of an event.

The entire force of Einstein's work, argued Bergson, was due to how it functioned as a "sign" that appealed to a natural and intuitive concept of simultaneity. "It is only because" Einstein's conception "helps us recognize this natural simultaneity, because it is its sign, and because it can be converted into intuitive simultaneity, that you call it simultaneity," he explained.[17] Einstein's work was so revolutionary and so shocking only because our natural, intuitive notion of simultaneity remained strong. By negating it, it could not help but refer back to it, just like a sign referred to its object.

Bergson had been thinking about clocks for years. He agreed that clocks helped note simultaneities, but he did not think that our understanding of time could be based solely on them. He had already thought about this option, back in 1889, and had quickly discounted it: "When our eyes follow on the face of a clock, the movement of the needle that corresponds to the oscillations of the pendulum, I do not measure duration, as one would think; I simply count simultaneities, which is quite different."[18] Something different, something novel, something important, something outside of the watch itself needed to be included in our understanding of time. Only *that* could explain why we attributed to clocks such power: why we bought them, why we used them, and why we invented them in the first place.

The philosopher described how change was all around us, yet—paradoxically—most scientists downplayed this aspect of the world. Even the theory of evolution (in the standard interpretation due to Herbert Spencer) considered the production of new evolving forms in terms of the recombination of old material. In so doing, it ignored the emergence of novelty into the scene of life. By focusing exclusively on the real world as eternally fixed, one could lose sight of new possibilities: "Let us say that in duration, considered as a creative evolution, there is the perpetual creation of possibility and not only of reality."[19] What would happen if thinkers across the world embraced the radical change described by Bergson? Bergson was well aware of the consequences. For one, scholars would have to temper their attempts to know the world solely through its material make-up. Materialism, a philosophical outlook closely associated with the philosophy of René Descartes, appeared at risk. Einstein, argued the philosopher, was blindly following Descartes's footsteps. Bergson ended his controversial book with the stark sentence: "Einstein is the continuator of Descartes."[20]

Bergson knew fully well that the Cartesian metaphysics, which he saw underlining Einstein's work, was extremely attractive, yet he believed it led to a number of paradoxes and contradictions both technical and ethical. He was hardly the first (Bergson compared his work to the seventeenth-century criticisms by the theologian Henry More) or the last to point out its deficiencies. Descartes had inspired many to adopt a seductive mechanistic philosophy that considered our bodies to be just like machines. His own philosophy, centering on the connection between material and immaterial realms and detailing how differences between body and spirit were not absolutely fixed but rather shifted throughout time and history, solved some of its problems. He had been widely celebrated for advancing a new solution. Was Einstein oblivious of Bergson's successes? Was he even aware of them?

You might think, argued the man who would confront Einstein, that the words written on this page are simple material stimuli that set off ideas in your mind. Well, you are wrong: if we change or omit most of the letters here, you will most likely still be able to read the text. Readers typically recognize only eight or ten letters from about thirty or forty of them, filling in the rest from memory, he explained. Reading this very sentence would not be possible without an "exteriorization" of your memories, which arguably mix in with the letters on this page. Mind and matter, he explained, join right here and now on this page, on every page. They join every time we read a clock or recognize an image, no matter how simple. But if matter could not be separated from mind, on what grounds could scientists distinguish their work from those of humanists?

Enlightenment notions of selfhood appeared equally fragile. "Our personality," he explained, "changes without ceasing." "We are creating ourselves continually," and that produced new opportunities.[21] The past—including our memories and historical sense—as well as the future and our sense of it, changed with time. So did ethics. What could prevent us from revising some of our sturdiest taboos some time in the future? Bergson combated the tendency to think of the past and of our memories as something that could never be changed by reversing the way they were commonly understood. If our idea of the past was rethought as that which we could no longer act on, then by changing our actions, we could reach back into the past. "The past," he insisted, "is essentially that which *no longer acts*." Our perception of the world was not, as commonly thought of, merely contemplative and disinterested, rather it was already shaped by our memories. Both were defined by our sense of what we could act on. Bergson warned his readers that unless they acknowledged the active role played by memories, they would inevitably come back to haunt them: "But if the difference between perception and memory is abolished . . . we become unable to really distinguish the past from the present, that is, from that which is *acting*."[22] The distinction between the past, the present, and the future was determined physically, physiologically, and psychologically. By tugging away at the concept of time as a thread running through physical astronomy to moral philosophy, Bergson questioned some of the most standard tenets of his era. Even the very possibility of determining how old we were getting could not be absolutely determined for "aging and duration belong to the order of quality. No work of analysis can resolve them into pure quantity."[23]

Einstein's theory of time, argued the philosopher, was particularly dangerous because of how it treated "duration as a deficiency." It prevented us from realizing that "the future is in reality open, unpredictable, and indeterminate." It eliminated real time; that is, "what is most positive in the world."[24]

EINSTEIN'S TIME: PSYCHOLOGY OR PHYSICS

Months after the debate, Einstein wrote a letter to Lord Haldane, who had authored an important book on the theory of relativity. He had "received the Bergson book and read part of it but have not yet been able to make up my mind about it finally."[25] Later that fall, he finally found time to examine it more carefully.

Einstein brought Bergson's book with him to read during his long trip from Marseille to Japan. The day the ship exited the harbor he started reading. A "major racket" woke him up early the next morning because the crew was scrubbing down the vessel. He jotted down some quick notes about it in his travel journal: "The philosophers constantly dance around the dichotomy: the psychologically real and physically real, and differ only in evaluations in this regard." He considered them as cyclically caught in an eternal debate between idealism and materialism. Acknowledging that Bergson had fully "grasped the substance of relativity theory," he failed to see how Bergson's views about time were precisely designed to move beyond the very two categories into which he pigeonholed the philosophers' métier. Bergson's contribution "objectivized" *psychological* aspects of time, he wrote.[26]

Bergson did not think that Einstein had ever understood him. He once confided to another friend that Einstein could not comprehend him because "he is not that familiar with philosophy and especially with the French language."[27] Einstein had perhaps not even "read my book," relying on secondhand accounts of "this or that French physicist who did not understand me, and who, not having the philosophical background needed to understand me, would remain impervious to my explanations."[28]

During the debate, Einstein explicitly stated what he held to be the purpose of philosophy and why it should *not* play a role *at all* with respect to time. In the face of his contradictor, he gave to philosophy a very limited role. He proceeded to explain himself. He mentioned two common ways of thinking about time, psychological and physical. Psychological time was the time perceived by a person, while physical time was time as measured by a scientific instrument, such as a clock. Time as measured by an instrument was often different from time perceived by a person. Factors such as boredom, impatience, or simple physiological changes affected psychological perceptions of time. With the spread of timekeeping devices, the difference between time *felt* and time *measured* became increasingly noted. We know, for example by reading the diary of Franz Kafka, that in intimate accounts of that period, an "inner clock" often seemed to disagree from an "outer one."[29]

But in most cases, physical and psychological conceptions of time did not have to differ too much. Most people could estimate time in a manner that accorded pretty well with that of a clock, determining very precisely the time for breakfast, lunch, and dinnertime. Most people could also judge if two events were simultaneous in a way that accorded pretty well with simultaneity as measured by instruments. But the opposite was true when dealing with very fast events. In these cases (such as during the finish of a horse race), the deficiency of perceptions of simultaneity when compared to simultaneity as determined by an instrument was clear; these determinations differed significantly from those determined with instrumental aides. In a universe marked by events occurring close to the speed of light, the difference between the two was extreme.

According to Einstein, philosophy had been used to explain the relation between psychology and physics. "The time of the philosopher, I believe, is a psychological and physical time at the same time," he explained in Paris.[30] But relativity, by focusing on very fast phenomena, had shown just how off-the-mark psychological perceptions of time really were. Psychological conceptions of time, Einstein insisted, were not only simply in error, they just did not correspond to anything concrete. "These are nothing more than mental constructs, logical entities."[31] Because of the enormous speed of light, humans had "instinctively" generalized their conception of simultaneity and mistakenly applied it to the rest of the universe. Einstein's theory *corrected* this mistaken generalization. Instead of believing in an overlapping area between psychological and physical conceptions of time (where both were important although one was admittedly less accurate than the other), he argued that they were really two distinct concepts: a mental assessment (the psychological one) that was wholly inadequate when compared to the "objective" concept: physical time.

Bergson and Einstein accepted that an essential difference existed between psychological and physical conceptions of time, yet they made different deductions from this. For Einstein, this led him to conclude that "the time of the philosophers does not exist, there remains only a psychological time that differs from the physicist's."[32] For Bergson this lesson—that psychological and physical assessments of time were different—made, on the contrary, the philosopher's task even more interesting, especially because no one, not even physicists, could avoid the problem of relating time back to human affairs.

QUESTIONING THE STATUS OF SCIENCE: PHENOMENOLOGY AND THE "CRISIS OF REASON."

"I do not believe that there are any longer any *philosophical* problems about Time," explained the philosopher and mathematician Hilary Putnam in 1967.[33] Scientists would most likely still agree with his assertion. Philosophers would not. As Bergson was increasingly sidelined, the authority of science with respect to questions of time ascended to new heights.

More than three decades after the debate, the philosopher Maurice Merleau-Ponty wondered whether "it is to science alone we must go for the truth about time and everything else."[34] The founder of French phenomenology was secure in Bergson's chair at the Collège de France. Following the tradition of paying homage to his predecessors, the new professor discussed Bergson during his inaugural lecture. As a close friend of Jean-Paul Sartre and Simone de Beauvoir and part of a group of intellectuals who gathered at the café Les Deux Magots in Saint-Germain-des-Prés, and who coedited the influential journal *Les Temps modernes*, he continued to refer to the philosopher throughout his life.

Merleau-Ponty took us once again to that meeting: "On April 6, 1922, Einstein met Bergson at the Philosophical Society of Paris." Merleau-Ponty explained how "Bergson had come 'to listen.' But as he was arriving the discussion flagged."[35] Einstein's science, if seen through the lens of Bergson's philosophy, he argued, led to a veritable "crisis of reason." The philosopher's lectures of 1955–1956 centered on Bergson's challenge to Einstein's theory of relativity.[36]

The by-then-common interpretation of the debate, where Einstein appeared victorious over Bergson, had led to a dangerous outcome, according to Merleau-Ponty. Now an all-pervading scientism overruled experience: "The experience of the perceived world with its obvious facts is no more than a stutter which precedes the clear speech of science."[37] Merleau-Ponty continued to write about Bergson. In 1959 he delivered a talk at the end of the "Bergson Congress," which included presentations by Gabriel Marcel, Jean Wahl, and Vladimir Jankélévitch.[38] Throughout his career, he sought to reintroduce embodied perception back into theories of knowledge, inspiring a generation of scientists, writers, and artists to come. While scientists often spoke of lines and circles, in actual life we never encountered these shapes in that perfect geometrical way, he argued. The same was true for measurements of time. By excluding our actually perceived environment, he continued, modern science had lost touch with reality. What would science look like if it reintroduced the world as seen, heard, and felt? For decades, he dedicated himself to answering these questions.

While for many prewar authors Bergson's philosophy was dangerous, in the postwar period Merleau-Ponty saw a potent danger in the unbridled rationalism of his era: "Not counting its neurotics, the world includes a good number of 'rationalists' who are a danger for living reason." Regaining "living reason" for Merleau-Ponty did not mean abandoning science, but it meant giving a renewed place to philosophy within science: "And reason's vigor is on the contrary bound to the rebirth of the philosophical sense which will of course justify scientific expression of the world, but in its proper order and place in the whole of the human world."[39] Although he was never an enemy of science and frequently inspired scientists (especially neurophenomenologists), Merleau-Ponty nonetheless tried to put cold rationalism back in its "proper place."

Why, asked Merleau-Ponty, did everyone turn to physics and to physicists for answers? Why were they even consulted as public intellectuals about everything from fashion to government? Merleau-Ponty mocked the

> extravagances of journalists who consult the genius about questions which are most alien to his field. After all, since science is thaumaturgy, why should it not perform one more miracle? And since it was precisely Einstein who showed that at a great distance a present is

> contemporaneous with the future, why not ask him the questions which were asked of the Pythian oracle?[40]

In the 1950s and '60s the debate between Einstein and Bergson was as relevant as ever. "Today, as thirty-five years ago, physicists reproach Bergson for introducing the observer into relativistic physics, which, they say can make time relative only with instruments of measurement or a system of reference," explained Merleau-Ponty.[41] But the observer, according to him, should never be made irrelevant; instruments by themselves would never entirely demystify time.

The pendulum swung rapidly in the 1960s. The place of "reason" shifted from being narrowly associated with science to becoming a closer ally of philosophy, as many thinkers shied away from an initial fascination with Einstein and gravitated toward Bergson. In the *Phenomenology of Perception*, Merleau-Ponty insisted on the importance of our individual assessments of time. To stress how time was dependent on embodied consciousness and how it was not a mere physical quantity of a disembodied universe he exclaimed: "I am myself time."[42] He, after all, had learned from Bergson that "we do not draw near to time by squeezing it between the reference-point of measurement as if between pincers." On the contrary, "in order to have an idea of it we must on the contrary let it develop freely, accompanying the continual birth which makes it always new and, precisely in this respect, the same."[43]

Was philosophy limited to studying the "stutter which precedes the clear speech of science"?[44] Should philosophers accept the new role of post-war science? Einstein's supporters would largely answer affirmatively, but back in Paris a new generation of young writers would not accept this limited role. Phenomenologists were not the only ones concerned about the role of philosophy in a century marked by the rise of science.

PART 2

THE MEN

CHAPTER 4

The Twin Paradox

BOLOGNA, ITALY

When did Bergson first learn of Einstein's work? In the spring of 1911 "under the protection of the king of Italy," world-renowned scientists and philosophers, including Bergson, met in Bologna during the Fourth International Congress of Philosophy.[1] Bergson's reputation was at its peak.[2] Conference participants could hardly wait to hear him speak. But a competing session dealing with a "paradoxical fact" in Einstein's work—delivered by a junior and largely unknown scientist—captured the imagination of many of the attendees.[3]

The presenter, Paul Langevin, asked members of the audience if "anyone among us" would want to "dedicate two years of his life to find out what Earth would look like in two hundred years."[4] All that a willing volunteer needed to do, Langevin continued, was to travel to outer space at a speed close to that of light. Easy, right? Langevin delivered this question not as a peddler of dreams and fantasies, but rather as a pure and honest physicist. If someone did indeed agree to the quick trip, he argued, they would come back to find out that time on Earth had passed more rapidly. They would see the world two hundred years later. "The most definitely established experimental facts of physics can permit us to affirm that this will be the case," he confidently stated.[5] Bergson, we are told, was seething in the audience, already getting ready for a fight.[6]

For those who did not consider time travel exciting enough, Langevin offered something else. He promised his listeners eternal youth: "One could now say," he claimed, "that it is enough for us to get agitated, to become accelerated, in order to age less rapidly." When Einstein learned about Langevin's presentation, he at first considered it "the thing at its funniest."[7] But soon afterward, he started considering it very seriously and dedicated himself to exploring this aspect of his own work.

Bergson was not amused. In Bologna, the philosopher delivered "L'intuition philosophique," one of his most famous lectures, but the direction of his thought would soon change in light of Langevin's presentation of Einstein's work. It would take him almost a decade to craft a response.

Langevin's presentation was simply brilliant. With a mix of philosophy and science, and with key references to popular science fiction stories (of Jules Verne), it captured the imagination of an eager public. It was even more successful than the earlier talk by the eminent Henri Poincaré, another contributor who refused to see in the theory of relativity the same revolutionary potential.[8] Langevin's talk, which did not even refer to Poincaré, was quickly published in *Scientia* and a summary of it appeared in the *Revue de métaphysique et de morale.*

The international congress at Bologna that year was a complete success. Its numbers had ballooned from a mere 150 when it started in 1900 to between 500 and 600. The most brilliant minds of the era were there, exchanging "personal introductions" and engaging in opportune, although "informal discussions in the corridors."[9] Both scientists and philosophers attended. The philosophers were proud to be able to meet in specialized congresses as scientists had long done and to greet them as esteemed colleagues. "Philosophers," noted one attendee, "can meet together as men of science have long been accustomed to do, and can regard philosophy as a body of knowledge which, like science, is advanced, grows and progresses."[10] In this way, they might be able to jump on the bandwagon of "scientific progress." Yet they were also proud to do even more than scientists. At that congress, Bergson agreed with his teacher Émile Boutroux, who gave a talk "The Relation of Philosophy to the Sciences," that "while science regards things as purely objective, as dehumanized . . . , philosophy insists on regarding them in connection with the aspiration and will of man." For this reason, added Bergson, science was typically "defiant" toward reality whereas philosophy was "sympathetic" toward it.[11]

Langevin's presentation stole the show. After Langevin's presentation in Italy, Einstein's work suddenly appeared much more interesting and amusing than it ever had been—even by Einstein's own standards. Einstein was thrilled. His affinity with Langevin was such that in a letter written to him shortly before his 1922 trip, he could not conceal his excitement: "I rejoice like child to think that soon I will be able to stroll in the streets of Paris once more with you."[12]

LANGEVIN RETURNS FROM BOLOGNA, 1911

After learning about Langevin's success, philosophers in Paris immediately called on the scientist as soon as he returned home. They wanted to scrutinize his words on *their* ground, at the Société française de philosophie, the selfsame forum that would host Einstein and Bergson a decade later.

Einstein's 1905 paper was hardly a reason for concern for anyone in France during these years, but the presentation that Langevin had just delivered in Italy caused a significant stir. Many in the public and beyond started to speculate how the physicist's theory would affect Bergsonian philosophy.

Langevin was one of the first scientists in France to espouse Einstein's theory. After learning about it, he quickly became "the apostle of the new gospel."[13] His involvement with relativity theory was so thorough that Einstein, at the time of Langevin's death, even claimed that his friend would have in all likeliness developed it himself had others, including Einstein himself, not done it: "It seems to me certain that he would have developed the special theory of relativity if that had not been done elsewhere, for he had clearly recognized the essential points," he explained.[14] Throughout his life, Langevin defended Einstein with what critics considered a "bitter zealousness."[15]

Langevin and Einstein were close friends and had close political affinities. During an unforgettable trip to Paris in 1913, Einstein became close with others in Langevin's circle of friends, most notably with Jean Perrin and Marie Curie. Einstein owed the invitation to come to Paris in 1922 to Langevin. One year later, to reciprocate Einstein's acceptance, Langevin went to Berlin to demonstrate at a pacifist rally in Berlin, posing prominently for photographers alongside Einstein.

Langevin and Bergson worked at the prestigious Collège de France, where Langevin remained under the shadow of the towering philosopher.

Langevin entered the Collège de France in 1902 as a substitute and earned a permanent position there in 1909. One student of the Collège de France during the years that Langevin taught there referred to the institution as "the house of Bergson," while commenting that even those who were "studying science did not know the name of Langevin."[16] The differences between Langevin and Bergson eventually became as stark as those between Einstein and Bergson. The Langevin-Bergson conflict underlined the Einstein-Bergson clash.

Alongside Einstein, Langevin became a member of the International Committee on Intellectual Cooperation led by Bergson. Langevin allegedly had an affair with Marie Curie after she became a widow, and together they formed part of a select group of politically engaged French scientists who became increasingly attracted to leftist politics.[17] Langevin was close to Georges Politzer, who wrote the most virulent book-length attack against Bergson, titled *La fin d'une parade philosophique: Le Bergsonisme* (1929), using for a pseudonym Voltaire's real name, François Arouet. One of the driving messages of Politzer's book was to show the incompatibility between Bergsonian philosophy and science: "One could confront against science the Bergsonianism that pretends to be based on scientific results, and to show how distant scientific bases and Bergsonianism are from each other."[18] Langevin and Politzer founded the journal *La Pensée*, advocating what they called modern rationalism to cure the excesses of Bergsonian and other forms of philosophizing. Both men suffered dearly during the German occupation of France and under the Vichy government. Politzer was shot in 1942. His wife and Langevin's daughter, Hélène Solomon-Langevin, were sent to the Auschwitz concentration camp in the famous "convoi de 23 janvier." None of this silenced Langevin, who in 1944 joined the Communist Party.

While in Bologna, Langevin discussed the consequences of relativity theory for a "voyager on a rocket ship." His account is widely considered to be the first original formulation of the thought experiment initially known as Langevin's paradox and later baptized as the "twin paradox." The common way the paradox is usually presented can be summarized as follows. According to the theory of relativity, two twins (one which traveled in outer space at speeds close to the speed of light, and another one who remained on Earth) come back to Earth to find that time slowed down for the twin who had traveled. The twin who stayed would have aged more rapidly; the traveling twin would be younger. Their clocks and calendars would show different dates and times. Although it was formulated as a thought experiment, many scientists started to ask if the various effects described and predicted by Einstein's theory showed that the paradox would indeed take place.

Langevin's original publication did not talk about twins or use the common names later given to them by Bergson of Peter and Paul; rather, he simply described a single "voyager" taking off from Earth in an imaginary rocket and only imagined what would happen when he returned. Einstein's "special" theory of relativity initially only dealt with a situation in which a "voyager" would depart and did not completely explain his return. The twin paradox took time to reach this traditional, often-repeated form.

What would happen if the voyager eventually returned to Earth? This question took years to be resolved. At first its full implications escaped even Einstein. After Langevin's intervention Einstein increasingly started to ask if the delay in the time marked by a clock would also affect biological—and not only physical—processes. In an unpublished manuscript written after Langevin formulated the paradox, Einstein started to take it more seriously.[19] He started to consider these delays as more than just affecting clocks, claiming that relativity theory described "the temporal course of no matter which process."[20]

IN PARIS

A few months after his brilliant presentation in Italy, Langevin was called to the Société française de philosophie to speak to an audience composed mainly of philosophers but also attended by illustrious scientists. The philosopher Abel Rey was the first to comment after listening to Langevin's presentation in Paris.

Rey immediately speculated, what would Bergson think? He explained how Bergson in his original theory had overturned "the tradition of Aristotle, Descartes, Leibniz, and even Kant," whose theories of time all (according to him) considered it in terms of spatial distances. Einstein's theory had taken "much farther than had ever before been done, the notion of a parallelism between space and time." Bergson, most likely, would probably reject this parallelism. "True, Bergson has the right under his system to not see anything there [in Einstein's theory] other than a new effort from the part of science to spatialize time." But maybe Bergson would draw another conclusion as well. Perhaps, speculated Rey, the philosopher would be even more bothered by Einstein's claim that absolute space and time did not exist—no matter how useful these concepts could be and how sought after they were by scientists. Since Bergson's philosophy considered science as an extraordinarily successful technique for dealing with the "material world," how would the philosopher then explain the claim in Einstein's theory that one could never know which one of the times of competing systems was correct and thus should consider them both as equally valid? Bergson "could be reluctant to see in the Universe, as in knowledge, unconquerable levels [*plans irréductibles*]," surmised Rey.[21]

In Paris, Langevin's presentation to the Société seemed preposterous to many of those who were listening. The attendees' incredulity centered on Langevin's claim that the effects of relativity described by Einstein would also affect biological beings *and* psychological processes and that, therefore, they would affect "the common conception of time." Langevin was bold at first, insisting that the effects on time predicted by relativity theory affected both mechanical and biological processes: "The principle of relativity consists in admitting that even if other means (mechanical, biological, etc.) could be brought to a level of precision comparable to the first [measures of time and space made by optical and electromagnetic means], they would also furnish the same results."[22] He firmly believed "it is therefore necessary, from the point of view of the principle of relativity, that all mechanical, electrical, optical, chemical and biological processes employed for measuring . . . time lead to concordant results."[23]

Langevin's choice of words was also controversial. He described time dilation by saying that "of two clocks, one grows older than the other," arguing that the equivalence between physical and biological processes he described were "very probably" accurate. After hearing some of his colleagues balking at the seamless expansion of the theory from the realm of the physical to the biological, he concluded with the evocative statement, "but we are ourselves clocks."[24]

But wait, are we really just like clocks? The philosopher Brunschvicg was one of the first to disagree.[25] He reminded the attendees that for

Langevin's hypothesis to be correct, scientists still needed to prove that biological processes underwent the same temporal transformations as physical ones: "It still must be established that the life of the clockmaker is linked to the movement of the watch, and that biological or physical phenomena depend on physical phenomena that are used to measure time."[26] And even if one could accept a certain link between clock processes and biological ones, scientists should not forget that humans themselves made clocks in the first place, and that these would simply not exist if humans had not made them: "You are not only one of those clockmakers linked to a clock, you are a maker of clocks."[27] We are not clocks, argued Brunschvicg. We are clock *makers*. Brunschvicg underlined the inventive, productive, and ever-changing qualities of humans over their predictable, mechanical ones. He also stressed their will to power.

If stubborn, self-centered, or chauvinistic enough, one traveler could still claim that only *his* clock showed the real time. Could power dynamics between the two travelers play a role? Brunschvicg brought into the discussion the issue of "domination," remarking that physicists should not forget that the "observer" of relativity theory "would *want to dominate* the diverse groups of observers, who were incapable of bringing the clocks into agreement, instead of being confused among them."[28] Brunschvicg, in short, did not admit that Einstein had started a revolution.[29]

After Brunschvicg's comments, others jumped in, criticizing how Langevin described clocks as "aging" and "growing old." "But so be it! Call it aging, if you want, the acceleration of the hands of a watch," expressed one exasperated member. The physicist Jean Perrin, a strong supporter of Einstein and a friend of Langevin, added with irony: "When physicists say 'aging,' that is one word I especially like."[30]

ÉDOUARD LE ROY: SCIENCE, RELIGION, AND MAGIC

The "Bergsonian" philosopher, Édouard Le Roy, a man who would later be responsible for bringing in Bergson into the discussion with Einstein, was already listening attentively.[31] Ending his long silence that day, he offered to help the attendees move beyond their disagreements. "Permit me to adopt for a moment the role of interpreter," he politely intervened.

Le Roy had a brilliant idea to solve the impasse: Why not simply use different terms for what physicists were referring to and for what philosophers meant when discussing time? Why not use "hour" for the time of physics, and "time" for that of philosophy? In this way, Le Roy aimed to set boundaries and curb the philosophical pretensions of physicists such as Langevin.

After listening attentively to the objections of Le Roy, Brunschvicg, and others that day, Langevin retreated. He qualified some of the most dramatic conclusions of his Bologna lecture. Modestly, he admitted that he did "not have the pretension of speaking from the point of view of a philosopher." These issues, he explained, were really up to them to sort out: "It is up to the philosophers to say which are the elements of the notion of time that must be modified."[32]

The differences between Langevin and Le Roy widened with time. Langevin's relationship to Einstein strengthened in direct proportion to his distance from Le Roy and other "Bergsonian" philosophers. Le Roy and Bergson, in contrast, grew closer. After years substituting for Bergson at the Collège de France, Le Roy obtained his chair in 1921. What role did Le Roy play in creating and cementing further divisions between Einstein and Bergson? One important author to become embroiled in the Einstein-Bergson debate described Le Roy as being "at the origin of Bergson's error."[33]

Le Roy was more Catholic than the Church and more Bergsonian than Bergson. His fascination with Bergson started when he read *Matter and Memory* (1896). He compared Bergson to Socrates, arguing that both had revolutionized our theory of knowledge to the same degree. In works that followed, he attributed to science an enormous, but not an entire, role in human affairs. Science gave us "the schematic pattern of the world and its elements," but it was also important not to lose sight of "the specific, the concrete, and the living." He considered himself to be a person who had a "love of positive science, but who could not resign himself to sacrifice the richness of thought, the representation of the unity of knowledge, and the mutual relations between different orders of inquiry."[34] He became a key member of Catholic Modernism, an influential reform movement that started within Catholicism but that was eventually condemned by the Church as too radical. The article that first turned the Church against Le Roy was titled "What is a dogma?"

In that article, he argued that reason itself sufficed for understanding Christ. Because of this article, he was eventually placed on the margins of Catholicism. Le Roy started to diverge from Catholicism by focusing even more intently on life and the living. Official Catholic philosophy at that time was dominated by Thomist rationalism, whose main principles were derived from St. Thomas Aquinas' reinterpretation of Aristotelian philosophy. Against St. Thomas, Le Roy invoked St. Paul, who focused intently on love and life; against Descartes, he backed Pascal, a critic of cold rationalism; and against Einstein he defended Bergson.

In *Duration and Simultaneity*, Bergson developed one of the ideas brought up by Le Roy during his first meeting with Langevin at the Société française de philosophie a decade earlier. He found extremely agreeable the option of finding other words for the time concepts used by scientists.[35] When discussing how Einstein used "this common word [simultaneity] in both cases," Bergson described it as a trick used by the physicist so that science could "operate magically." It was his duty as a philosopher to note the difference between Einstein's use of the term and its everyday meaning. He urged scientists to "invent another word for it, any word."[36] Exasperated by how physicists used concepts related to time in tricky, dual senses, he asked: "Does not science act on us like ancient magic?"[37]

In the 1950s Le Roy compiled Bergson's collected works. He decided not to include *Duration and Simultaneity* in the collection, leaving the impression in the eyes of many that Bergson himself did not endorse the book. But that was hardly the case.[38] Bergson never recanted a single word he had written or said about Einstein's theory of relativity.

CHAPTER 5

Bergson's Achilles' Heel

When Bergson first entered into the fray in 1922, he insisted that only one of the two clocks' time was "real" while the other was "fictional." The two times, he argued, *could not* be compared because one of them was the exact mirror image of the other. They were "absolutely reciprocal."

The philosopher's refusal to understand both times in equal terms would become his Achilles' heel—the reason why legions of readers would brand him as not having understood the theory of relativity. Time dilation can be explained solely by using special relativity.[1] Yet the case of a returning twin was much more complicated, because scientists needed to introduce a change in direction, and therefore acceleration, into their theories.

Einstein's theory of relativity did not at first deal with acceleration or with a change of direction. It covered only movement that was uniform and linear. It was only after the "special" theory of relativity was expanded into the "general" theory of relativity that Einstein started to seriously consider the case of a returning twin.

In 1905 Einstein imagined what would happen if a traveling clock and a stationary one would meet again. One of them, he surmised, would be behind the other. "The clock that has been transported from A to B will lag $\frac{1}{2}tv^2/c^2$ sec," he explained.[2] Yet Einstein soon realized that this would require one of them to be accelerated. His theory, before it was expanded into the general theory, did not account for acceleration.

Einstein was ambitious and quickly started working on the problem of acceleration.[3] At first, this new work faced the risk of undermining his early work on relativity, but Einstein continued to plow ahead with the hope of developing a new, more comprehensive theory. His efforts paid off, and by 1915 he had a system that included both acceleration and gravitation. What is more, this general theory did not undermine his earlier work. It confirmed it as a perfectly valid, albeit simpler, incomplete case. His special theory was still correct, even when considered separately from the general theory. But the general theory was even more controversial than the special one. In the general case, there was no difference between gravitation and acceleration. The effect of being pushed to the floor when an elevator ascended was of the same nature as the effect of being pushed to the ground everyday by the Earth's gravitational pull. If a person was inside a spacious chest resembling a room hooked to a string which could be raised or lowered, what would they feel? They would feel the force of acceleration in the same way as they felt that of gravitation. When movement was perfectly smooth and constant, they would feel absolutely nothing, but when it accelerated or slowed down, they would feel a pressure on the soles of their feet comparable to that which would be sensed given a change in the gravitational field.

Bergson, for the most part, focused on the special theory of relativity. "We remain therefore within the frame of the special theory," he explained initially, investigating movement only when it was "straight and uniform."[4] The philosopher knew fully well that this approach had its limitations, "because in the special theory there is something that demands the general theory."[5] How would he deal with the general theory? The question of a returning clock was particularly complicated because the equations for time were different in the earlier special theory than in the later general one. Which of these two equations was most relevant? When readers were asked to consider clocks not only traveling away from each other, but meeting once again, how should they understand their time? Should they use the algebraic equations of the special theory or the differential equations of the general one?* Yes, both were correct, but philosophically they each had very different meanings. Bergson asked for clarification, since they each seemed connected to two radically different notions of time.

*The equation for dilated time in the special theory is: $t_1 = t_2/\surd(1 - v^2/c^2)$. For the general theory, the equations for time are: $ds^2 = c^2dt^2 - dx^2 - dy^2 - dz^2$ and $d's^2 = c^2dt'^2 - dx'^2 - dy'^2 - dz'^2$.

In his popular work, Einstein mostly focused on the simpler theory based on algebraic equations that could be understood algebraically. The other one, requiring knowledge of advanced differential calculus and geometry, was glossed over. The time dilation equation for the special theory $t_1 = t_2/\surd(1 - v^2/c^2)$ clearly showed how one time variable decreased with an increase in the velocity of the other system. The equation for the general theory was much different, as it showed a differential for time squared (dt^2) sitting next to those for space x, y, z and amounting to a result (ds^2) that could not be easily associated with a familiar, measurable quantity. "A formula such as $ds^2 = -dx^2 - dy^2 - dz^2 + c^2dt^2$ seems to takes us outside of any reference system, to the Absolute, facing an entity comparable to a Platonic idea," explained Bergson.[6] This was quite different from how time was described by reference to the special theory. Which of the two referred to time? The first was simple and logical and referred to concrete, widely used concepts; the second one much less so.

SHOULD SCIENTISTS CARE?

Einstein's general theory remained a difficult work well into the second third of the twentieth century. By 1922 it counted a number of successes, yet it was not wholly embraced or even understood by most physicists. Certain problems remained with it that would only be solved until decades after the Einstein-Bergson debate. In contrast, Einstein's efforts at popularizing the special theory of relativity had paid off generously. Most physicists by 1922 considered it to be sound, logical, and coherent. The public at large was thrilled by it.

After Bergson was sharply criticized for considering the two travelers in relativity theory as completely interchangeable, he explained himself more fully. He had, on various occasions, described specific "conditions" under which the two times would differ in comparison to each other: "Under these conditions, the time of Paul is one hundred times slower than that of Pierre," he wrote.[7] But the discrepancy between the twin's times still did not necessarily imply that they should both be considered on equal terms. After receiving numerous criticisms from Einstein's allies, Bergson focused intensely on the implications of having the twin's clock times differ. This was most clearly stated two years after his first encounter with Einstein in "Les Temps fictifs et les temps réel" (May 1924). There he insisted that even if the twins' clocks differed, his major point still held: that philosophy had a right to study these differences. So what if one of the twins' clocks showed a different time than the other's, asked Bergson? This discrepancy did not necessarily mean that time itself became dilated and should be understood in the way that Einstein proposed.

Bergson argued in a manner consonant with how Einstein's general theory of relativity was understood at that time: that acceleration was the essential difference that produced a difference in times. Acceleration created a dissymmetry. To Bergson, this in turn proved that the two times were not equal in every sense: "So, if one wants to deal with real Times then acceleration should not create a dissymmetry, and if one wants for the acceleration of one of these two systems to effectively create a dissymmetry between them, then we are no longer dealing with real Times."[8] Acceleration was an inescapable mark of a difference in the clocks' travel itineraries. Since a difference existed, one connected to a difference in time, then their times were not equal in *every* sense. After all, one would have the extraordinary experience of having done something different—being propelled to outer space and jolted back to return to Earth—while the other one comfortably remained at home. These differences were extraordinary, he argued, and physicist had no right to brush them aside and consider both observers as dealing with one single, precious, and contested entity, time.

If the dissymmetry due to acceleration was ignored, then Bergson was ready to concede to Einstein: "one could naturally say that [clocks traveling at different speeds] cannot run in synchronicity." In these cases "in effect Time slows down when speed increases." But for Bergson the introduction of acceleration proved that the times described by Einstein were not all equally real. "But what is this Time that slows down? What are these clocks that are not in synchronicity?"[9] These clocks were not equal in every way because one of them had gone through something that the other one had not.

When observers or clocks disagreed because they traveled in different ways, how could one claim—with full certainty—that one was right and the other wrong? Could one disregard the fact that they disagreed because they traveled in different ways? Could one prevent them from being judged in terms of their different histories? Could their different trajectories, memories, and experiences be neglected? Bergson would reply no. If considered from "the social point of view," he insisted, these distinctions mattered profoundly.[10] Einstein would answer yes.

Should scientists care about these differences? To understand the case of a returning clock only under the frame of the special theory, scientists and philosophers would think up new scenarios for decades to come. The first way of thinking about it involved a case in which the two separating twins were connected to each other via electromagnetic waves. In this way the two twins exchanged time signals every step of the journey. The differing times of the twins could be compared side-by-side and step-by-step. Alternatively, they laced the journey of each clock with other coordinated clocks. Or they completely automated a whole set of clock comparisons. One of the most popular scenarios involved the introduction of a third clock—now called the three clocks paradox. Two twins could compare their clock against a third, unaccelerated traveling clock and ascertain a time difference. These new scenarios solved some problems but opened others. The plot only thickened and the debate intensified.[11]

Bergson's limited focus on the *special*, instead of on the *general* theory of relativity, could be easily seen as his main weakness. For this reason, the philosopher expanded his work to account for the problem of acceleration. Yet his clarifications on this topic were largely ignored. Because Bergson was largely considered to have been "mistaken," readers forgot one of his central messages: that philosophy had the right to study the processes that lead us to infer certain conclusions from directly ascertainable (yet limited) observations, in science and in general.

But one aspect of Bergson's critique of relativity continued to resonate for decades: that experimental results did not lead directly to Einstein's conclusions. Scientists listened to him. Influential physicists who worked on relativity theory agreed with him, including Paul Painlevé, Henri Poincaré, Hendrik Lorentz, and Albert A. Michelson. Accepting "the invariance of the electromagnetic equations" did not necessarily lead to Einstein's interpretation.[12] Admitting that no experiment could be designed to decide if one of the two times could have a special status, all of these scientists refused to accept that one of them could not be *chosen* over the other for a specific purpose, nor did it mean that nobody would *ever* be able to find a way to single out one time as unique and special sometime in the future.

THE PREVIOUS WINTER

During the winter of 1921, neither Bergson nor Einstein had met, yet some scientists and philosophers in France were already setting the stage for the debate that was to take place in the spring. Langevin and Édouard Le Roy—one the defender of Einstein, the other of Bergson—sparred publicly once again, crafting arguments that would resurface in the spring. Objections to Einstein were launched at a "great gathering" at the Sorbonne. Bergson presided over one of the sessions. The most interesting session, however, was not his, but one about Einstein, who once again was defended by his friend and disciple Langevin.

That winter, philosophers and scientists alike were already "planning a rendezvous at the lectures which Einstein himself was soon to conduct in Paris." Two clear camps, one for and the other against the scientist, were shaping up. "The meeting which attracted the greatest attention on the part of the public, and caused the most 'excitement,'" was Langevin's defense of Einstein. He argued against Paul Painlevé, one of the most renowned mathematicians in France, who used some of the same arguments that Bergson would later bring up in *Duration of Simultaneity*. Their disagreement "was for the auditors a fascinating spectacle" mainly because of the contrast between the two men involved: Painlevé's "ardent, brilliant eloquence" against "the smiling and tranquil simplicity of M. Langevin."[13]

The debate between Painlevé and Langevin enthralled the participants: "Perhaps the most animated of the general meetings held in the afternoon was that at which Relativity was the theme. Here two famous French mathematicians, Langevin and Painlevé, crossed swords, the former finding more of significance, the latter less, in the theory of Einstein."[14] An attendee described how *science* was the topic of most interest during this meeting of *philosophers*:

> Of particular interest, however, was the *séance generale* for the section of logic and philosophy of science presided over by Monsieur Painlevé

> of the Institute . . . and debated with extraordinary power and vivacity by Professor Langevin and Monsieur Painlevé. Langevin arguing in defense of the relativity theory and Painlevé arguing without compromise against it. A more brilliant occasion of this sort can hardly be imagined than this general session.[15]

Einstein followed these events in Paris from faraway Berlin. He was so concerned about them that at a dinner he attended before leaving for Paris he talked to Count Harry Kessler, a well-connected man-about-town and diplomat who had just come back from Paris. Kessler had spoken to Painlevé, and Einstein was interested in hearing all about it. He asked him "to repeat more than once and verbatim" any comments Painlevé had made to him.[16] The physicist needed to be prepared.

PAINLEVÉ AGAINST LANGEVIN

Painlevé, in addition to being a renowned mathematician, had an illustrious political career as minister of war and prime minister of France. He remained an active politician for most of his life, trying unsuccessfully to stem the depreciation of the franc.[17] He insisted that he admired Einstein as a colleague and person but did not agree with his interpretation of relativity theory. "Apropos of Einstein, Painlevé is credited with the saying that 'one may admire the skill of a diver even though one attaches no great value to the pearls which he brings to the surface.' "[18] Einstein was similarly gracious and critical at the same time.

The meeting in Paris between Christmas of 1921 and the New Year was to be followed by the one next spring, when alliances and arguments would shift once again. When the director of the Société française de philosophie learned that Einstein had accepted Langevin's invitation to lecture at the Collège de France, he saw a clear opportunity for continuing "the discussion about the most recent approaches to relativity theory . . . that were initiated during the extraordinary Christmas session," in which "our colleague and friend Painlevé" played such an important role. While Painlevé had been extraordinarily successful that winter meeting, Einstein's visit succeeded in reversing that outcome. "The attitude that one has here with regard to your theories is now completely different from how it was before," explained his friend Maurice Solovine to Einstein, a few weeks later.[19]

The Christmas meeting at the Sorbonne, although at face value a professional meeting of philosophers, had clear political overtones. It was strategically planned to strengthen the links between France and its allies. "Invitations to a meeting, to be held in Paris in December, 1921, were extended not only to the English and American societies, but also to those of the other countries which were the allies of France in the war," explained a commentator.[20] Painlevé, who by then already multitasked as mathematician, politician, and philosopher, was principally responsible for drawing links between the philosophical discussions that day and the political tasks at hand. During the closing speech, he used the occasion to defend the role of France during the Great War. "The response to these expressions of appreciation and friendship was made by M. Painlevé, member of the Institute [Institut de France], who is a gifted speaker." Painlevé wanted peace. He "made an eloquent plea for moral effort in the task of peace now confronting the world as no less imperative than was the military effort of the war."[21]

Bergson, who had defended the role of France during the Great War, supported these collaborative efforts. He presided over the afternoon session, using his fluent English to reach out to Anglo-American participants: "As the audience was overwhelmingly French, Bergson gave for their benefit a *résumé* of the points made by each speaker. He did this with an ease and precision which showed his mastery of English, which he has spoken since his childhood."[22] Bergson was active throughout the day and night. He "attended some of the social gatherings and entertained a number of the delegates at his home."[23] Einstein and Langevin, in contrast, were adamant critics of the political situation at the time and of the role of France and Germany during the war. Together they used Einstein's trip to Paris as an opportunity to publicize their views.

Painlevé had recently published two important articles on the general theory of relativity in the prestigious *Comptes rendus*. One of his followers described his contributions as accepting all of the results of relativity theory but giving it a different interpretation: "Painlevé (1921) has given alternative forms of relativity theory which agree with Einstein's in those portions which can be tested by observation, and yet disagree altogether." His reinterpretation could have important consequences, even to the point of reducing Einstein's to a fad: "If Painlevé's views can be sustained, the preference of mathematical physicists for Riemannian geometry may be only a passing phase."[24] The mathematics developed by Bernhard Riemann were useful for studying the world in terms of four dimensions and were essential for Einstein's general theory. But perhaps Painlevé might save physicists the effort of having to master them.

When Einstein finally arrived in Paris, Painlevé once again played an important role. He intervened before Einstein's debacle with Bergson and then again during the debate itself. At that time, Painlevé was deeply immersed in politics: he headed a leftist Christian political party and social movement (the Ligue de la République) and was organizing an alliance between radical socialists (the Cartel des gauches) and the International Workers' movement.[25] For this special occasion, and to the "great joy of his friends," they saw how "he abandoned politics for a few hours" to discuss relativity theory.[26]

ACCELERATION

In one of the meetings with Einstein at the Collège de France, Painlevé asked what would happen "if the train goes in reverse." Bergson was in the audience listening attentively. Among many other attendees "whose names escape me," recounted the astronomer Charles Nordmann, "and who were modestly lost in this assembly of the French intelligentsia," the "profile without end of Monsieur Bergson stood out."[27]

"What time will the clock on the train mark when it comes back to the departure station?" Painlevé asked. His question was tricky. Einstein, backed by Langevin who was "whispering answers" behind him, replied that "it will be behind."[28] Painlevé, according to one witness, thought differently: "It should mark the same time."[29]

Painlevé accepted the special theory of relativity in its entirety. He acknowledged its bulletproof logical coherence, but he was not so sure that one should accept all the premises of the general theory. "It is certain that one cannot find a logical contradiction in the special theory of relativity, but considerable difficulties arise when one passes from one inertial system to another," he told Einstein.[30]

Painlevé's comments focused on these, as yet not completely solved, difficulties. Even Langevin agreed about the problems of acceleration and return, pointing out that he had noticed them first. But for Painlevé these difficulties also proved that the special theory had to be considered in a different way. They showed that a fixed observer and a returning one were not necessarily speaking about the same thing. It showed that there was no necessary "univocal correspondence" between the two. For him, this lack of correspondence created a "fundamental dissymmetry" between the two observers that rendered all talk of "reciprocity" invalid.

Langevin replied immediately to Painlevé's comments, taking priority for himself. "I have to insist that this lack of symmetry," he told the audience, "was brought out by me in 1911 at the Bologna congress of philosophy and in my course at the Collège de France."[31] But although Langevin and Painlevé agreed about the nature of the difficulties, they disagreed about their significance. When two observers compared their clocks, and when they saw these disagree, it was because they were comparing apples and oranges, insisted Painlevé. They were simply not comparing the same thing, time. Langevin agreed there was a "lack of symmetry" between them, yet still considered both as marking time.

The question of a returning clock would remain at the center of Bergson's critique. Bergson thought that the reason why acceleration—and how it was responsible for creating the differences in times—was kept out of most presentations of relativity theory stemmed from a deeper problem. It was a kind of trick, designed to "dissimulate the difference between the real and the virtual" in Einstein's work. Its reintegration into discussions of relativity—which would open up questions about the relation of the special to the general theory—was "superfluous for the physicist, yet capital for the philosopher."[32]

In the decades that followed, Einstein's general theory received ever more support. A group of scientists working from new American large-scale observatories became "Einstein's jury," ruling in favor of the physicist.[33] This generation of researchers, who understood time in terms of the general relativity equations, had no use for concepts developed in discussions aimed at clarifying the relation of the special to the general theory. Painlevé's "univocal correspondence" and Bergson's "lived experience" were, in contrast, designed to show how differences in travel conditions created differences in time. When clocks went their separate ways, attempts to understand their differences in ways that were not directly tied to the set of concepts used by this new generation of researchers were increasingly irrelevant.[34] But Einstein's victory did not come easily—some of the most prominent scientists to work on relativity continued to side with Bergson.

CHAPTER 6

Worth Mentioning?

UNIVERSITY COLLEGE, LONDON

On May 4, 1912, two months before his death, Henri Poincaré, a renowned scientist and philosopher of science, went to London to deliver a lecture on the theory of relativity. It would be his last significant statement on the topic. Poincaré did not even mention Einstein. He did, however, mention Bergson. "The time of scientists comes out of Bergsonian duration," he explained.[1]

The relation between Einstein and Poincaré is one of the most intriguing episodes in the history of science, one shrouded in mystery and controversy.[2] Poincaré was so involved with relativity theory that many have claimed that he deserved credit for it. Einstein read Poincaré's work avidly before and after writing his famous 1905 paper, perhaps missing only a paper or two.[3] By 1921 a close friend warned Einstein how a French colleague was "claiming everywhere that your [Einstein's] discoveries don't even belong to you. Poincaré purportedly invented everything; you just had to develop his ideas."[4] Poincaré's earliest work on the topic preceded Einstein's by many years, but it differed from Einstein's in important ways. Why would such an expert on relativity theory explain Bergson's views of time and space while he considered Einstein's name not even worth mentioning—even in a lecture on those topics? Why do certain scientists come so close to major breakthroughs but fail to push them through?

Poincaré's books and numerous articles in renowned journals were read widely. He came from an elite family accustomed to occupying top-level appointments in prestigious institutions. His cousin Raymond Poincaré was prime minister of France on five separate occasions and president of France from 1913 to 1920. Poincaré moved easily from public service roles (mainly as a consultant for the mining, transportation, and telecommunications industries) to fundamental research, contributing across applied and theoretical sciences in engineering, mathematics, physics, astronomy, and even philosophy.

Although Poincaré was one of the most prominent scientists to work on relativity theory, he did not accept Einstein's conclusions. He gladly accepted all of the theory's experimental achievements but made the decision to stick to "ordinary mechanics" instead. In London, Poincaré asked: "What will be our position with regard to these new conceptions?" The physicists who decided to adopt them were not "constrained to do so," he answered. Those who did surely acted because for them it was "more comfortable, that is all." The position of those who rejected them was equally "legitimate." Poincaré believed that in the long run most scientists would opt against Einstein's system.[5] He was wrong, but he would not live to see it. A few months after his trip to England, the great mathematician died from complications related to prostate surgery.

Bergson deeply admired Poincaré, to whom he referred as "the great mathematician and philosopher."[6] Ten years after his death, the presence of this man was still felt the evening that Einstein met Bergson. After introducing Einstein, the organizer of the meeting reminded the audience that "the Société française de philosophie had among its founding members another scientific genius: he was named Henri Poincaré."[7] In the discussion that ensued, references to Poincaré resurfaced again and again. France, after all, had a strong tradition of collaboration between scientists and philosophers. The Société française de philosophie even used the word "philosophical sciences" to describe its goal of bringing about "agreement and regular meetings between scientists and philosophers."[8] Bergson backed this mission: "The close interaction between philosophy and science is a fact so constant in France that it could suffice to characterize and define French philosophy."[9]

Poincaré's main difference with Einstein (and this is why Einstein received and deserved credit for revolutionizing physics) was that he did not believe that these relativistic effects were that revolutionary. He did not think that the concepts of time and space should be overhauled. He worked closely with Hendrik Lorentz, who had developed the relativity equations that Einstein would later use and who called the changed magnitudes predicted by the theory local time and apparent length. Yet for Einstein there was nothing unique, let alone "local" or "apparent," about them. Einstein's genius centered on his reinterpretation of the notion of time, a contribution that was essential and novel to his work.

Historians have consistently considered that Poincaré, like Bergson, failed to fully understand relativity theory.[10] But the story of his relationship to Einstein and to the theory of relativity is much more complex. The problem was not that he did not understand it; the problem was that he did not want to accept it. In this respect, he would soon be left in the minority and labeled as retrograde.

Years before Einstein, Poincaré had made a particularly radical point, reminiscent—although different—of the claims that Einstein would later make: "Of two watches, we have no right to say that the one goes true, the other wrong; we can only say that it is advantageous to conform to the indications of the first."[11] In 1900 he gave a clear physical interpretation of the Lorentz transformation equations in terms of the slowing down of clocks and shortening of measuring rods.[12] He also considered the idea of redefining standards of length in terms of the time taken by light to travel a certain distance. In Poincaré's widely read book *The Value of Science* (originally 1905), he clearly described a "new mechanics" where "no velocity could surpass that of light" and where "inertia increasing with velocity, the velocity of light would become an impassable limit."[13] These sentences described concepts that were very close to those in Einstein's work, which appeared that same year. In all of these texts, he developed "strikingly similar" ideas to Einstein—sometimes years before.[14] But for Poincaré, they had an entirely different significance: this "new mechanics" would never be the be-all and end-all of physics.

Independently of Einstein, Poincaré explained that if one changed how physicists traditionally conceived of time, a cataclysm would follow, comparable to that which "befell the system of Ptolemy by the intervention of Copernicus."[15] These lines show just how clearly he foresaw the potentially revolutionary lessons that could be drawn from Lorentz's relativity equations. Yet, in contrast to Einstein, he did not want to espouse such a radical theory. At the same time that Einstein was pushing for a revolution, Poincaré was fighting against one.

POINCARÉ AND EINSTEIN

After the organizer of the debate mentioned the mathematician who had been dead for almost a decade, another commentator introduced Poincaré's views into the discussion once again. When weighing in on the benefit of relativistic versus nonrelativistic approaches, he concluded: "The question concerns which of the two languages is most comfortable."[16]

Poincaré's philosophy is generally summarized as conventionalism (or *commodisme*, in French). It was driven by the idea that scientists could choose among various ways of describing the same phenomena and that their choice was more conventional than necessary. A *conventionalist* perspective, rather than aiming to describe how things really were (as a *realist* would), maintained instead that scientific descriptions arose from the particular needs of different professions and the individuals who espoused them.

Einstein disagreed with those who described his theory as one possible "language" out of many others. Throughout the meeting, he forcefully fought against the view, frequently associated with Poincaré but also present in Bergson's work, that his theory was only one of other viable options. "One could always choose the representation one wants if one believes that it is more comfortable than another for the work to be done," he admitted. Yet he concluded firmly: "but that does not have an objective meaning." Einstein criticized the philosophical view that allowed for the same phenomena to be described in various ways and which claimed that choice between two alternative theories should be left up to the individuals involved. "But there are objective events that are independent of individuals," he protested that evening.[17] In other conversations, he aligned himself with the position associated with the physicist and philosopher Ernst Mach, who had argued that if two competing theories were available, the most "economical" one should be adopted.

Einstein and Poincaré met each other only once in 1911, at the Solvay Conference, a prestigious event for scientists in Brussels. There they argued about the behavior of molecules in gases at low temperature.[18] Afterward, Einstein described the old man as "simply negative in general, and, all his acumen notwithstanding, he showed little grasp of the situation."[19] Even though they were working on similar research topics, and even though Einstein followed Poincaré's work closely, the younger scientist mostly ignored the elder: "Einstein continued to pass over the older scientist in utter and unbroken silence," citing him only once.[20] The silence was reciprocated ("From Paris, Poincaré echoed back Einstein's silence") and it stretched "for seven more years."[21]

Poincaré knew Einstein's work well enough to recommend the physicist for a job at the Swiss Federal Institute that same year, clearly showing a certain noblesse oblige toward his junior colleague. The recommendation, however, was hardly a full-fledged endorsement. More than half a decade had passed since the publication of Einstein's seminal papers, but Poincaré's letter included the caveat that Einstein, at present, did not have much original work to claim as his own, although it speculated that future accomplishments would surely come.[22]

By then, Poincaré had accepted some of the most revolutionary implications of relativity, which he nonetheless attributed to Lorentz—not to Einstein. Poincaré wrote a report on Lorentz's work in 1910, repeating some of the same claims he had made earlier when he nominated Lorentz for the Nobel Prize. Poincaré explained how in the case of traveling clocks, Lorentz had shown that it was impossible to claim one as correct and the other one as delayed. He explained how it was "impossible to detect anything other than relative velocities of bodies with regard to one another, and we should also renounce the knowledge of their relative velocities with regard to the ether as much as their absolute velocities." He concluded clearly: "This principle must be regarded as rigorous and not only as approximate."[23] That year (1910) in a lecture in Gottingen, he framed the choice between Einstein's and Lorentz's interpretations of the theory as mainly a matter of taste.

GEOMETRY AND EXPERIENCE

After his sojourn in Paris, Einstein embarked on a months-long trip around the world that included a stop in Japan. In Kyoto, the physicist continued his polemic against the ghost of Poincaré. He told audiences that "the foundations of geometry have a physical significance."[24] Einstein's assertion was extremely radical. It went against the usual view, espoused by Poincaré, that considered mathematics as a tool for understanding the physical world and that did not consider mathematical equations as representing the universe itself. At stake in his lecture stood one of the pithiest questions pertaining to the nature of mathematical knowledge and the shape of the universe.

Einstein attacked Poincaré most forcefully years after the Frenchman had died. *Geometry and Experience* (1921), one of Einstein's most celebrated texts, was a clear rebuttal of Poincaré's insistence that scientists would never be able to prove that the universe had a specific geometry. In that essay, Einstein explicitly rejected the conventionalist philosophy of the "acute and profound thinker" Poincaré, searching for reasons other than bare convenience to justify why scientists should adopt a particular scientific explanation over an alternative one.[25]

What is the shape of the universe? Does space have a certain shape? The specific topic of Einstein's attack against Poincaré was Riemannian geometry, which Einstein used in developing his general theory of relativity, but also at stake was the status of Euclidean geometry over non-Euclidean geometry, pertinent because Einstein's theory of relativity was closely associated with the latter. Do we live in a non-Euclidean universe?

Non-Euclidean geometries were strange: the shortest path between two points could be curved, parallel lines could intersect, and the angles of triangles do not always add up to the same number. Traditional Euclidian geometry typically assumed that only one line could pass through two points, that a straight line was the shortest distance between two points, and that through one point only one parallel line could be drawn to a given straight line. Non-Euclidian geometries denied these basic postulates. Euclidean geometry was three dimensional and was often described as based on simple principles. But in relativity theory, time was a fourth dimension on the same footing as the three traditional spatial ones. The basic postulates of Euclid, argued Einstein, were no longer adequate.

Was mathematics a tool used by scientists, or did it reveal the basic structure of the universe? Einstein tried to show that the question of which geometrical conception was true was "properly speaking a physical question that must be answered by experience, and not a question of a mere convention to be selected on practical grounds."[26] His argument could not be farther from Poincaré's view about the relation of mathematics to physics. According to Poincaré, mathematics was *not* an experimental science. "Geometrical axioms," Poincaré insisted, "are conventions" or "*definitions in disguise*."[27] Poincaré was an expert in the mathematics of both Euclidean and non-Euclidean geometries and saw the merits and disadvantages in each. But according to Einstein, something other than sheer practicality should be used to weigh in on the general question of which particular geometric construction of the universe was correct.

For Poincaré, the question of which geometrical system should be used was completely the same as which measurement standard should be used. Asking about the validity of Euclidean geometry over non-Euclidean ones was simply the same as asking if one use the yard or the meter: "What, then, are we to think of the question: Is Euclidean geometry true? It has no meaning. We might as well ask if the metric system is true, and if the old weights and measures are false."[28]

Poincaré's philosophical outlook can be understood in the context of the time-standardization efforts in which he participated. He saw France, Germany, and Britain engaged in a bitter debate about which country's time and timekeeping methods would prevail and proposed a solution based on determining what was most convenient for the parties involved. Time coordination (of traveling clocks or stationary ones) was, for Poincaré, mainly about finding a conventional agreement that would be useful for each case. In contrast to Poincaré, Einstein did not have firsthand experience of the long, painful negotiations undertaken by many nations in their effort to reach an agreement on standards of time.[29] For him, procedures for time coordination should not be understood as matters of convention.

While the topic of standards of measurement could seem far from debates about relativity theory, they intersected in key ways. Discussions about measurement standards stood at the center of debates about the validity of Einstein's theory of relativity.

EINSTEIN'S ENEMIES

Einstein had powerful enemies. Attacks against him due to his Jewish background were most prominent in Germany, but in France his background was also relevant. A distinguished historian at the Sorbonne compared the controversies surrounding Einstein to the infamous Dreyfus affair, a scandal that divided the French nation when an innocent Jewish captain was unfairly accused of treason and sentenced to life imprisonment in a penal colony in a case that took over a decade to unravel. The Einstein controversy, he argued, divided France across the very same lines: "The Dreyfus adherents claim that he [Einstein] is a genius, whereas the Dreyfus opponents say he is an ass. . . . The same groups line up and face each other at the slightest provocation."[30] Earlier in the century, Poincaré had sided with Jewish causes as a vocal defender of Captain Dreyfus and played an important role in securing the exoneration of the wrongly accused man. But Poincaré's criticisms and those of many others who sided with Bergson differed from those who attacked Einstein for being Jewish.

In Germany, Einstein's work was viciously denounced on August 24, 1920, in a series of anti-relativity lectures delivered at the Berlin Philharmonic Hall.[31] After that day, Einstein claimed that anyone who opposed his theory did it for political reasons.[32] By 1921 he did not consider the possibility of any legitimate scientific or philosophical critique: "'No men of science,' he replied, emphasizing the last word, 'were against my theory.'" When asked about the opposition to it he insisted: "that was purely political." Asked to elaborate more about the specific "political reasons," he referred to anti-Semitism: "Because I am a Jew."[33]

But despite Einstein's claim linking all criticisms to anti-Semitism, not *all* attacks were motivated by anti-Semitism or by political differences. Many prominent participants at the Collège de France and the Société française de philosophie (such as Brunschvicg, Langevin, Paul Lévy, Charles Nordmann, and Bergson himself), as well as the organizer, Xavier Léon, were Jewish. Bergson's critique and those of many of Bergson's followers differed from anti-Semitic attacks in important ways: Bergson never categorized Einstein's science as Jewish, and he never contested it for that reason. Their criticisms focused on specific claims present in Einstein's work.

BERGSON AND POINCARÉ

Poincaré's philosophy represented, in the eyes of many of his followers, a salutary entente cordiale between scientists and philosophers of his generation. Other thinkers were more radical, such as Édouard Le Roy and Pierre Duhem, who wanted to go even farther than he in stressing the *constructed* nature of many scientific claims. While they at times perceived him as more conservative than they, for many others he was too radical. Bertrand Russell even labeled him a nominalist, and in France, Louis Couturat hailed this assessment. The label of nominalist charged Poincaré with the claim that scientific truth was so particular and so dependent on individual cases and practical situations that absolutely no general, let alone universal, lessons could be drawn from it.[34] Poincaré resisted this characterization of his own work and distanced himself from any form of nominalism.

In *The Value of Science* (1905), Poincaré described Bergson's influence negatively, as dangerously anti-intellectual.[35] But later, Poincaré and Bergson came to be, in fact, quite close. Poincaré's younger sister Aline married the philosopher Émile Boutroux, who was one of Bergson's teachers.[36] Boutroux, in his *De la contingence des lois de la nature* (1874), had stressed the role of contingency over determinism in the laws of nature, contrasting his philosophy to a Cartesian theory of eternal truths. Bergson shared with Poincaré and Boutroux the same aversion to Descartes and eternal truths, even associating Einstein with that philosophical position.

Poincaré's famous *Science and Hypothesis* (originally 1902) opened with a clear attack against Le Roy, one of Bergson's closest allies. He associated Le Roy's philosophical position with nominalism and targeted an article by the young author entitled "Science et philosophie," which had claimed to "follow" Bergson "*pas à pas*."[37] In *The Value of Science*, published a few years later, Poincaré wrote that the worst aspects of Le Roy's philosophy were due to Bergson's influence on him. "Le Roy's doctrine," he explained, "has another characteristic that it doubtless owes to M. Bergson, it is anti-intellectualistic."[38] But, with time, his initial animosity against Le Roy and Bergson attenuated. Poincaré became a supporter of Le Roy.

By 1910 Bergson found that Poincaré and Le Roy had "arrived at analogous conclusions" in their respective philosophies.[39] Bergson sketched a rough map of the main contemporary philosophical schools in France in which he clarified his relationship to Poincaré's philosophy. He considered him part of a French tradition in which "mathematicians wrote the philosophy of their science, and even of science in general." "Today," he explained, that school is represented by "our great mathematician

Poincaré," whose philosophy, in a nutshell, showed the "symbolic and provisional character" of scientific knowledge.[40]

Bergson believed his philosophy and the one presented by Poincaré were no doubt "distinct," but he also believed that "they could rejoin" since "they also had points in common." What drew them together was that both "felt a strong repugnance toward a philosophy that wants to explain all reality mechanically."[41] In this camp Bergson included prominent psychologists (he cited Théodule Ribot, Pierre Janet, Alfred Binet, and Georges Dumas) and sociologists (he mentioned Emile Durkheim and Lucien Lévy-Bruhl). Bergson criticized how these thinkers adopted philosophical stances without knowing enough of that discipline. Poincaré, who excelled in both areas, was an exception. Bergson also supported Durkheim's enemy Gabriel Tarde, who extolled Bergson in his *Les Lois de l'imitation* (1890). As to the others, they held implicit and deeply flawed mechanistic, reductionist, and materialistic stances in which science was uncritically and unjustifiably above it all.

Not only did Bergson think highly of Poincaré, both as a scientist and as a philosopher, he believed that they had enemies in common. The enemy of his enemies was his friend. What was at stake in stressing the similarities between Bergson's and Poincaré's philosophy against others? The attacks launched by the other camp were simply too vicious and unfair, explained Bergson. They characterized his philosophy as "a return to the mentality of primitives."[42] What could be more insulting? By 1915 Bergson was even clearer about the points in common between his philosophy and Poincaré's:

> The great mathematician Henri Poincaré has arrived at conclusions of the same type as ours. He has shown what is due to man and what is due to the needs and preferences of our science by analyzing the conditions that bind the construction of scientific concepts in the web of laws that our intellect extends on the universe, by following a different route—much more direct.[43]

Both Bergson and Poincaré shared another similar mission. They insisted that relativity theory as conceived by Einstein *could* be adopted but that it did not *have to be* adopted. In a letter to Lord Haldane written soon after his debate with Einstein, Bergson explained his position using the language made famous by Poincaré, that of conventionalism: "I see, for my part, in the space time of Minkowski and of Einstein a very convenient [*commode*] form of representation, but that it is perhaps not essentially tied to the Theory of Relativity." What he protested in *Duration and Simultaneity* was Einstein's attempts "to elevate a mathematical representation into transcendental reality" and to refuse to see that his redefinition of time was "nothing but a convention, a convention nonetheless necessary for preserving the integrity of physical laws."[44]

Readers noticed that both Bergson and Poincaré also held similar positions with regard to their views on measuring time. Measuring time, in their view, destroyed parts of it. "At the same time as Bergson, Poincaré thus reopened the ancient refutation about the possibility of measuring time," explained a professor of philosophy. He remarked how Poincaré, in a manner quite similar to Bergson, insisted that scientists "did not" measure time, "but cut it up into pieces that they declare to be identical so that their equations be as simple as possible."[45]

In 1916, four years after Poincaré died, the mathematician and Bergson were considered to be close comrades in the general fight against materialism and mechanistic philosophies. In the Parisian lectures entitled "Faith and Life," which had clear religious overtones, the philosopher Paul Desjardins grouped both men together.[46]

POINCARÉ'S WORK

Poincaré was intimately acquainted with the problems of measuring time and determining simultaneity. In 1898 he published an important essay entitled "The Measure of Time" in the *Revue de métaphysique et de morale*. Why would an article on time measurement be fit to appear in a magazine devoted to metaphysics and morals?

The article was then reprinted as the second chapter in his famous book *The Value of Science*. In that essay, Poincaré remarked that there was no single procedure for time coordination that could be considered an absolute method for determining time. From 1898 onward Poincaré often wrote about how to deal with different times shown by two separated clocks.[47] The most convenient procedure should be adopted for each specific situation. It was not—as for Einstein—a problem that should be solved by revolutionizing the existing laws of physics.

Another important point in Poincaré's essay was that "we do not have a direct *intuition* of simultaneity, and also do not have it of the equality of two time intervals."[48] By criticizing our intuition of time, was he attacking Bergson? Bergson, after all, was known to many as a defender of intuition. He once described his philosophy as one that "appealed to sentiment, to intuition, and to interior life."[49] In denying that simultaneity could be directly intuited, would not Poincaré be taking a position against Bergson?[50] Yes and no.

Within "the realm of consciousness," Poincaré admitted that "the notion of time was relatively clear," but it was not clear enough to base scientific measurements on it.[51] His view was similar to Bergson's, who used the concept of intuition in a way that differed from how many psychologists used the term when they used it to refer to an estimate, guess, or hunch. Poincaré's point was not that intuitions about time did not matter for science, but rather that they were not consistent enough to become sturdy references for measuring time precisely and quantitatively. His claim against "direct intuition" was aimed at the psychological concept—not the philosophical one.

When Bergson stressed the importance of intuition, he did not claim that it could be used to quantify time. The intuition of time that he wanted to stress was precisely the aspects of our temporal sense that were the least repetitive, cadenced, and homogenous. It was the intuition of the "moving character of reality" that he stressed—not an intuition of simultaneity or clear succession of events.[52] Our intuition of duration, according to Bergson, was of its ever-changing character. One commentator explained it clearly: "This duration, which we perceive immediately, is made manifest in us and around us by an incessant rejuvenation: there are not two instants like each other."[53]

"AN AMAZINGLY SIMPLE SUMMARY"

Bergson's objections to Einstein's theory were disquieting for physicists. In light of his critique, many started to wonder that perhaps Einstein's theory should not have the importance that it did, that perhaps the physicist was offering only one way of interpreting reality, and that perhaps this interpretation was only one out of many other possibilities. Perhaps it was not even revolutionary but, rather, conservative. Einstein fought against these interpretations harder than ever after his meeting with Bergson.

Much has been written about Einstein's philosophical outlook. Scholars debate just how much of a realist Einstein was.[54] Evidence varies, because Einstein often offered contradictory views on this matter. His debate with Bergson, however, reveals that *for a delimited period of time* (starting with the attacks against him at the Berlin Philharmonic Hall and followed by Bergson's confrontation), he adopted a view where only his interpretation of relativity theory was deemed "objective."

Einstein's bold defense of his theory during these years as the only scientifically viable option stood in sharp contrast to both earlier and later interpretations of his own work and that of his colleagues. Einstein repeatedly extolled the dual virtues of simplicity and generality when evaluating scientific theories. He first argued that his theory should be adopted because it was simple and comprehensive, not because it was absolutely and definitively the only right one. He described it "as an amazingly simple summary and generalization of hypotheses which previously have been independent of one another."[55] His seminal paper on general relativity (1916) had only modestly claimed that his theory was "psychologically the natural one."[56] The fact that he included the caveat of a *psychological* benefit was important, marking an important difference with later claims that it was *necessarily* the natural one. In that paper as well, when he referred to time, it was with the additional clarification that he was referring to a certain "definition" of time, and even put the term "really" in quotation marks when writing about altered clock times. One clock "'really' goes more slowly than the [other] clock," he wrote.[57] In 1918 he again implied that a choice for or against his theory was legitimate: "Only utilitarian reasons can decide which representation has to be chosen," he explained.[58] A year later, after the eclipse expeditions had confirmed his theory, he offered one of his strongest statements describing how scientific theories rarely imposed themselves. Scientists usually had choices:

> When two theories are available and both are compatible with the given arsenal of facts, then there are no other criteria to prefer one over the other besides the intuitive eye of the researcher. In this manner one can understand why sagacious scientists, cognizant of both—theories and facts—can still be passionate adherents of opposing theories.[59]

But after the attacks on him in Berlin and in Paris, Einstein did not describe the availability of other theories as alternative options.

Bergson and most of his supporters did not deny the results of the theory of relativity. Most merely disputed the assertion that they led *necessarily* to Einstein's conclusions. Poincaré and others had often argued that scientists usually had a choice of which theories were better suited for particular cases. They sometimes argued that theories rarely corresponded exactly with reality, but they nonetheless considered them adequate and useful representations of it. Initially, it seemed that Einstein would have agreed with these characterizations, but he later took a stance against them. For a moment, he forcefully fought against the common view, made prevalent by Bergson's critique, that "whether 'simultaneity' can be given an absolute meaning" is a matter "which the physicists have been unable to decide."[60]

CHAPTER 7

Bergson Writes to Lorentz

HAARLEM, HOLLAND

In 1928 Einstein traveled to Holland to deliver a moving funeral speech. His colleague and friend Hendrik A. Lorentz had just died, at the age of seventy-four. On a Friday, exactly at noontime, the state telegraph and telephone services of the country were suspended for three minutes as a "tribute to the greatest man Holland has produced in our time."[1] Despite being a close friend, an illustrious scientist, and the author of the relativity equations Einstein used, the Dutchman never accepted his conclusions. Why not?

Lorentz received the coveted Nobel Prize for physics in 1902. In 1895 he proposed that lengths could contract under motion. A few years later, he proposed that clocks could slow down under these same circumstances. The reasons why scholars are fascinated with Einstein's relationship to Lorentz are similar to those fueling interest in his relationship with Henri Poincaré. Why did these two scientists fail to espouse the theory that would revolutionize physics? Einstein had a close and affectionate relationship to Lorentz, who was somewhat of a father figure for the younger physicist.[2] Their work on relativity was initially so similar, and it was so difficult to tease apart the contributions of each man, that it was often called the Einstein-Lorentz theory. But their relationship was not always smooth, particularly when it came to certain interpretations of relativity.

Abraham Pais, a colleague turned Einstein biographer, described him as not being able to "let go" of the old theories, much like Poincaré.[3]

Could Bergson have played a role in Lorentz's view of Einstein? A few years after the Paris meeting, Bergson started badmouthing Einstein to Lorentz—behind Einstein's back. In a letter to Lorentz, he explained why his philosophy was being received with such animosity on the part of physicists. He speculated that Einstein, along with numerous other physicists, simply did not understand him. To Lorentz, he offered a very negative view of the physicist:

> In general, relativity physicists have misunderstood me. They, by the way, frequently do not know my views except through hearsay, by inexact and even completely false accounts. This is perhaps the case of Einstein himself, if what they say about him is true.[4]

Bergson's argumentation differed from Lorentz's in various ways, but he wholly accepted and used Lorentz's equations. Like Lorentz, he defended a view for which Einstein's interpretation of relativity was not the only viable interpretation.

Bergson would also meet with Albert A. Michelson, the very author of the experiment that became central to relativity, and studied his work carefully. Historians have often puzzled why Lorentz, Poincaré, and Michelson—the three men whose research was closest to Einstein's—failed to embrace the theory of relativity wholeheartedly. The role of Bergson as an individual, colleague, mentor, friend, and confidant—in addition to the general role and impact of his philosophy—was key. How did Poincaré's, Lorentz's, and Michelson's views fit with Bergson's? These three men agreed that relativity theory as conceived by Einstein *could* be accepted yet did not *have to be*. Accepting it or not was a choice that, in the end, should be left up to individual scientists themselves.

Evidence suggests that Lorentz and Bergson saw eye to eye with respect to many pertinent issues of their time. Bergson considered his arguments in *Duration and Simultaneity* to be compatible with Lorentz's (although they were hardly identical). While Lorentz hoped that they would lead back to a traditional belief in a universal and absolute notion of time, Bergson held no such expectations. Bergson's support of Lorentz did not go this far. It did not amount to a defense of the controversial contraction hypothesis, a belief in the ether, or in the hope that the fixed-star hypothesis could be used a stable reference point for the laws of physics.

Discussions about the physics of time were particularly tense because of the involvement of all three men in the International Committee on Intellectual Cooperation (CIC), one of the most prestigious branches of the League of Nations. Lorentz was called to replace Einstein after the physicist angrily resigned from the committee. Later Bergson, who was president of the CIC, supported the Dutch scientist to such an extent that he nominated him as his successor. Lorentz succeeded him as president in 1925. Only a year earlier, Bergson had sent a letter inviting Lorentz to dinner and included the following promise: "We will be absolutely alone." That same letter thanked Lorentz for having sent Bergson a note on "the two clocks": "My first impression," wrote Bergson, "is that your argument is irreproachable." He then added a concluding clause explaining how both of their arguments were compatible, "and *for that reason* mine remains correct."[5]

LORENTZ AND EINSTEIN

Lorentz was a resolute admirer and supporter of Einstein personally and professionally who, for the most part, sympathized with Einstein's political outlook except with respect to the League of Nations. In contrast to Einstein's nearly nonexistent relationship with Poincaré, Lorentz and Einstein had an intimate and personal relationship that stretched across many years. The two men corresponded amply about both science and politics and were mostly friendly, although clear tensions surfaced on occasion. When Lorentz died, Einstein stood up to the occasion, speaking ceremoniously and generously at his funeral. "I revere you beyond measure," Einstein once wrote to Lorentz. Throughout his life, Einstein would continue to praise his colleague and his close relatives.[6] The admiration was mutual. Lorentz used all his influence to try to bring Einstein to work in Utrecht, although Einstein declined. After the Great War broke out, Einstein and Lorentz became even closer, sharing many political aims. Lorentz led various antiwar initiatives from Holland, and when the war started, Einstein immediately reached out to him to convey his international stance and opposition to the war.[7]

But behind these public forums lay an, at times, tense relationship. Lorentz's closeness to Bergson complicated their relationship. So did his affinity with Poincaré. Lorentz and Poincaré seemed so close to each other that a French scientist chose to describe Lorentz simply as "the Dutch Poincaré."[8] When in one of his last speeches, Poincaré talked about the theory of relativity, he attributed some of its most important characteristics to Lorentz.[9]

Both Poincaré and Lorentz continued to strongly support each other. In 1902 Poincaré nominated Lorentz for the Nobel Prize, which he indeed was awarded. In his nomination, Poincaré credited Lorentz for introducing the concept of time dilation. Poincaré's letter to the Nobel Prize committee supporting Lorentz described his "ingenious invention of 'reduced time,'" in which "everything happens as if the clock in one place slows down in comparison to the other." Lorentz, according to Poincaré, had also noted that there "is no conceivable experiment that can lead us to discover" a difference between a traveling and a stationary clock, with the additional result that no experiment could help decide on one its time over the other. He also credited Lorentz with changing our concept of simultaneity. Lorentz's "surprising" discovery explained why "two phenomena occurring in two different places can appear simultaneous even though they are not."[10] Lorentz himself later famously claimed that "Einstein simply postulates what we have deduced, with some difficulty and not altogether satisfactorily."[11]

Their assessment of Einstein's work was hardly anomalous. After the latter was published in 1905, it received almost no attention. One of the few notes that mentioned it drily included the caveat that it "leads to results which are formally identical with those of Lorentz's theory."[12] Einstein at first dutifully attached Lorentz's name to the theory, referring to it as "the theory of Lorentz and Einstein."[13] He only separated himself from Lorentz's position in 1907, referring separately to "the H. A. Lorentz theory and the principle of relativity."[14] That same publication described his differences with Lorentz in more detail. Einstein claimed that he took Lorentz's "local time" to be time in general: "Surprisingly, however, it turned out that a sufficiently sharpened conception of time was all that was needed to overcome the difficulty discussed."[15] Lorentz had referred to one of the changed magnitudes as local time (and in the similar length equations he called it apparent length), but Einstein eventually came to believe that there was nothing unique, let alone "local" or "apparent" about it.

Einstein's new "sufficiently sharpened conception of time" received additional backing from Hermann Minkowski in a famous September 1908 lecture.[16] Minkowski had been Einstein's mathematics professor at the polytechnic in Zurich where Einstein completed his doctorate. He would later become one of Einstein's most influential supporters, playing a key role articulating the differences between Lorentz's and Einstein's work in terms of Einstein's reinterpretation of local time.[17] In the famous speech trumpeting his student's accomplishments, Minkowski did not even mention Poincaré's.[18] Minkowski was also essential in combating one of the main accusations against the theory. He argued that relativity was not an artificial hypothesis but one that necessarily followed from this new conception of time.[19]

Poincaré and Lorentz never attributed the same meaning and importance to these conclusions that Einstein or Minkowski did. Why? And what was the role of Bergson in these discussions?

ARTIFICIAL OR NECESSARY?

From 1907 to 1911 Einstein tried hard to differentiate his work from Lorentz's. The distinction between the two approaches became so clear that eventually the senior scientist agreed to give the junior physicist full credit for a particular way of interpreting relativity. Although in his famous 1905 article Einstein clearly referred to "time," his claim was initially more modest than it would later be. Then, it was closely associated with clock time and did not even include all clocks. He was careful to say that his theory applied to balance-wheel clocks as well as to light-signal-coordinated clocks. A footnote added later when the article was published in a collection dedicated to the theory (probably by the editor of the collection, Arnold Sommerfeld), explicitly warned that it would "not" apply to "a pendulum clock, which is physically a system to which the Earth belongs. This case had to be excluded."[20] How did a paper about clocks, and only certain kinds of clocks, become a paper about time in general?

Einstein's initial caution about referring to time started fading in 1911. In a lecture given in January, Einstein claimed that the effects of time dilation that he had considered for moving clocks would affect "a living organism in a box."[21] When that same year Paul Langevin developed the famous twin paradox, he referred to clocks but also to biological aging.

Poincaré, and Lorentz himself in self-criticism, had found Lorentz's explanation to be ad hoc—a pejorative label for a scientific theory. During the 1900 International Congress of Physics, Poincaré commented on the artificial character of Lorentz's theory.[22] A few years later Lorentz recounted how "Poincaré has objected to the existing theory of electric and optical phenomena in moving bodies that, in order to explain Michelson's negative result, the introduction of a new hypothesis has been required, and that the same necessity may occur each time new facts will be brought to light." Lorentz accepted the criticism entirely, noting that "surely this course of inventing special hypotheses for each new experimental result is somewhat artificial."[23]

Lorentz continued to work hard on the topic, trying to combat the apparent artificiality of his theory by incorporating new experiments on the deformation of the electromagnetic mass of the electron. He now optimistically claimed that his work was based on "fundamental assumptions . . . with better results."[24] But even with this new research, he was not entirely successful in convincing his Parisian colleague. In *The Value of Science* (1905), Poincaré insisted that if Lorentz had succeeded "it is only by piling up hypotheses."[25] Yet he also lauded it as the "least defective" of all available options.[26]

Extending the accusations that had first originated with Poincaré, Einstein portrayed the advantage of *his* theory over Lorentz's as not being artificial. The reason he gave for the superiority of his interpretation was that, in contrast to Lorentz's, it was not "an artificial means of saving the theory." Einstein included a history of the development of this research where he differentiated his position from Lorentz's.[27] Minkowski reinforced this argument by arguing that Lorentz's theory appeared as "a gift from above," whereas Einstein's alternative allegedly emerged naturally from the observed facts.[28]

What distinguishes a theory that naturally reflects the order of the world from one that is artificially designed to match experimental results? These publications by Poincaré, Lorentz, and Einstein contained mud-slinging accusations of artificiality and ad hoc construction that ensued among the three men. Lorentz disputed Einstein's accusations that his theory was ad hoc, arguing instead that it was Einstein's solution that was most artificial. By 1909 Lorentz conceded in print that Einstein could "take credit" for "the manifestation of a general and fundamental principle" of relativity—something he himself had not done—but only if one recognized how "artificial" Einstein's contribution really was.[29] In a text that (seemingly) generously gave credit to his colleague for the relativity theory, he added: "Einstein simply postulates what we have deduced, with some difficulty and not altogether satisfactorily." The result was "rather complicated and looks somewhat artificial."[30] In lectures that followed (in 1910) he used the term "Einstein's Relativitätsprinzip," clearly marking his difference.[31] He accepted that Einstein's theory could be considered the best available explanation of certain experimental results, but he kept searching for an alternative theory.

TENSIONS DURING THE WAR

In six lectures delivered in Gottingen (1910), Lorentz accepted that Einstein was right but claimed that he was right too: "Which of the two ways of thinking you would like to join, is a decision that depends entirely on each individual."[32] In the years that followed, Lorentz would be clear about the benefits of each approach.[33] This status quo, in which both interpretations were considered as equally viable, changed after signs of tension between Einstein and Lorentz resurfaced at the end of 1911, when Einstein refused a generous job offer to move from Prague to the Netherlands (an offer made to him largely through Lorentz's efforts). Einstein negotiated instead to go to the Zurich Polytechnic. "But if they were to learn the day after tomorrow, or at some other date in the very near future," that I am not interested in the Utrecht job, "they would lose their fervor at once and keep me forever in suspense," he wrote to a friend.[34] Negotiating simultaneously with two institutions was risky personally and professionally, since for the position in the Netherlands he was dealing with "the greatest man in our field, who is also a personal friend."[35]

In three lectures given in Haarlem in 1913, Lorentz was even clearer that the decision as to who was right, Einstein or him, could not be left up to experiment and certainly not to Einstein himself, for that matter. He criticized Einstein's "short and quick" dismissal of these "questions" and his claim that they were simply "nonsense."[36] The issues at stake, Lorentz insisted, were epistemological: "The evaluation of these concepts belongs largely to epistemology, and the verdict can also be left to this field." Scientists were free to chose between them depending on "the mindset to which one is accustomed, and whether you feel most attracted to one or the other view."[37] The presenter of Einstein's Nobel Prize restated (almost verbatim) the view which Lorentz had expressed years earlier, that the validity of Einstein's theory of relativity "*pertains to epistemology* and has therefore been the subject of lively debate in philosophical circles."[38]

In an article published for a general audience in the influential *Kultur der Gegenwart*, a multivolume "encyclopedia of the present" commissioned by the German Reich, Einstein responded to the view that anyone could chose which of the two interpretations was better. The article included a stinging criticism of Lorentz's interpretation: Einstein explained how "Lorentz's theory arouses our mistrust."[39] When Lorentz read these lines, he quickly accused Einstein of "raising the impression that only 'fictitious' things were involved here and not a real physical phenomenon."[40] In January 1915, Lorentz personally complained about this article in a personal letter to Einstein: "In your article of the *Kultur der Gegenwart*, I find in the discussion . . . the remark: 'This manner of thinking up *ad hoc* hypotheses to cope with experiments . . . is very unsatisfactory,'" noticing that it referred directly to him.[41] He also chastised Einstein for "presenting a personal view as self-evident" by claiming that it was "hopeless" to continue to search for a distinguishing factor that might lead scientists to select one time in relativity theory over the others. Lorentz then proceeded to describe all that was wrong with Einstein's interpretation, settling largely on two main points.

The first problem had to do with the equivalence of time and space. Space and time were simply not interchangeable, argued Lorentz: "An unmistakable difference exists between spatial and temporal concepts, a difference which you also certainly cannot remove entirely. You cannot view the time coordinates as totally equal in status with the space coordinates." The second issue had to do with the equivalency of the status of the different times. Lorentz was ready to concede that imperfect spirits could not assign a different status of t_1 over t_2, but a "universal spirit" could. But, what is more, Lorentz argued that there was something of a "universal spirit" in all of us: "Surely we are not so vastly different from it."[42]

What did Lorentz mean by "universal spirit"? Traditional conceptions of time in connection to the perceptual qualities of a hypothetical or real universal spirit had a long history going back to medieval times. In Newton's time, this universal spirit was directly taken to be God. Were these scientists still debating, albeit in secularized terms, questions of a theological nature? To Lorentz, the issue of a "universal spirit" and the ability to disentangle the equivalency of the status of t_1 and t_2 was not a question of physics properly, but one that went "beyond the bounds of physics."[43]

Einstein replied to Lorentz, apologizing to some extent: "Although I had 3 years to compose it, I had completely forgotten and was reminded of my commitment . . . one week before the delivery deadline." He pleaded: "So please, do not punctiliously weigh every word!" He explained that his assessment of the equal status of t_1 and t_2, was based on the fact that so far there were no "physical grounds (accessible in principle to observation)" for distinguishing them—a fact that Lorentz himself knew and acknowledged. Scientists could have other reasons for wanting to distinguish between them, but Einstein cited Ernst Mach as responsible for leading him to believe that "a worldview that can do without such arbitrariness is preferable, in my opinion." Einstein finished his response by saying that agreement on these issues was probably not forthcoming, at least not through correspondence: "Finally, as far as the question of *time* is concerned, we are scarcely going to be able to debate this effectively by letter. I shall be glad to come to Holland again to discuss this and other matters, when the sorry international entanglement is finally overcome."[44]

Even though their disagreements about the nature of time intensified during the Great War, the conflict, in many other ways, drew the two men closer. During those years Lorentz and Einstein corresponded frequently about the horrors of the war, lamenting how it hurt relations among scientists.[45] They also corresponded about their current work. In 1919 Lorentz was among the first physicists to explain Einstein's theory to the general public. His short, popular book, *The Einstein Theory of Relativity*, called it "a monument of science" and extolled the "indefatigable exertions and perseverance" of Einstein.[46] Lorentz nonetheless cautioned that "in my opinion it is not impossible that in the future this road [research on the ether], indeed abandoned at present, will once more be followed with good results."[47] Lorentz continued to search for a stable background that could serve as an anchor for an absolute concept of time, be it the ether, a concept of space that could serve as reference point, or the fixed-star hypothesis. Either of these would all do what he wanted: to serve as absolute reference points. If scientists had not found these references to date, it did not mean that they never would.

In the 1920s Lorentz continued to support Einstein personally, despite their other differences. His support peaked after Einstein suffered anti-Semitic attacks at the Philharmonic Hall. Yet tensions remained. In 1922 he still insisted that "one may, in all modesty, call true time the time measured by clocks which are fixed in this medium [space], and consider simultaneity as a primary concept."[48]

By then, it was clear that the two men differed in their interpretation of relativity but that they agreed on many aspects of international politics. Both scientists lamented the exclusion of German nationals from scientific forums after the war, yet they each had different ways of protesting this exclusion. For Lorentz, it entailed reaching out to German scientists, including Einstein. For Einstein, it meant boycotting certain forums perceived to be exclusionary. Sometimes these efforts clashed with each other. Lorentz had been president of the Solvay Conferences since 1911, and Einstein later boycotted these meetings. Einstein raised a complaint against the exclusionary policies of the International Committee on Intellectual Cooperation (CIC), first led by Bergson and later by Lorentz. Einstein claimed that the Committee was excluding all Germans, an assertion that provoked Lorentz to respond with an angry letter (on September 15, 1923) explaining to Einstein that it was not true that German scientists were excluded on principle. He offered clear evidence, including the repeated invitations that had been made to *him*. As Einstein continued to launch criticisms and miss meetings of the CIC, Lorentz as its new president continued to distance himself from Einstein's theory.

During these at-times tense discussions, Lorentz continued to consider Einstein's theory as one of many other possible options. Special and general relativity were undoubtedly correct: "I do not mention that, also in my opinion, not only the theory of relativity but also your gravitation theory can remain valid in their entirety." But they were not the only way to see things: "They will just not impose themselves on us so much as the only possible ones."[49]

In a series of lectures delivered at the California Institute of Technology (and published later) Lorentz again explained the differences between a "physicist of the old school" and the "relativist." Both agreed that nobody could "make out which of the two times is the right one." But the old-school physicist was ready to acknowledge that he "preferred" one of them, whereas for the relativist, "there cannot be the least question of one time being better than the other." Lorentz's personal preference was to maintain "notions of space and time that have always been familiar to us, and which I, for my part, consider as perfectly clear and, moreover, as distinct from one another." Why give up these clear advantages? "My notion of time is so definite that I can clearly distinguish . . . what is simultaneous and what is not." Why give up this clear benefit?[50]

A science writer of that period explained that "to ask which of these durations are the real one is equivalent to the question of which is the true color of a piece of opal. It can be yellow if we look at it from a certain angle; red when we move toward the left, green or blue if we move ourselves toward the right."[51] Lorentz nonetheless continued to defend his position. In 1928, prompted by new experimental results, he again described the option he advocated, giving Einstein full credit for relativity. While he had introduced the concept of local time, in contrast to Einstein he "never thought that this had anything to do with the real time. The real time for me was still represented by the old classical notion of an absolute time, which is independent of any reference to special frames of co-ordinates. There existed for me only this one true time."[52] Einstein, Lorentz insisted, could take full responsibility for what he had done:

> So the theory of relativity is really solely Einstein's work. And there can be no doubt that he would have conceived it even if the work of all his predecessors in the theory of this field had not been done at all. His work is in this respect independent of the previous theories.[53]

He granted to Einstein a lot: "To the experimental evidence which we already had, the charm of a beautiful and self-consistent theory was then added."[54] Nonetheless Lorentz continued to believe in his hypothesis: "Asked if I consider [my hypothesis] a real one, I should answer 'yes.' It is as real as anything we observe."[55] And he continued to believe that Einstein "simply postulated" what he had laboriously deduced.[56]

CHAPTER 8

Bergson Meets Michelson

PARIS, FRANCE

Test it. Measure it. Why not solve the debate between Einstein and Bergson by simply measuring time? That way, scientists and philosophers could potentially determine if it behaved in the manner described by Einstein or in the way that Bergson understood it. Just measure the time of clocks traveling at different speeds and determine if one slowed down in comparison to the other. Couldn't Einstein simply prove—experimentally—that Bergson was wrong?

Together with the chemist Edward Morley, the American physicist Albert A. Michelson devised the famous Michelson-Morley experiment, which is usually considered to be central to Einstein's special theory of relativity. Einstein himself claimed that it was. Historians and philosophers have been doubtful, debating the actual role played by the experiment in the formulation of his theory.[1]

Bergson knew Michelson well. He even understood that the notion of time he had proposed in *Duration and Simultaneity* adhered more closely to the results of Michelson's experiment than Einstein's theory did. On March 19, 1923, almost one year after the debate, Bergson met Michelson and had an important conversation with him. He found that the physicist was "completely shocked at having enacted such a revolution by a simple experiment that appeared, even to himself, susceptible to a completely different interpretation."[2] Bergson's interpretation of the experiment was consonant with Michelson's views about it: because the results of the experiment could be understood in an entirely different way, it did not directly prove Einstein's theory.

By measuring time, scientists destroyed some of it, argued Bergson. Time measurements were surely valuable, but they referred to something different from those aspects of time that so fascinated him. Clocks were ideal for studying time in physics, but how they related to our general sense of time would still need to be settled, explained the philosopher. Physicists could respond: "If you cannot in principle measure something, it surely does not exist." But Bergson had evidence to the contrary—evidence that time existed and was not entirely measurable. What made something measurable? How did changes in measuring techniques affect our sense of reality? The possibility of measuring certain things, and our desire to measure new ones, changed radically throughout human history. For that reason, scientific arguments based solely on measurement *results* were sorely deficient descriptions of nature.

The debate between the two men soon engulfed one of the foundational concepts of modern science: experiment. What proof did Einstein have against the propositions of Hendrik Lorentz, Henri Poincaré, and Henri Bergson? Could the debate between Einstein and Bergson be decided experimentally? The experimental method, after all, was exactly what set science apart from other endeavors, humanistic or practical.

Let us travel to the Société française de philosophie, the same forum where Einstein encountered Bergson, but arrive one year earlier. During the spring of 1921, the philosophers of the Société française de philosophie invited Michelson to discuss his recent research.[3] Michelson was by then an established experimentalist who had received the Nobel Prize in physics for 1907.

Why was Michelson invited to the Société française de philosophie? What could be of interest to philosophers? In the days preceding Michelson's visit, Bergson met with him. "I had a conversation with him that interested me greatly," he explained to Xavier Léon, who would later host Einstein.[4] Bergson would continue to meet with the experimental physicist and converse with him, using what he learned to build a case against Einstein.

THE CRUCIAL EXPERIMENT

The Michelson-Morley experiment is usually considered the paradigmatic example of a *crucial experiment*—one that blindly and justly helps decide between competing hypotheses; one that settles debates and lets the facts of nature speak for themselves; one that is the arbiter between right and wrong, truth and error; one that becomes a model for all sciences, even the social sciences, that aim to be experimental.

The concept of the crucial experiment is much older than the Michelson-Morley experiment. It is usually traced to Francis Bacon, who used the term *instantia crucis* to refer to an experiment that would prove one of two hypotheses right and disprove the other. Yet many of our ideas about what scientific experiments are stem from a widespread misunderstanding of Michelson and Morley's experiment. Einstein differed with Lorentz, Poincaré, Bergson, and even Michelson himself in how they interpreted it.

Michelson's shock at seeing his experiment be used as proof for Einstein's theory of relativity can be understood in part because he did not create his experiment as a test for it. In fact, he undertook it more than a decade before the theory even appeared, in 1881, and then repeated it many times, often hoping for a different outcome.[5] From the moment of its first formulation to the end of his life, Michelson would not believe many of the conclusions that scientists drew from his own experiment—especially not Einstein's. He identified relativity theory mainly with Lorentz's work and his equations more than with Einstein's particular interpretation of them. "These [the famous Lorentz transformation equations] contain the gist of the whole relativity theory," he explained.[6]

In 1907 Einstein started claiming that the famous experiment was central to his theory although in his famous 1905 work he had not drawn a connection to this experiment at all.[7] Two years later, he thought it was essential. From that day onward, Einstein frequently referred to the Michelson-Morley experiment as offering unambiguous proof for his interpretation of the theory of relativity. Was it?

Did Einstein have all the experimental proof he needed? Bergson did not think so. He criticized Einstein for not giving enough importance to experimental results. Ironically, the philosopher wanted even *more* weight placed on the results of the Michelson-Morley experiment than even Einstein himself. In some of the first sentences of *Duration and Simultaneity*, he complained that "the theory of relativity is not precisely based on the Michelson and Morley experiment."[8] Bergson insisted that one should go back to those experimental results and to the formulas attached to them in order to properly understand Einstein's contributions. This exercise, furthermore, would lead readers to understand the role of experiment in the sciences more generally. Because of his no-nonsense reliance on experimental facts, Bergson found that his argument was most easily understood and accepted by scientists with a practical bent: "The only ones who seem to have understood my book are some polytechnicians and engineers."[9]

Einstein initially cited the Michelson-Morley experiment as proof that the speed of light was "constant" since even the fast velocity of the Earth (approximately 1600 km/h at the equator) did not seem to affect it. Some of the most dramatic consequences of relativity theory and its paradoxes followed from the constancy of the speed of light. Light, scientists had noticed, did not seem to behave like anything else known to us. If a traveler on a train threw a ball toward the direction of the train's movement (say at 20 mph), when seen from the outside the train, the ball would be traveling at its initial speed (20 mph) *plus* the speed of the train. But if the same observer sent a ray of light—instead of a ball—in the same direction, its speed would not have to be added to that of the train. The velocity of light would be the same for the person on the train as outside, since it did not need to be included as in the classic Galilean case exemplified by the train and the ball. According to Einstein, the experiment also proved that the ether—a substance believed to fill empty space—did not exist. In consequence, no stable, single background could be used as a reference point to regulate moving phenomena. In other words, any frame of reference was just as good as any other—none could have a privileged status.

Initially, however, the experiment was designed for an entirely different purpose. Michelson conceived of it as a way to reveal the effect of the Earth's velocity on light waves. Why did this experiment become so important for Einstein's theory years later? Why did it become so crucial for discussions about the nature of time?

Michelson's experiments about the speed and behavior of light, it turns out, were undertaken in efforts to find *better measurement systems* than science could offer at the end of the century. They were part of broader efforts to find better ways of measuring time and length. Einstein was profoundly aware of initiatives to find absolute standards of measurement. His knowledge of current measurement techniques and limitations was so thorough that in 1901, when looking for a job, he described himself as "a mathematical physicist familiar with absolute measurements."[10]

A VIEW FROM NOWHEN: SEARCHING FOR A "NEW CLOCK"

By the end of the nineteenth century, scientists had pointed out too many problems with the traditional way of determining time: a system based on the rotation of the Earth against the stars. Scientists referred to this method as the sidereal clock. They had noticed, in comparing the Earth's rotation against mechanical clocks and in following through all the implications of thermodynamics, that the Earth's velocity was slowing down. So if the rotation of the Earth against fixed stars could only keep time imperfectly, what new clock could take the place of the standard yet defective sidereal clock system?

Archimedes is said to have exclaimed, "Give me a place to stand outside the Earth and I will move it." The need for such a place was felt not only for the purpose of accumulating forces through a smart use of a fulcrum, but it was important in seemingly more modest ways: for measuring. To measure with precision, something had to be compared against something else, but this something else had to be different from the object measured. Ideally, this something else should be unchanging, so that subsequent measured values could remain comparable. Ideally, it should not be damageable by human or natural actions, revolutions or cataclysms, so as to withstand the vicissitudes of time and history. Ideally, it should be mobile, easily accessible, and reproducible, so that the standard, or a copy of it, could be used over and over again. Ideally, everyone should agree on the same standard. Some of these ideals, it turns out, conflicted with each other.

To properly measure time, scientists needed such an absolute standard, something unchanging, unaffected by the stresses and strains of the universe as we know it. We can nearly say that what they needed was something divine, for god and divinity were frequently understood by these very same qualities. They needed a view from nowhere.[11] But they also needed a view from nowhen.

The quest of finding a natural standard of measurement had consumed scientists since the time of the French Revolution to the moment when Michelson's standards were proposed in the first decade of the twentieth century. Standards of measurement were so important that determining them was the main source of scientists' income after modern nation states replaced previous royal patronage systems.[12]

Was light a better clock than the rotation of the Earth? Could one second of time be defined as the time taken by light to cover a certain fixed distance? Since the late 1870s, many scientists asked if the speed of light could be used to measure time more accurately. If scientists could accurately determine a certain distance, and if they could find something that covered this distance at a constant speed, they could define a unit of time as the time needed to cover this particular distance. What about length? Could it also be used to define a distance? Michelson and others thought so. He worked hard to create such a system.

To implement an alternative measurement system based on the properties of light, scientists needed to measure lengths and the speed of light as exactly as possible. In 1879 Michelson measured its value to be 299,944 ± 50 km/s.[13] These experiments were state of the art—it was extremely hard to measure something as quick as light. Only a handful of previous researchers had attempted this feat. But when Michelson compared his number against theirs, he found that it differed significantly with previous measurements. Why was his result so different?

LIGHT AS A STANDARD OF LENGTH

Perhaps the problem lay with the unit of length used by Michelson and others. Many scientists had good reasons to believe that the length of the meter—a length that was ostensibly related to precise astronomical constants—had been mismeasured.

At the end of the eighteenth century, French scientists defined the meter as a 1/10,000,000th part of the Earth's quadrant circumference. They constructed a platinum prototype of this length, kept it safe in an underground vault, and devised a system for making copies so that all other measures of length would derive from this unique standard. Yet this platinum meter bar was much maligned. Despite cover-up efforts by certain scientists intent on promoting a different view to the public, uncertainties and doubts about its value (and that of other standards, such as the yard) remained strong.[14] Up to the first decade of the twentieth century, the "international" meter bar kept in Sèvres could not hide its imperfect origin, which seemed more Gallic than universal.

Improving on current standards of length was particularly urgent because of the criticisms they faced. Some of the debates pertaining to the meter bar were due to uncertainties in previous measurements of the Earth's quadrant circumference. Estimates of this value were undertaken in ancient times by measuring the difference between the angles of shadows at two different places, but these measurements were only approximate because the Earth was not perfectly round. An alternative technique was proposed at the end of the eighteenth century when a portion of the circumference was measured step-by-step by two traveling astronomers who embarked on "the most important mission that any man has ever been charged with."[15] Armed with portable theodolites used for cadastral surveys, they determined the difference in longitude between Dunkerque and Barcelona by going there and measuring the distance they covered. Extrapolating extensively from these measurements, they determined the length of the Earth's circumference.

The measurements they brought back were so confusing, and so filled with errors, that they led scientists to classify the errors into three main types: random, systematic, and personal errors. But despite prompting advances in the mathematical theory of errors, uncertainties (in the determination of the Earth's circumference) spread to those of the meter bar based on it. Napoléon did not accept defeat easily, and this was one of his pet projects. In 1800, he unveiled a platinum meter bar based on these measurements, crowning it as a natural standard against which all measures of length should be compared.[16] Almost a century later, scientists would ask if instead of basing it on the Earth's circumference, they could use something much smaller: wavelengths.

Michelson took on the quest of redefining the length of the meter by reference to wavelengths. His work on this topic became a key reference for a growing number of scientists who considered light waves a better standard for measurements of length and their frequency as a better standard for time.

The idea of using light waves as standards had been seriously considered by the eminent physicist James Clerk Maxwell, known as one of the founders of electromagnetism. In his famous *Treatise on Electricity and Magnetism* (1873), Maxwell asked how the sharp frequency lines in the spectrum of chemical substances could be used as ruler marks. He proposed to leave behind determinations based on the size of the Earth's circumference and also to abandon the controversial measures of the Earth-Sun distance, to focus instead on the spectrum of the sodium atom.

By the end of the nineteenth century, scientists had greatly improved their techniques of spectral analysis. The German experimenters Robert Bunsen and Gustav Kirchhoff pioneered these investigations in the 1860s. Soon thereafter, spectral analysis became widely used for determining the chemical makeup of a variety of substances. Could spectra be used for something else? If the vibration frequency of molecules was constant, scientists could use the distance between vibrations as a standard of length, and the frequency of the vibrations could be used as a standard for time. Standards based on these molecular spectra were better, argued Maxwell, because molecules were "imperishable," "unalterable," and "perfectly similar."[17] Maxwell saw in them—and in their peculiar stability—the undeniable mark of God.

At least since the times of Democritus, the idea that atoms could be absolute units was seriously considered. Perhaps these indivisible building blocks of nature could be used as units on which to base physical constants. Critics of Maxwell did not believe that molecules and atoms were as unchanging as he claimed. Thomas Huxley, Darwin's famous champion, doubted that "atoms are absolutely ingenerable and immutable," insisting that "the supposed constancy of the elementary atoms" was as baseless a belief as "the constancy of species."[18] In an evolutionary universe, nothing seemed unchanging.

Michelson did not give up. He explored the possibility of basing standards of length on vibrating light waves by trying to make the waves themselves into a standard. He described his interferometry experiments as designed for "making a light wave a practical standard of length."[19]

Metrologists were particularly enthusiastic about Michelson's research, and they decided to bring the American pioneer to France. From 1892 to 1893, he worked at the Bureau international des poids et mesures to determine standards of length. Michelson's stay in France culminated with the construction of a costly apparatus designed to compare the length of the meter against wavelengths from various light sources. While in France, he famously measured the meter against the wavelength of cadmium light.[20] A few years later, this standard received further confirmation with the work of Alfred Pérot and Charles Fabry, who built another interferometer to determine the length of the standard meter by comparing it against fixed light waves.[21]

According to its supporters, Michelson's light standard for length completed the eighteenth-century dream of finding a standard that was based on a natural constant—and one much easier to measure than the circumference of the Earth. Michelson's boss, the director of the Bureau international des poids et mesures, could not be happier. With modifications, Michelson's method continued to be employed in the twentieth century. In 1967 the meter was redefined as 1650763.73 times the wavelength of the orange-red light emitted by ^{86}Kr, a natural krypton isotope. The basic premise of light-based standards remained the same: the use of either the frequency or wavelength of well-defined chemical sources produced under fixed conditions.

Michelson was showered with accolades. The man who first sponsored Michelson's work in France explained how at the fin de siècle scientists had finally found an ideal natural standard—one sought intensely by the Jacobins after the French Revolution and eluding scientists until the twentieth century. The next director of the Bureau international des poids et mesures, Charles Édouard Guillaume, expressed the importance of Michelson's work for metrology: "The classic research of Michelson and Morley . . . has placed in the hands of physicists rays that can be standards of length . . . and, at the same time, measurement instruments."[22] In 1907 Michelson was crowned as the man who was able to prove that an absolute standard of measurement existed, and he received the Nobel Prize for his work related to this endeavor. In an address to celebrate his award, a colleague explained: "It is to Michelson's eternal honour that by his classical research he has been the first to provide such proof."[23]

TESTING THE LIGHT STANDARD

Physicists were largely delighted with light-based standards of measurement, but another group of scientists were not so happy, to put it mildly. Astronomers found that when they used the value for the speed of light as it had been calculated from experiments on the Earth's surface in their formulas, the results did not match previous calculations or previous observations. Even the basic dimensions of the universe and the velocities of the planets appeared to be different. Calculations involving lengths were a problem, and so were those involving time.

The lack of consensus surrounding basic astronomical standards only got worse after Michelson's research. Soon after the American scientist published his work on the speed of light, astronomers quickly calculated the value of the Earth-Sun distance using his result instead of their own suspect observations (based on observations of the transit of Venus). The two did not match. This new way of calculating the coveted constant only increased uncertainty and fueled competition to determine the true value. By the end of the nineteenth century, disagreements about the exact value of the Earth-Sun distance remained enormous: about 1,500,000 miles.[24] Some sided with the physicist's calculations, sidelining those of the astronomers. By 1881 Michelson's value on the speed of light was a strong contender among the available alternatives that could be used for determining the Earth-Sun distance.[25]

Initially, measurements of the speed of light had been deduced from astronomical observations. In the seventeenth century, Ole Rømer, while working at the Paris Observatory, famously calculated the speed of light from observations of Jupiter's satellites.[26] Rømer's measurements therefore depended on the value of the diameter of the Earth's orbit around the Sun. Because it depended on the Earth-Sun distance, it was subject to the changing dimensions of an unwinding and slowly but surely shrinking solar system. In contrast, speed-of-light measurements done on Earth, such as Michelson's, which used a beam of light and bounced it off mirrors on the Earth's surface, would be largely unaffected by any changes in the solar system.

When the two values for the speed of light were entered into the classic formulas of celestial mechanics, the results of many calculations were slightly but significantly different. Which ones were correct? Why did they differ? "In the terrestrial methods of determining the velocity of light, the light comes back along the same path again," noted Maxwell, thinking that the light's back-and-forth trajectory might be the cause of such differences.[27] To determine if measurements of the speed of light or other measurements based on light waves could serve as absolute standards, it was important to investigate how the speed of light would change in different situations and directions.

Michelson created his famous Michelson-Morley experiment as part of his efforts to obtain a better standard for measuring time and length that both physicists and astronomers could use. What did Michelson find from his work with Morley? Nothing. Did he find what caused the difference between the values found by astronomers and those preferred by physicists? No. He looked and looked, moving his instrument from laboratories to mountain peaks to the quiet deserts of California, and—in disbelief—he did not find any effect of the ether or of the velocity of the Earth on light waves. The result of his experiment was simply null.

But if the ether was not the culprit, why did the speed of light determined physically differ from that determined astronomically? Michelson was not sure why. He, like so many others, started pointing fingers. They found many potential culprits and many possible sources of error.

Perhaps there was an error in the number physicists were using for the Earth's velocity? If this error was corrected, perhaps their numbers would now agree. In narrow terms, Michelson explained his motivation for creating the famous experiment simply as offering another measurement of the Earth's velocity: "In 1880 I conceived for the first time the idea that it should be possible to measure optically the velocity w of the Earth through the solar system."[28]

A popular technique for measuring the velocity of the Earth was based on the concept of stellar aberration. When looking at the stars, scientist had to tilt their telescopes to compensate for the Earth's movement in much the same way that someone walking rapidly with an umbrella under the rain has to tilt it slightly forward to gain more protection. The most reliable late nineteenth-century measurements of the Earth's speed using measurements of stellar aberration attributed to it a velocity that was 1/10,000 the speed of light. This number depended on the concept of a stationary ether as developed by the scientists Augustin Fresnel and George Stokes. Michelson concluded from his investigations that there was something wrong with the current theory of stellar aberration and that this error spread to values of the velocity of the Earth derived from it. This could explain the null result, but there were other possible explanations as well. Michelson considered at least two of these. One could be due to how the Earth dragged the ether along with it, so that no "ether wind" could be found at the Earth's surface, where the apparatus was firmly located. Another explanation was called the Lorentz-Fitzgerald contraction hypothesis, which argued that the length of rigid bodies changed according to their motion through the ether.

In 1892 Lorentz explained the null result of the Michelson-Morley experiment through a controversial contraction hypothesis. He argued that a shortening in the length of the apparatus in the same direction of the Earth's movement countered the ether's effect on light waves. Because of this contraction, the effects of the Earth's velocity on light waves could not be found. Soon after this explanation was first introduced, Michelson himself considered this hypothesis a "rather startling though not at all improbable explanation of the negative result."[29] A few years later, once he had wrapped his head around this possible explanation, he considered his experiment as providing potential evidence for Lorentz's hypothesis where "the length of all bodies is altered (equally?) by their motion through the ether."[30] Michelson and Lorentz were soon joined by Poincaré and many others who tried to answer these questions.

Einstein's portrayal of Michelson's experiment as one that unambiguously proved that the ether did not exist and that previous notions of simultaneity, time, and space needed to be completely overhauled seemed to Michelson and others far-fetched, at the very least. Since its inception in 1800, scientists could not fully trust the current definition of the meter bar. By midcentury they could not trust the stability of the solar system and of the sidereal clock based on it. A quarter of a century later they could no longer trust the Earth-Sun distance.[31] Soon they would admit that they could not even determine, through electromagnetic means, that the Earth moved through space. What had they learned from all their experiments, calculations, and observations? As electromagnetic communication networks increasingly crisscrossed the globe (both telegraphic and wireless), scientists became increasingly sure about one thing: the behavior of light on the surface of the Earth.

Was the speed of light a better clock? Was it, as well, a better measure for length? "How do we go about measuring?" asked Poincaré in the article that advanced ideas that would later be considered as most similar to Einstein's. "The first response will be: we transport objects considered to be invariable solids, one on top of the other. But that is no longer true." In cases involving electromagnetic phenomena, "two lengths are equal, by definition, if they are traversed by light in equal times."[32] As scientists became increasingly unsure about other standards, particularly astronomical ones, they gained confidence in the possibility of defining the meter in terms of light waves. Part of the success of this standard was the result of its constant velocity in a vacuum: Michelson had been unable to find an effect from the Earth's velocity on the speed of light. Was its use as a measurement standard justified? If so, could it be used to determine other coveted physical constants that were increasingly hard to determine due to the changing dimensions of the universe? Scientists could finally have a better standard for measuring not only length but also time, and what is more, for using the same standard for both.

CONSTANT VELOCITY, OR RECIPROCAL CHANGES IN TIME AND LENGTH?

Einstein's first successful entry into the contemporary debates pertaining to time measurement standards appeared in his famous 1905 paper. In that epoch-making work, he simply considered what would happen to the laws of physics *if* the velocity of light was considered to be constant.[33] Could it be used as a standard for measuring other astronomical values? If it were constant, scientists would have in their hands a potential solution to the century-long quest for setting and determining some of the most important values of celestial mechanics.

Bergson noted in his controversial book how problems with "the sidereal clock lead to consequences that impose the adoption of a new clock." He agreed that the "light clock—that is to say, the propagation of light"—was the best available alternative.[34] Yet he insisted that using the speed of light as a way to measure time was equivalent to agreeing on a new clock or agreeing on a way to measure of time, and not, as Einstein proposed in Paris, the most "objective" way of understanding time—and certainly not a reason to force scientists and philosophers to completely change their current understanding of it.

Bergson repeatedly reminded his readers how debates about *time standardization* and *measurement units* were essential for understanding the theory of relativity. Its paradoxes, he claimed, arose because of how time-measurement units were defined in it. The speed of light was the same for an "extraterrestrial onlooker" traveling close to the speed of light as it was in a system at rest. Bergson agreed with that essential fact. But he stressed that it was only true because of reciprocal changes in the units for time and length. "My units of time appear to him as longer than his own," he insisted. An observer would determine that light was propagated with the same speed in every direction, but only because the units used to measure its velocity where changing too: "I am measuring with a ruler whose length he sees changing."[35]

The question for Einstein's critics who doubted the conclusions he drew from the claim advanced in his 1905 paper that the speed of light was "constant" was, how would one know? The problem was particularly pertinent because many philosophers (such as Poincaré) and many scientists felt that there was no practical way to determine that standards—any standards—were absolutely constant and unchanging. To measure a constant velocity one needed an absolute standard of length *and* an absolute standard of time. Some scientists of the time felt they had neither. If Michelson's light waves were used for length and time, any reciprocal changes would cancel each other. Could this be a positive quality for a standard?

Poincaré, who actively participated in debates pertaining to the standardization of time, along with many other scientists of his time, was convinced that standards (especially standards of time) were conventional and not absolute. Some of the most important scientists had lost hope in this quest, believing that absolute standards could never be found. The astronomer Charles Nordmann concluded firmly: "We will never find in nature a standard, a fixed unit." This was simply an "impossibility" leading to the dramatic conclusion that

> In the eternal swell that jolts us, carries us and at times drowns us, not only is there not a rock for us to tie our forsaken boat, but the buoys themselves that we lay to measure our traveled path, are naught but

> floating mirages. And at the mysterious depths of things, our anchors slip without ever grasping the floor.[36]

It would be tempting to think that scientists could measure time straight from their laboratories, without reference to the stars and without needing to deal with the perturbations of the unwinding universe. Of course, scientists would have to agree on a particular standard and procedure—a formidable task given the competing interests of the different parties and nationalities involved. But the problem of time standards was even more complicated than was suggested by the already difficult task of bringing different nations and groups into agreement. Since midcentury, scientists had not only debated whether one standard was better than another for some or other purpose (such as for traveling in the desert or timing a harvest), or whether the decision should come from some person or other in power (such as a Roman emperor or a Holy Pontiff), or whether it benefited one group over another (such as railroad engineers over astronomers), they also increasingly grew doubtful that absolute standards existed in nature.

How important were debates about standards of time to relativity? The issues at stake were very different from previous debates about the standardization of weights and measures—scientists were not simply debating the use of a particular standard, such a yard or a meter. Einstein's solution to the scientific problems of his era involved considering that both time and length units dilated under certain circumstances. It did not matter how these units were expressed because the units *themselves* were changing. Scientists and philosophers were well aware that they could adopt a different system to define time and length units in ways that would compensate for these dilation effects. Or they could accept Einstein's brilliant offer. For the most part, they did.

Even after Einstein's theory of relativity was expanded significantly, criticisms pertaining to its relation to light-based standards continued to linger to the century's end. In the 1970s, the Harvard physicist Léon Brillouin described the problems in determining the constancy of the speed of light given that time and length were *both* defined using light waves: "The unit of length is based on the spectral line of krypton-86 . . . and the unit of time is based on the frequency of a spectral line of cesium. . . . Hence the same physical phenomenon, a spectral line, is used for two different definitions: length and time." Under this system, any change in the velocity of light could go undetected because the changes would cancel out when length was divided by time: "It should be stated, once and for all, whether a spectral line should be used to define a frequency or a wavelength, but not both!" he wrote with complete exasperation.[37] Since length and time were both defined through light waves, scientists were left with no other standard with which they could measure the velocity of the waves themselves.

Brillouin protested that "with the legal definitions of length and time it seems rather difficult to check experimentally" some of the claims of relativity theory. Brillouin blamed metrologists: "This raises a very real problem of metrology."[38] It also raised a very real problem for science and for the philosophy of science, leading thinkers to reconsider the effectiveness of experimental "proof" in one of the most successful theories of modern science and in science in general. Brillouin, who had grown up in Paris, emigrated to America after the war. Toward the end of his brilliant career, he wrote *Relativity Reexamined* (1970), which addressed some of these questions. He remembered very well being present at the conferences at the Collège de France during Einstein's legendary visit to Paris.[39]

CHAPTER 9

The Debate Spreads

GENEVA, SWITZERLAND

When the philosopher Léon Brunschvicg first spoke on April 6, 1922, he said something in a different tone of voice. He started his comments generously, remarking on "his excitement at greeting among us a man who, by his work as scientist and philosopher . . . aggrandizes our idea of humanity." But he also mentioned another reason. Brunschvicg's enthusiasm at meeting Einstein was not limited to the physicist's scientific work. The transcriber of the meeting placed his words in parentheses: "by his work as scientist and philosopher (and for other reasons as well, on which I will not insist, but which he surely knows are on our minds)."[1] What could these "other reasons" mentioned by Brunschvicg be?

A few days after Einstein and Bergson met in Paris, the two men were to meet again in Geneva at the inaugural meeting of the International Committee on Intellectual Cooperation (CIC), a forerunner of UNESCO and one of the most prestigious branches of the League of Nations. Five days before (on April 1, 1922), Bergson had been unanimously confirmed as its president. Einstein would later join as a member.

How did Bergson's and Einstein's involvement with the League of Nations affect discussions about the nature of time both within and outside the institution? The League of Nations was one of a host of competing institutions that tried to forge agreement about time standards and distribution protocols. The CIC was founded on the hope that scientists and intellectuals could lead nations by showing them how cooperation and agreement across borders could be forged, peacefully and civically, so that senseless wars and conflicts would be avoided. It initiated a wide range of diplomatic activities pertaining to elementary and secondary education, the creation of bibliographic databases, faculty and student exchanges, intellectual property, nomenclature, scientific organization, dissemination and access to knowledge, classified research, and so on. Besides its obvious political mission, the League of Nations spearheaded initiatives to standardize the calendar and to set and coordinate world time.

Which standards of time would ultimately prevail would not only be a political victory for a particular international body, they would also reflect scientists' views about the properties of time and space and how they should best be measured. What is one second of time? The answer to this question was complicated, especially after scientists realized that the common way of defining a unit of time by using the rotation of the Earth against the stars (the sidereal clock) was subject to numerous errors. Einstein was directly concerned with time standards, understanding them in terms of how they were measured by a "light clock." In good clocks, time intervals need to be as identical as possible. If certain properties of light (including its frequency and propagation speed) were constant, light would be an ideal clock. Time intervals could be determined in terms of light traversing a certain distance and lengths by the distance covered in a certain amount of time (as defined above). When time and length were defined in this way, the strange dilation described by relativity were readily apparent.

By the time of their debate, both Bergson and Einstein were very active outside their specific disciplines.[2] Bergson was perhaps the most politically engaged intellectual to emerge in France since Émile Zola. He was a major political asset during World War I, when he was enlisted by the French government to participate in two high-level "diplomatic" missions. In 1916 he went to Spain to try to persuade it to side with the allies. The next year he traveled to the United States to help convince Woodrow Wilson to enter the war against Germany. His vocal endorsement of the war against Germany won him many enemies and a few friends. He gained a reputation as a forceful intellectual leader, becoming the president of the prestigious Académie des sciences morales et politiques.

Einstein's involvement in politics started after Bergson's. He publicly took a pacifist stance during the First World War and started to participate in the New Fatherland Association, a group advocating early peace and the formation of a federal structure for Europe. Soon after the war broke out, many prominent German intellectuals signed the famous Manifesto of the 93 (including Max Planck) to express their adherence to the German nation and army. Einstein, who had recently moved to Berlin, and three other colleagues, drafted their own countermanifesto but managed to convince only one graduate student to sign.[3]

After his 1919 launch to stardom, his energy and that of his wife was directed toward helping his fellow Jews. In the spring of 1920 he started offering accredited university courses to Jewish students unable to register at German universities because of (unofficial) quota limits.[4] During those years he became a strong supporter of cultural Zionist causes, advocating "large-scale Jewish colonization" of "Palestine" and the creation of a university in Jerusalem.[5] His visit to the United States in April 1921 was explicitly designed to raise funds for the Hebrew University of Jerusalem.

By 1922 Einstein was deeply involved in various political causes for more than half a decade already. He was a symbol for science and for politics. "I would quite like to say a word to one or the other serious politician, if the opportunity presented itself," Einstein explained to Langevin, in a letter preparing his trip to Paris. He wished to do "something," he explained "against the harm that is being borne out into the world from your fine city."[6] Einstein wanted to use his trip to Paris "to have occasion to speak with scientists about the possibility of restoring international relations in the scientific world, but not just with well-wishers and pacifists."[7] The initiatives of the League of Nations were not enough for him.

Einstein continued to advocate political causes throughout his life. After moving to America, he became famous for drafting a letter to the president alerting him of the potential dangers of nuclear weapons. He was active in the civil rights movement, taking personal risks during the McCarthy era and earning a hefty FBI file.

A political Bergson-Einstein debate was waged alongside the scientific and philosophical debate. Bergson was the single most politically committed French intellectual of his time. Einstein's participation in the CIC was equally crucial. Bergson needed him. Acknowledging that the CIC "will only succeed, will only impose itself by the prestige and the authority of its members," he worked hard to include Einstein.[8] But disagreements between the two men plagued the committee, which finally failed in 1939 in the face of the upcoming Second World War.

THE POLITICS OF TIME

Politics affected discussions about time directly, particularly in connection to the success or failure of institutions vying to set and maintain time standards. From the mid-nineteenth century to the early twentieth, the task of standardizing time internationally seemed so daunting, so passionately fought, and so difficult that many scientists placed their hopes in the League of Nations, which had been created in Paris in the aftermath of the peace conference of spring 1919 to regulate disputes among nations—including time. The standardization of time had progressed tremendously after many nations in the late nineteenth century agreed to set the prime meridian in Greenwich, England. But some prickly issues remained, particularly with respect to how time related to the calendar. Gaining agreement on these two topics was considered to be a contribution to—and in fact in some instances, even a prerequisite for—world peace. Uncertainties in the determination of time led to uncertainties in the determination of longitude. The transmission and standardization of time had clear geographic consequences not only for mapmaking but for fixing national borders. Both could easily lead to diplomatic conflicts and even to wars. Settling border disputes was an extremely pressing need, particularly the one between France and Spain concerning the border in Morocco, the disagreement between France and Germany over the Congo-Cameroon border, and the territorial disputes between France and Italy in Tunisia.

Most members of the League of Nations wanted all nations to adopt one calendar and agreed that it should be the Gregorian calendar. They hoped to integrate religious holidays within this calendar, so that the latter would finally be fixed and not change from year to year. The date of Easter, for example, varied as much as thirty-five days and changed every year. But the question of Easter was extremely complicated. The need to adjust time according to time zones had been extremely controversial, but attempts to adjust religious holidays was even more divisive. The commission that was dedicated to this problem resolved to be joined by representatives of the Holy See and the Orthodox and Anglican churches.

Charles Nordmann, the astronomer from the Paris Observatory who greeted Einstein at the border during his visit and who rode with him into the city, believed that "the general acceptance of the Gregorian calendar can and must be part of the condition imposed on the world by the League of Nations."[9] Nordmann, however, hoped that questions touching on religious issues, like that of the date of Easter, would not be part of the discussions. "From a strictly dogmatic point of view, in what concerns the date of Easter feasts and other more general questions, the reform of the Gregorian calendar will not take on difficulties of a nature which it considers insurmountable," he recounted.[10] The benefit of the Gregorian calendar, according to him, resided in that it lay "at the center of gravity of civilization."[11] The success or failure of the League of Nations hinged on the question of calendar reform: "The outcome of these projects will permit us, in every case, to judge the League of Nations."[12] Nordmann was optimistic: "We count heavily on the League of Nations to reform the calendar. We are right."[13]

When suggestions to adjust clocks back or ahead for a full day or longer were brought up, they inevitably caused anxiety and controversy. Nordmann described how when Bulgarians adopted the Gregorian calendar they suddenly "aged officially by 13 days."[14] And if the astronomers of the Paris Observatory agreed to join the Greenwich standard, a small consolation would be that they would rejuvenate: "Some have been consoled in thinking that rejuvenating 9 minutes and 21 seconds by the force of law is not a trivial blessing."[15] The Russians would not agree to change the calendar because "the mass of ignorant muzhiks" would not want to "age" those days and a single "ex-beautiful woman" from "down there" could singlehandedly bring the reform to a halt.[16]

A group of astronomers presented a new perpetual calendar for "consideration *without recommendation*" that boasted various advantages: it would only "need to be printed but once," every new year would begin on a Monday, and "holidays [would] *always* fall on the same day of the week."[17] But ambitions to reform the calendar were curbed by its explicit commitment to avoid "far-reaching reforms touching the every-day life of the people, *at the present moment*, even though such reforms seem in themselves desirable."[18] Reluctance to push for bolder reforms was due to "this very unsettled time in the world's history" when "all sorts of reformers, wise and otherwise, are springing up advocating changes in all sorts of things."[19] Many scientists hoped that the League of Nations would help them push these reforms along. The hope of its members was that nations could start agreeing on bigger issues once they settled on the more pragmatic ones.

Scientists working within international organizations played important roles in initiatives to standardize instruments, methods, procedures, and publication and bibliographic practices. The organization of the League of Nations (and of the CIC within it) was modeled after previous scientific international commissions that had tackled questions of standardization. Global sciences (such as geodesy and meteorology), global industries (such as electric, telegraph, and rail), and global standards (time, longitude, weights, and measures) depended centrally on international agreement across nations.

TIME AND THE LEAGUE OF NATIONS

The League of Nations (and the CIC within it headed by Bergson) faced direct competition for questions related to scientific standards from the International Research Council (IRC).[20] The IRC constituted itself as an umbrella institution for many others—including the committee responsible for determining the very unit of time. In a decision that would have important consequences for the theory of relativity, in 1919 the Astronomical Union of the IRC set up the Committee on Standard of Wave-Length.[21] The decision to eventually "place metrology in the British Empire on a wave-length basis" was eagerly accepted in the United States by the Board of Trade and the National Physical Laboratory.[22] Other standards, such as those based on the traditional meter bar at Sèvres, would now have to compete against those proposed by this new Anglo-American front.

Einstein was deeply concerned about any intervention from the League of Nations on topics that could potentially affect the definition of time. Could the recommendation to use light waves as standards be at risk? One of Einstein's biographers described his work in the League of Nation's CIC branch as centering on topics "which were technical and belonged to his own domain." Einstein "intervened in the course of successive sessions in favor of proposals . . . such as, for instance, the project for a universal synchronization of comparative astronomic measures, as well as proposals for the needs of telegraphy, etc.; or the proposal for the creation of an international meteorological office."[23]

After the Great War, a number of debates pertaining to the definition, determination, and distribution of time gained a certain closure, at least in practical terms.[24] A worldwide time distribution and timekeeping system had been set in motion during the Congrès international pour la réforme du calendrier, which had met in Liége May 27–29, 1914, with the express purpose of reaching agreement with respect to time and the calendar, but the Congrès had been unable to complete its work because World War I had started later that summer. The task of reforming the calendar had to wait until after the war, when the need for reform was even more pressing than before. The postwar period presented a unique opportunity for gaining consensus with respect to time—the Allied nations could legitimately exclude the others.

In 1919 the IRC urged the formation of the International Astronomical Union (IAU) to regulate and promote scientific standards. In July, during a meeting in Brussels, they started the actual process. Participants determined that the IAU should "adopt" the central authority on time, the International Time Commission (ITC). The ITC would be in charge of distributing time. After extended negotiations, they agreed that the bulk of its work should be carried out by the Paris-based International Time Bureau (ITB), which was located at the Paris Observatory, and as a compromise, that time be "expressed in Greenwich Time."[25] To avoid appearing as simply a Franco-British alliance, the ITB was charged with "harmonizing" time sent to it by various other "National Centers" located at international observatories worldwide and with calculating a mean value from them. From there, it could be sent back to the world as "universal time." The ITC included as its task the regulation of "ordinary" time, seeking to reach "individuals, etc." and stating as its mission the process of "unifying time" signals "whether these are scientific signals of great precision or ordinary signals adapted to the needs of navigation, meteorology, seismology, railways, postal and telegraphic services, public administration, clock-makers, individuals, etc."[26]

How did this new institutional organization affect science? The IRC, through the IAU, recommended that standards should eventually be determined by reference to light waves, asking "that the question of a primary standard of wave-length be held in abeyance—that is not to be reopened." Time and length standards, they concluded, were "a matter belonging entirely to the Committee on the International Bureau of Weights and Measures."[27] With this decision, the path leading to the eventual adoption of light waves as a privileged unit of measuring time and length was cleared. This decision, more importantly, favored scientists, such as Einstein, who analyzed space and time concepts in terms of the behavior of light waves.

AGAINST THE LEAGUE OF NATIONS

Einstein initially portrayed his political and scientific outlook as fitting within the collaborative efforts of the league. "The theory of relativity does not topple Newton's and Maxwell's theories, just as the League of Nations does not annihilate the states that join it," he explained.[28] But his initial support for the League of Nations was lukewarm and would eventually completely wane.[29] Einstein missed the CIC's second meeting (in September 1922) because he was in Japan.[30] At that time, the members of the CIC were divided between those who believed that scientists should make any research that could be have potential military applications available to all and those who believed it should remain classified.[31] That day, the committee discussed scientists' research into toxic gases "in order to reduce to a minimum the chances that they would be used in a future war," concluding that making this research public would constitute a grave danger to all.[32] The CIC agreed to leave the issue up to the personal morality of each scientist, a position that greatly disappointed Bergson.

Einstein was explicitly considered as someone who could counter the "Latin element" of France, Spain and Italy, Portugal and Latin American nations in the League of Nations. To try to get him to participate more fully in its activities, a colleague explained to him how "this committee, like all the organizations of the League of Nations, is in danger of having the Latin element overrepresented."[33] Yet Einstein did not feel he could adequately represent Germany in these forums because of his "condition as Israeli, on the one hand, and on the other because of his antichauvinistic feelings from the German point of view do not permit him to truly represent the intellectual milieu and the universities of Germany."[34] He also considered resigning because of prevailing anti-German sentiments. Nonetheless, Einstein was invited as a representative of German science. Einstein was in a difficult position. Marie Curie and others pleaded with him to remain a member, and he agreed.

The physicist resigned on March 1923, publishing a sharply worded public statement against the League of Nations. Einstein protested that it had watched helplessly as the French, refusing to send the problem of Germany's war reparations to arbitration, occupied the German Ruhr region. The government of Raymond Poincaré, Henri Poincaré's cousin, had sent the troops to the Ruhr industrial area to seize control of production. "The League of Nations," argued Einstein, "fails not only to embody the ideal of an international organization, but actually discredits it."[35]

His highly public resignation only made the work of the League of Nations and the CIC more difficult. His behavior appeared paradoxical to many of his colleagues. How could a scientist who preached about internationalism refuse to take place in these outreach activities? After all, he was being invited (indeed, they had pleaded) as a German-born member. Had not Einstein repeatedly protested the exclusion of German scientists?

His colleague Max Born immediately wrote to him upon hearing the news of Einstein's resignation. Could it be true? "The papers report that you have turned your back on the League of Nations. I would like to know if this is true. It is, indeed, almost impossible to arrive at any rational opinion about political matters, as the truth is systematically being distorted during wartime."[36] But Einstein had indeed resigned. He was replaced by Lorentz, who would play an increasingly important role at the CIC, becoming its president after Bergson's retirement.

During this tumultuous period, Einstein considered his theory of relativity in both *political* and *scientific* terms. In a letter to his friend

Maurice Solovine, he connected his decision to resign directly to Bergson's reception of relativity:

> I resigned from a commission of the League of Nations, for I no longer have any confidence in this institution. That provoked some animosity, but I am glad that I did it. One must shy away from deceptive undertakings, even when they bear a high-sounding name. Bergson, on his book on the theory of relativity, made some serious blunders; may God forgive him.[37]

In four short sentences Einstein referenced God, politics, and relativity. As if these topics were not complicated enough, Einstein dated his letter to Solovine by reference to a religious holiday, Pentecost 1923, instead of to the calendar date of May 20, 1923. When months later he was forced to explain his decision to resign from the CIC and to combat views that he was being pro-German, he again explained his decision by reference to his scientific work, stating that his position was consistent with the theory of relativity. In a letter to Marie Curie, he explained: "Do not think for a moment that I consider my own fellow countrymen superior and that I misunderstand the others—that would scarcely be consistent with the theory of relativity."[38]

EXCLUSION AND BOYCOTT

Since the First World War, Bergson had criticized the policies of exclusion of various so-called international forums and academies. As president of the Académie des sciences morales et politiques, he was pressured by a group of members of the Institut de France who demanded the expulsion of foreign associates of German nationality. The philosophers of the Institut, as a group, condemned this initiative. Bergson, during his tenure as president of the Académie drafted a declaration that condemned the war but did not go to the extreme of expelling German nationals.[39]

Bergson's compromise with regard to German scientists was exceptional during those years. After the war, the politics of exclusion intensified as Germans were consistently excluded from most "international" scientific forums. In 1922 the Académie de sciences refused to invite

Einstein to lecture while he was in Paris because Germany did not belong to the League of Nations.[40] An antagonistic newspaper questioned their logic: "If a German were to discover a remedy for cancer or tuberculosis, would these thirty academicians have to wait for the application of the remedy until Germany joined the League?"[41]

The politics of exclusion was more complicated when it came to Einstein, who did not consider himself as adequately representing Germany yet continued to protest against the exclusion of Germans. Einstein boycotted these forums *despite* being invited. After resigning from the CIC, he explained to Marie Curie: "I have requested, furthermore, that I not be invited to Brussels [to attend the Solvay Conference]."[42] Einstein nonetheless constantly referred to the exclusion of German scientists as a reason for his unforthcoming support. Despite the efforts of Curie, Lorentz, and others, Einstein continued to be extremely critical of the CIC and the League. To his friend Michele Besso, Einstein explained how he felt "proud not to have been duped by the League of Nations." That would have been a complete "waste of time and energy," due to the complete "hypocrisy" of the institution.[43]

On March 6, 1924, crisis hit the CIC. Gilbert Murray, a scholar of ancient Greek literature, world peace advocate, and vice president of the CIC, sent a letter to the *Times* accusing the commission of being anti-German. The CIC responded by publishing numerous invitations to German scientists and institutions as evidence of how *they* had turned down these invitations. Bergson, who could not disguise his anger toward Murray and those who were claiming that the CIC was exclusionary, needed Einstein more than ever.[44] When asked how they should best respond to the crisis and to these accusations, Bergson responded: "It would be extremely useful if Prof. Einstein joined the commission again."[45] Bergson quickly sent him a second invitation.

Should Einstein be asked to rejoin, even after he had sent to the press insulting remarks about the League of Nations? The question of reintegrating Einstein into the CIC was also affected by pressure from the British, who sought to profit from two weak points in France: the diplomatic isolation that France's occupation of the Ruhr area had brought on and the concurrent devaluation of the franc. Gilbert Murray asked Einstein if he would join, and he accepted on June 25, 1924.[46] Reintroducing him was awkward for the president. Bergson saluted him as a member who "was at the same time old and new" and whose "reputation is universal."[47]

THE MEETING BREAK

New articles on the conflict between the theory of relativity and Bergson's philosophy appeared in the July 1924 issue of the *Revue de philosophie*. In that volume, the alleged equality of the different times in the twin paradox was again debated. The philosopher Isaac Benrubi, among others, decided to attend the CIC's meeting in Geneva (July 25, 1924) only after learning that both Einstein and Bergson would attend.[48] The fate of the CIC was now colored by the Bergson-Einstein debate. For its participants, the debate was at least as important as the meeting itself.

For the *meeting proper*, Bergson reintroduced Einstein with a flattering introduction, but during the meeting *break*, their differences once again became evident. Benrubi approached Einstein to ask him what he thought of *Duration and Simultaneity*. Einstein offered his official response—that Bergson had not understood the *physics* of relativity; that he had made a mistake. Asked if he would continue the fight against Bergson, Einstein responded: "No, I do not intend to do that, unless Bergson himself provokes a polemic. But that would not help anybody."[49] Was Einstein willing to let bygones be bygones? Reporting to the *Frankfurter Zeitung* right after the meeting, Einstein expressed discontent about the dominant role of the French participants, who were led and chaired by Bergson as president: "It is true that the French mentality may unwittingly have dominated the proceedings to some extent."[50]

Einstein and Bergson did not manage to work together at the CIC. Passions again flared when the French government offered the CIC the option of building the International Institute of Intellectual Cooperation in Paris. Einstein (among others) expressed his concern that the CIC was international only nominally and, in effect, nationalistically French. But Bergson thought he should not turn down the government's generous offer. When Bergson accepted, Einstein became more and more suspicious of the CIC's underhand nationalism. He did not attend the next meeting, which was held in Paris, instead of in Geneva.[51]

Einstein continued to support alternative political forums, even hoping to enlist Bergson in some of them. Together with the Zionist leader Chaïm Weizmann, he invited Bergson to participate in the inauguration of the Hebrew University of Jerusalem. Bergson immediately declined, explaining that although he was "touched" by the invitation he was simply too busy supporting *other* causes. Bergson then asked Einstein once again about his participation in the League of Nations, urging him to please attend the next session.[52]

Einstein's support of Zionist causes seemed, in the eyes of his critic, to be more important than his defense of international causes. Einstein responded by arguing that "my Zionism does not preclude cosmopolitan conception." When pressed to explain himself, he responded that he was "*against* nationalism but *in favor* of Zionism."[53] Einstein's Zionism became increasingly controversial as relations between Jews and Arabs worsened in the Middle East. As he continued to advocate cultural Zionist causes and criticize the League of Nations, he was reproved by some of his fellow German Jews, who did not yet see a necessary tension between German nationalism and Judaism. Fritz Haber, a prominent Jewish German chemist and friend of Einstein who had played an important role in developing and using toxic gases during the Great War, disagreed with Einstein's activism. He worried that it would only fan the flames of anti-Semitism in Germany.[54]

Einstein was totally "bored" by the League on Nations. During the meeting of July 1925, he killed time by thinking about physics. "This letter," he wrote to Besso, "was written during a boring meeting of the League of Nations."[55] In August 1925, Einstein once again criticized the CIC for its two-facedness. Bergson resigned, citing an illness.[56] He completely retired from public life. Lorentz assumed the presidency. But even after Bergson's resignation, Einstein did not increase his attendance; from 1926 to 1930, Einstein attended only three meetings. He even asked if Langevin could go to them in his stead.[57]

When he did attend the CIC, Einstein advocated Germany's participation in the League of Nations. Even after the country was accepted in 1926, he continued to protest the excessive dominance of the French. His complaints escalated after the CIC inaugurated its institute in Paris with Einstein publicly complaining of the "impression of French predominance," especially since "the chairman of the committee has so far also been a Frenchman," that is, Bergson.[58] Einstein believed that the domination of the CIC by the French (through Bergson) was "a fact which is not conducive . . . to international solidarity."[59] Elsewhere the physicist described the initiative as a "keen disappointment" and a "weak and imperfect instrument" that has "by no means fulfilled all the expectations that accompanied its founding."[60]

To the mathematician Jacques Hadamard, he described the league as "impotent."[61] And to the *New York Times*, he described himself as "rarely enthusiastic about what the League of Nations has done or has not done."[62] Einstein explained to Paul Painlevé that "I have always regretted the fact that the institute was established in Paris and financed exclusively by French funds."[63] He advocated instead "to move the institute in toto to Geneva and have all countries contribute to its financial support under a quota system."[64] Together with Paul Langevin, his lifelong ally in the defense of relativity theory, he worked on the prospect of creating an alternative international association of leading intellectuals to exert political influence. His opinion of the institution worsened even more when Alfredo Rocco, the Italian minister of education, who was of a clear fascist persuasion, joined as member. In July 1930 Einstein criticized the committee yet again and resigned for good. Murray, one of the CIC's most active members, wrote about the woes that Einstein's resignation brought to the institution and pointed his finger to a quick fix: "The best solution of all these difficulties is obvious! It is that you should remain with us, but perhaps that is too much to hope for."[65] It certainly was.

THE DISAGREEMENTS CONTINUE

As they disagreed about world politics, the physicists continued to disagree on pertinent scientific issues. Einstein decided to resume attending the Solvay Congresses, which were organized by Lorentz, on April 1926. During the following year's Solvay Conference, Einstein started a notorious debate with one of the founders of quantum mechanics, Niels Bohr. During the meeting, Einstein "expressed . . . a deep concern" over physicists' disagreement about causality in physics, uttering his argument that God did not play dice with the universe: "ob der liebe Gott würfelt."[66] From then on the Solvay Conferences "took quite a dramatic turn" plagued by disputes pertaining to the relation between relativity and quantum mechanics—two areas of science based on radically different conceptions of how the universe worked.[67] Bohr's challenge profoundly concerned Einstein, who "expressed a feeling of disquietude as regards the apparent lack of firmly laid out principles for the explanation of nature, on which all could agree."[68] Defenders of Bohr and quantum mechanics increasingly turned to Bergson for support.

The physicists continued to disagree, but meeting together became increasingly difficult because of the deteriorating political situation. Einstein was unable to attend the Solvay Conference of 1933. What was worse, the tone of their discussions appeared increasingly unfit for scientists, who risked further discredit in the eyes of the public. Bohr explained the urgency of having additional international scientific meetings so that participants would not leave with a negative impression of science. He had in mind Einstein's remarks about what God would or would not do: "Utterances of this kind would naturally in many minds evoke the impression of an underlying mysticism foreign to the spirit of science." Bohr explained how at the "Congress of 1936 I therefore tried to clear up such misunderstandings."[69] Despite his best efforts, the International Congress for the Unity of Science (1936) was yet another display of disunity.

In 1938 the CIC tried to forge some consensus on the main issues splitting the physics community, in particular the quantum mechanics–relativity debate, under the auspices of the League of Nations. "These aspects of the situation were especially discussed at a meeting in Warsaw in 1938, arranged by the International Institute of Intellectual Cooperation of the League of Nations," recalled Bohr.[70]

We now know that the League of Nations did not fulfill its much-hoped-for expectations: agreement, either political, or scientific, or of any other kind. The question of what kind of reality should be given to times marked by the two clocks continued to be debated. And the Einstein-Bergson debate was now entangled with the larger problem of how relativity could be reconciled with quantum mechanics.

By then, Bergson had given up active political involvement. He dedicated the rest of his life to writing his last works, including *The Two Sources of Morality and Religion* (1932), a book whose tone regarding war, peace, and cooperation was influenced by his own experience at the CIC. In particular, it was colored by the CIC's failure with respect to arms control. Bergson explained how the failure of the League of Nations was *not* due to its powerlessness or to its lack of a means of enforcement, as many (including Einstein) believed: "Even if the League of Nations would take an armed form sufficient in its appearance, . . . it would collide with the profound war instinct that covers civilization."[71] The war instinct, according to Bergson, would be better tackled by rethinking the idea of instinct itself and of the role of biology in determining behavior, that is, by not attributing to them such dominant roles.

Einstein thought differently. In 1932 he was asked to publish a volume of correspondence between himself and another leading intellectual on the challenges facing mankind. Einstein considered Langevin first, but later settled on Freud. Freud, in contrast to Bergson, was much more pessimistic about the inevitable direction of history and the role of destructive instincts in humans. While Bergson continued to argue in his book that these impulses could and should be overcome and that philosophy could lead the way toward a better world, Freud considered civilization as inevitably imperiled by them. He argued that the instinctual drive of humans toward destructiveness could never be overcome, basing his argument on evidence from biology, anthropology, and psychology. Between 1933 and 1935, the CIC published the correspondence between Freud and Einstein in *Why War*.[72]

Bergson's response to anti-Semitism and the growing horrors of Nazism was very different from Einstein's. Einstein publicly differentiated his views about Jewish identity from the mainstream Central Association of German Citizens of the Jewish Faith. He did not agree with those who considered Judaism simply as a faith or as a religion, arguing instead that being Jewish was first and foremost about "ties of blood and tradition" and therefore a distinctly racial category.[73] "The way I see it, the racial peculiarity of Jews will necessarily influence their social relations with non-Jews."[74] Bergson did not espouse a racial definition of Judaism, especially not one that placed it in conflict with their national affiliations. In *The Two Sources of Morality and Religion* he firmly criticized the unwarranted extension of biological notions of race to the realm of human culture. The book contained a sharp statement against Friedrich Nietzsche, who was being read—including by Einstein himself—as an authority describing how power relations mapped unto racial ones.

The Nazis rose to power early in 1933. Einstein abandoned his pacifist stance, hoping that another institution "differing from the present League of Nations in Geneva, would have at its disposal the means for enforcing its decisions."[75] He started to advocate an international standing army and police force. The CIC lost momentum and had its last meeting on July 1939.[76] During the years that followed many of Einstein's supporters suffered the consequences of the deteriorating political situation.

CHAPTER 10

Back from Paris

BERLIN, GERMANY

When Einstein returned to Berlin from Paris, he found that the visit had been "unforgettable, but devilishly tiring."[1] Fatigue was the least of his worries. Einstein soon feared for his life.

Walther Rathenau, the man responsible for convincing him to go to Paris, was targeted and killed that summer. A band of men intent on assassinating prominent Jews shot him at close range and finished him off by throwing a hand grenade into his car. His assassination foretold a deteriorating political situation in Europe and the immense difficulties in improving Franco-German relations. Einstein's confrontation with Bergson did not help matters. In Germany, Einstein's work was widely attacked, often because of blatant anti-Semitism. Einstein and his supporters defended themselves bravely, but after his visit to Paris, they would have to deal, in addition, with Bergson's objections and those of his numerous supporters.

Einstein had discussed his theory with Rathenau, who had asked him what would happen if instead of light signals striking a moving train, an assassin would throw a stick of dynamite on the train carrying the czar of Russia. "What startles the czar twice is a single matter for the assassin," Rathenau concluded.[2] Rathenau attempted to extend to this violent scenario the claim that certain events that were simultaneous for one observer (seen once) were sequential for another (seen twice). But Einstein's work showed that a different logic applied to a light signal than to a solid body, be it a vehicle or a stick of dynamite. Relativity was not necessary for understanding the latter. Could he have helped Rathenau in other ways? Could he now help the wider world?

Einstein's trip was considered diplomatic by some but provocative by others. On his journey back, Einstein stopped at battle sites in Rheims and its surroundings to show "how ugly war is."[3] Were the French or the Germans to blame? Bergson unambiguously blamed Germany. Einstein, in contrast, did not point his finger to any nation in particular, feeling instead that all "horrid Europeans" were at fault.[4]

CONCEPTUAL HISTORY: ERNST CASSIRER

Einstein's politics and fundraising for Zionists causes made him more controversial than ever. Some of his closest friends warned him of the dangers, yet many sided with his causes. Einstein received the backing of some of the most important German philosophers, who simultaneously started to retreat from Bergson. By far the most important endorsement of relativity by a philosopher came from one of the most talented men in the profession: Ernst Cassirer. Cassirer was a few years older than Einstein and can roughly be considered to belong to the same generation. He came from a privileged, elite Jewish family and was well established as the chair of philosophy at Hamburg University, where he embarked on an ambitious and highly successful career.

In the spring of 1920 Cassirer sent Einstein the manuscript of a book on the theory of relativity that he hoped to publish. The letter accompanying the manuscript explained to him that his purpose was to "bring about agreement" between physicists and philosophers and to "avoid misunderstandings" among them.[5]

Cassirer's letter, by virtue of coming from a German philosopher who did not outright attack the foundations of relativity, offered Einstein the possibility of a rapprochement with the philosophical community, at least with part of that community in Germany. The offer of a white flag was rare from philosophers, since one of the most prominent philosophical schools in Germany, grouped under the name of the Kant Society, was clearly against some of Einstein's most important claims. A few days after Cassirer's letter arrived, Max Wertheimer, one of the founders of Gestalt psychology, wrote a letter to Einstein warning him how the Kant Society wanted to invite him to Halle "to uncover in public before a philosophical tribunal the elementary absurdities of Einst[einian] theories."[6] He recommended that Einstein not go. Wertheimer warned him that the philosopher Oskar Kraus was poised to attack him. Kraus and others would use Einstein's theory of relativity to prove the main points of the Philosophy of As If, which had been developed years earlier by the philosopher Hans Vaihinger.[7] According to its proponents, scientists worked under the assumption that theories matched perfectly well with reality, but this belief, in the end, was unwarranted.

Wertheimer thought that Cassirer was different from almost all of the other philosophers in Germany. Cassirer, explained Wertheimer to Einstein, was a real ally. Offering an otherwise almost wholesale indictment of German philosophers, he referred to Cassirer alone as a unique person of "earnest intention."[8]

Einstein paid attention and considered to "call Halle off because the blather would make me sick." Instead of going, he dedicated himself to digging into Cassirer's manuscript, which initially he found "less amusing" than playing with his friend's kids.[9] A few days later, he had made up his mind: "*I am not going to Halle.* It would be senseless."[10] And he thanked Wertheimer: "It is very nice of you not to let me fall into the trap."[11] His wife agreed: "How glad I am that you aren't traveling to Halle! All that fuming, for what purpose? You won't convince that sort anyway."[12] Thanks to Wertheimer's and his wife's warnings about "the trap" coming from "that sort" of people, Einstein succeeded in distancing himself from antagonistic philosophers and drawing closer to Cassirer, a person of "earnest intention."

That summer (1920) Einstein responded approvingly to Cassirer's request for comments, concluding that "I think your treatise is very well suited to clarify philosophers' ideals and knowledge of the physical problems of relativity."[13] Their disagreements were minor, but nonetheless existed. Einstein thought Cassirer should place less emphasis on "conceptual tools" than he had done, stating that "I do not think that our choice of these [conceptual] tools is constrained *by virtue of the nature of our intellect*." He also urged Cassirer to place more emphasis on experiment and measurement: "The theory of relativity stands and falls as a *physical* theory," he underlined. Cassirer obliged, making necessary revisions to that end: “In particular, I have now given stronger emphasis to the theory of relativity’s purely empirical point of departure, which, when set against the analysis of theoretical assumptions, certainly did get the shorter end of the deal.”[14] Cassirer had initially operated under the assumption that the nature of our thought processes and the thinking apparatuses shaped what was known—but retreated from this position after being chastised by Einstein.

Cassirer’s support of Einstein was clear from the very beginning of their correspondence. Even before Einstein’s work appeared in print, Cassirer considered the physicist as someone who could “guide us toward a return to critical rationality and staid, factual inquiry.”[15] Why was Cassirer so intent on defending “staid, factual inquiry”? The immediate context of Cassirer’s remarks appeared in light of the attacks that Einstein was facing from the “Syndicate of German Scientists,” who held an event in August 1920 at the Philharmonic Hall in Berlin to discredit the theory of relativity.

After the Berlin anti-relativity lectures, Einstein increasingly described all criticisms of his theory as politically and racially motivated. He explicitly claimed “that there are other motives behind this undertaking than the search for truth.” These attacks would not have been launched “were I a German nationalist, whether bearing a swastika or not, rather than a Jew of liberal international bent.”[16] Einstein explained that “to my knowledge there is hardly a scientist among those who have made substantial contributions to theoretical physics who would not concede that the entire theory of relativity is logically consistently structured and that it agrees with the experimental facts now available.”[17] Einstein was right: most scientists agreed with these two aspects, logical consistency and experimental verification, but many of them still did not agree that all of the theory’s conclusions were absolutely necessary.

Cassirer stood firmly by Einstein because of the dangerous political situation in Germany. He warned that “the more conditions here worsen,” the more “we need the men” like Einstein. He was convinced that attacks on the physicist’s theory were politically, rather than intellectually, motivated. In defending Einstein, Cassirer had a mission: “I hope to be able to do my part in contributing at least a little toward checking the mental confusion of these things that still seem to exist in so many minds and seem to be deliberately exploited by some quarters.”[18]

Despite doing "his part," conditions worsened tragically. In 1933, when the Nazis seized power, Cassirer would be dismissed from his professorship in Hamburg and forced into exile. In America, he would continue his defense of Einstein and fight against Bergson until his death.

In 1921, the year Einstein started to portray the acceptance or denial of his theory as a political matter, Cassirer published *Einstein's Theory of Relativity*, through the prestigious publishing house of his cousin Bruno Cassirer.[19] In publications that followed, Cassirer attacked Bergson.[20] By the time the third volume of *Symbolic Forms* appeared in 1929, it was clear that Cassirer did not think highly of Bergson, nor of the many other representatives of the Philosophy of Life movement in which he grouped him.

KANT'S GHOST

The enemy of his friend was also Cassirer's enemy. What most separated Cassirer from Bergson was the legacy of Kant. The two philosophers held distinctly opposite appreciations of Kant's philosophy. Cassirer was a neo-Kantian, whose work fit within the school of thought of his teacher, Hermann Cohen. Cohen, one of the first unconverted Jews to hold a university professorship in Germany, was an authority on Kant. He offered a highly influential interpretation of Kant's work that influenced generations to come, reading him as a science-friendly epistemologist rather than as an obscure metaphysician. His reading of Kant would soon be contested by Martin Heidegger. According to Cohen, the role of philosophy with regard to science was that of providing an epistemological framework for it. Read in this way, Kant was the perfect model for Cohen, and later for Cassirer, since he showed how science could be both empirical (touching on the concrete) *and* idealistic (connected to the eternal). Bergson was resolutely anti-Kantian and in more ways than one. "Kant's error has been to consider time as homogenous," he explained in *Time and Free Will* (1889).[21]

Cohen had just written a manuscript on the relation between reason, science, and Judaism that "violently critiqued Bergson." When Cassirer read it, he tried to appease Cohen and convince him to tone down his attack. In response to Cassirer, Cohen promised to "reexamine" his text.

As soon as Cohen returned to his house after these conversations, he telephoned Cassirer. In an "intimate and tender voice," he told Cassirer that he had attenuated his critique significantly, so that Bergson's message now appeared "as the most trivial, most insignificant, and most garrulous I have ever read." Cassirer's wife, who was listening in on the call from the second telephone on the line, could not help but burst into laughter. In the end, Cohen heeded his colleague's advice.[22] Cohen's preface to the book, translated as *Religion of Reason Out of the Sources of Judaism*, appeared without a direct reference to Bergson. Yet the claims he advanced throughout could not be more different from those of the French philosopher. Refusing to oppose intuition against reason, Cohen strove to show how science, reason, and Judaism strengthened each other.

What was Einstein's position with respect to these topics? According to Einstein, Cassirer initially went too far in emphasizing the role played by concepts in shaping science. For Cassirer, the reality revealed by physicists was "through and through mediated."[23] Cassirer's view on this matter was indeed very far from that proposed by Einstein during those years. In his early work, Cassirer was not content with simply defending a "neat formulation of the physical criterion of objectivity, that everything that can be measured exists," which Einstein (at a certain moment of his career) fully endorsed. He found an additional task for the philosopher to tackle, one that "involves the problem of discovering the fundamental conditions of this measurability."[24] But Cassirer would increasingly leave these questions behind as he allied his position more strongly with Einstein's.

"An experimental decision between Lorentz's and Einstein's theories was thus not possible," wrote Cassirer in his book on relativity. "Between them," he insisted, "there could fundamentally be no *experimentum crucis*."[25] To some, these were alarming claims. If not on clear experimental proof, then on what basis could one judge Einstein's merits? In his book on relativity, Cassirer endorsed Lorentz's view of relativity as much as Einstein's. What is more, he repeated the claim, made initially by Lorentz, that Einstein simply postulated what he had earlier deduced. With biting humor, he traced this strategy to Goethe, who had explained in a letter that "the greatest art in theoretical and practical life consists in changing the problem into a postulate; that way one succeeds." Cassirer stated that "this was the course which Einstein followed in his fundamental essay." Indeed, Einstein himself had initially described his contribution as one that "raised" a "conjecture . . . to the status of a postulate."[26]

Cassirer was nonetheless very generous to Einstein throughout the book. He lauded the physicist for taking science "a step further by freeing it still more from the presuppositions of the naïvely sensuous 'substantialistic' view of the world." Cassirer viewed the progress of science, and of civilization, as one that slowly moved away from a primitive and mythical focus on specifics and substance to a modern focus on symbols through abstraction. His view stood in sharp contrast to other philosophers who mourned the move of science toward increasing abstraction as a loss and who frequently gained inspiration from Bergson.

As the political situation worsened in Germany, Cassirer would abandon his early evenhanded approach, in which Einstein's contributions appeared side by side with Lorentz's and in which Einstein's theory appeared without the strong support of experimental evidence, and become one of Einstein's strongest supporters. *An Essay of Man* (1944), written in exile, was a panegyric statement for science and a tribute to his friend: "Science is the last step in man's mental development and it may be regarded as the highest and most characteristic attainment of human culture."[27] It stood at the pinnacle of culture precisely because of its relation to "spontaneity." Throughout the text Cassirer stressed the theoretical aspects of science, excluding from it any mundane or technological connections. Einstein was the most recent representative standing at the front of an illustrious lineage of scientists traced back to Galileo:

> The work of all the great natural scientists—of Galileo and Newton, of Maxwell and Helmholtz, of Planck and Einstein—was not mere fact collecting; it was theoretical, and that means constructive work. This spontaneity and productivity is the very center of all human activities. It is man's highest power."[28]

The tome simultaneously discredited Bergson, arguing that his philosophy was based on "a mode of receptivity, not of spontaneity."[29] The concepts of receptivity and spontaneity became the favored terms used by Cassirer to distinguish his philosophy from others. Cassirer associated spontaneity with engagement—political, social, and humanitarian—and optimism. He associated receptivity with passivity, inaction, and pessimism.[30] Bergson's philosophy "at first sight . . . would appear to be a truly dynamic more energetic philosophy," but in distancing the "vital impulse" from the realm of active life, it was forever doomed. Scientific work, in contrast, embodied the "spontaneity" that Cassirer dearly admired. These labels, and the philosophical stance associated with them, were clearly personal and political. "Bergson's ethics is a consequence and a corollary of his metaphysics," Cassirer concluded.[31] Cassirer's work fell in line with other work of that period focused on the history of ideas that considered science as the main accomplishment of the Enlightenment and that distanced it from technology, industrialization, the horrors of war, and global poverty.

After his death, Cassirer's strong dislike of Bergson became evident. In the fourth volume of *Symbolic Forms*, a work that remained unpublished during his lifetime, Cassirer accused Bergson of espousing and defending irrationality and, what is more, of overemphasizing the role and the weight of the past instead of stressing the possibilities of the future. He grouped Bergson and Heidegger together and found in them the roots of the intellectual decline of Europe. The third section of the volume, entitled "Time According to Bergson and Heidegger," sought to discredit both.[32]

CHAPTER 11

Two Months Later

FREIBURG, GERMANY

For some young scholars at the time, the next most intellectually stimulating event after the Einstein and Bergson encounter was a meeting in London led by a philosopher who was then spearheading what would become one of the leading schools of the century: phenomenology. The chief proponent of phenomenology at that time was Edmund Husserl. His student and personal assistant, Martin Heidegger, would emerge in the late 1920s as an inspiring young force behind the new movement. Citing Einstein and Bergson directly, he offered an option for moving beyond the impasse of April 6, 1922. In doing so, he attempted to take phenomenology into a different direction from what his teacher, Husserl, had initially proposed.

The summer after Einstein's visit to Paris, Husserl was scheduled to give four lectures at University College in London (in June 1922). A young aspiring philosopher, Alexandre Koyré lamented that he could not attend. In a personal letter to Husserl, he excused himself for his absence, noting how he considered the lectures in England "more important" than Einstein's visit to France.[1] Koyré, who was originally from Russia, had been part of Husserl's circle in Gottingen. He would move to France from Germany, where he would serve "as a bridge between Germany and France and between Husserl and Bergson."[2] When German forces marched into Paris, Koyré had to move again. He crossed the Mediterranean to relocate first in Cairo at Fuad University and then later, in America.[3] He took his Bergson with him, training numerous students, before becoming a leading historian of science in America.

By the time of Husserl's London lectures, the differences between Husserl and Bergson were already quite stark.[4] In 1917, a student of Husserl, Roman Ingarden, wrote an extremely critical dissertation about the French philosopher, claiming that Bergson's philosophy led to an ineffective skepticism rather than to purposeful action.[5] Ingarden published his dissertation in 1922—decisively marking off Husserlian phenomenology from Bergson's philosophy.[6] In the years that followed, Husserl and Heidegger emerged as attractive alternatives to Bergson. Each proposed different ways of understanding philosophy's relation to science and to Einstein's work and new—original and divergent—ways of understanding time.

HUSSERL AND HEIDEGGER

Heidegger's *Being and Time* (1924), a quickly written introductory volume to a proposed multivolume project, inspired philosophers for generations to come. What did the enigmatic title refer to? "As regards the title 'Being and Time,' 'time' means neither the calculated time of the 'clock,' nor 'lived time' in the sense of Bergson and others," he explained, years after the book appeared.[7] When he wrote that book, Heidegger was dissatisfied with the two dominant conceptions of time: Einsteinian and Bergsonian. For Heidegger, the two conceptions of time—"clock time" and "lived time"—one associated with Einstein and the other with Bergson, were symptomatic of the broader divisions of rationality and irrationality, where the first was associated with science and the second with experience. Heidegger criticized both of these divisions, investigating their emergence. "'The irrational' also appears and, in its wake, 'lived experience,'" he explained. These were strong words: "irrationality" thrived when "lived experience" was excluded from science.[8]

Heidegger's first confrontation with relativity theory occurred before he wrote *Being and Time*, while he was still a student in Freiburg. In his habilitation lecture, given in 1915, he confronted the theory of relativity directly, claiming that Einstein was not dealing with time but rather only with *measurements* of time. These two should not be confused. "We usually overlook the following," he explained. "As a theory of physics, the theory of relativity is concerned with the problem of measuring time, not with time itself." His judgment was severe, especially pointing out the limits of the theory: "The theory of relativity leaves the concept of time untouched," he boldly concluded. Einstein erroneously assumed that time was a "homogenous, quantitatively determinable character"—which it was not, argued Heidegger. The theory's limited notion of time was the result of how it spatialized time, assuming it to be a homogenous and geometrical concept. Einstein, continued Heidegger, claimed that time "can be placed alongside three-dimensional space." The philosopher disagreed with this point of view.[9] By treating time as space, scientists missed what Heidegger considered to be a much richer topic: an investigation into those aspects of time that could *not* be studied in the same way as space.

Heidegger's argument echoed Bergson's own critique of measurement that had been advanced in *Creative Evolution* (1907). The act of measuring time, according to both, destroyed much of it. "We, as it were, make a cut in the time scale, thereby destroying authentic time in its flow and allowing it to harden. The flow freezes, becomes a flat surface, and only as a flat surface can it be measured," wrote Heidegger.[10]

In his early work, Heidegger held certain views that were similar to Bergson's, namely that the time of physicists referred to something other than that of the philosopher's and that this time, was in essence, immeasurable. But as Heidegger's thought developed, he would increasingly distance himself from any notions of Bergsonian time, even as he continued to sharpen his critique of Einstein. The problem with Einstein's work was that it was based on a simplistic and inadequate notion of *measurement*. Unless this notion was clarified, Einstein's claim that he was dealing with real time remained suspect. To base a concept of time on measurements of time without noticing evident complexities in the concept of measurement itself was simply inadequate: "Any axiomatic for the physical technique of measurement must *rest upon* such investigations, and can never, for its own part tackle the problem of time as such."[11] Heidegger did *not* argue *for* or *against* the validity of the theory of relativity. He instead expressed the need to think about "the problem of the *measurement* of time as treated in the theory of relativity."[12] Measurement could not simply *give* answers about time, since it itself occurred *in time*. The "temporal meaning of measurement" itself had to be considered, and it had to be considered *first*, before anything else, since it was more basic and more essential than any derivative scientific results.

Heidegger's "The Concept of Time" lecture (originally 1924) diagnosed a damaging divide in the two dominant ways of thinking about time: the scientific notion of time and the lived notion. In this short lecture, Heidegger explained how a renewed interest in the concept of time was largely due to Einstein:

> Interest in what time is has been reawakened in the present day by the development of research in physics.... The current state of this research is established in Einstein's relativity theory.

The physicist, argued Heidegger once again, used clock time. And clock time, he repeated, was a grossly inadequate concept for understanding time: "Once time has been defined as clock time then there is no hope of ever arriving at its original meaning again," he warned.[13] A review article on the same topic also written that year (but it remained unpublished) directly cited Bergson's *Duration and Simultaneity*.[14]

The next summer, Heidegger gave a full course, "The History of the Concept of Time." His research was motivated by "the present crisis of the sciences," which Heidegger blamed largely on Einstein: "In physics the revolution came by way of relativity theory," he explained. Heidegger considered that most conceptions of time—including Einstein's—derived from Aristotle's: "Basically the concept of time as Aristotle conceived it is retained throughout."[15]

Bergson made similar arguments years before Heidegger, finding the origin of contemporary notions of time in Aristotle and proposing to go beyond it. Aristotle's error, Bergson argued repeatedly, was to describe time as analogous to space. Bergson proposed the concept of *duration* as an alternative to Aristotle's notion of time. In contrast to Aristotle, he revived a different antique authority, Plotinus, whom he pitted against Aristotle. Bergson tried to decouple the Aristotelian connection between time and space, restoring to time instead its connection to *duration*—a nonspatial, nonmeasurable, and indivisible concept.

Heidegger acknowledged Bergson's critique of Aristotle. "Bergson," he explained, "in fact makes an attempt to go beyond this concept to a more original one." For that reason, Heidegger set him apart from the majority of previous thinkers. "This justifies our treating him separately," he added. Yet Heidegger also wanted to improve on Bergson's critique, claiming that the French philosopher remained "traditional" in the way he approached time. Heidegger criticized how Bergson chose to focus on quality instead of quantity and on succession instead of instantaneity. In his opinion, Bergson's contribution merely shifted the debate to these other categories: "Basically, when we consider the categorical fundamental he [Bergson] presupposes, namely, quality and succession, Bergson does not advance the matters at issue and so remains traditional."[16] In his view, by proposing the opposite notion to Aristotle's, the force of Bergson's philosophy remained merely one of negation.

How would Heidegger improve on Bergson? Heidegger proposed to question the opposite terms, quality and succession, as well. His project would differ from Aristotle's, Einstein's, *and* Bergson's in this essential respect. Instead of continuing to debate about *what time is*, which was at the crux of the Einstein-Bergson debate, he proposed to take a step back and ask, what, after all, makes time? His stunning answer proposed that "human life does not happen in time but rather is time itself."[17]

Heidegger rarely did what he said he was going to do. He proposed to discuss his critique of Bergson in the last section of his four-hour-a-week course at Marburg University. Yet, by the end of the semester, it was clear that he was unable to complete the themes he had proposed in outline at the very beginning. The course ended before he could deliver the final lecture—the one slated for Bergson. The notes from this course nonetheless served him as an "early draft" for *Being and Time*—a book described as an introduction to a longer multivolume work that never materialized.[18]

In *Being and Time* Heidegger turned even more decisively against Bergson. In Bergson's philosophy, Heidegger continued to see a mere inversion of the Aristotelian conception of time. Bergson strongly and consistently criticized Aristotle's notion, and Heidegger seized on that fact. Although in his very early work Heidegger considered the Bergsonian critique of Aristotle as important and inspirational, by now he considered Bergson's duration to be a simple inversion of it. It was a significant improvement, but not a monumental one: "Every subsequent account of time, including Bergson's, has been essentially determined by it [Aristotle's concept]."[19] *Being and Time* was a clear statement "against Bergson's thesis." Heidegger's target was "the traditional concept of time, which has persisted from Aristotle to Bergson."[20] But it was also a statement against Einstein's simplistic conflation of time with the measurement of time.

Heidegger set out to do the contrary of what Einstein had done to time—in a radical way. Whereas the Aristotelian tradition from which Einstein drew (according to Heidegger) represented a culmination of a longstanding denial of differences between past and future, left and right, and up and down, for Heidegger these differences were now deemed essential. In "everydayness," he argued, they mattered substantially. Details as mundane and frequently overlooked as having something next to you or laying close to your dominant hand could have important consequences for what would happen next. They spoke volumes about the organization of the world. Even the future, ostensibly unknown to us, was primed to follow a certain course in connection to these frequently overlooked details.

HEIDEGGER AND CASSIRER

In 1928 Einstein gave a lecture to more than four hundred persons in the Hochschule in Davos about the theory of relativity. In that same location one year later, Heidegger confronted Cassirer.[21] Cassirer, a defender of Einstein and enemy of Bergson, was now pitted against a man ready to leave *both* Einstein and Bergson behind. By then, Heidegger was not only ready to confront Cassirer; he was also poised to abandon his teacher, Husserl.

Heidegger edited and published Husserl's *The Phenomenology of Internal Time-Consciousness* (in 1928) based on lectures given by his mentor between 1905 and 1910, with some supplemental material from later years. In it, Husserl explained the dilemma of time motivating his investigations in clear Bergsonian terms: "Time is fixed, yet time flows."[22] The published text made it easier for Heidegger to compare his approach against Husserl's, permitting him to highlight the differences between his own philosophy and that of his mentor. In contrast to *Being and Time*, which proposed to take philosophy in a new direction, Husserl's *The Phenomenology of Internal Time-Consciousness* seemed a comparatively conservative text, proposing to provide a descriptive phenomenological account of time according to a strict method. It suggested looking at how "objective" time arose from the subjective perception of flow: "In the flow of time . . . a non-flowing, absolutely fixed, identical, objective time becomes constituted."[23] By then, the approaches of Husserl and Heidegger were clearly different and, at times, even at odds.

During these years, the philosophical community noticed a clear falling out between the two men, both personal and professional. Husserl's work accepted a division between internal and external assessments of time. He agreed with a standard division of labor in which the philosopher focused on the "internal" aspects. Heidegger's work, on the other hand, proposed to take a step back in order to go forward. He aimed to analyze time even before it was divided into internal and external qualities. Husserl's method remained a "scientific" investigation into how time appeared to us before it emerged as a clear concept (such as that of traditional psychological investigations). Heidegger's method, in contrast, proposed to be "pre-scientific." Heidegger accepted the conclusions of the exact and mathematical sciences, but—in contrast to some of Einstein's defenders such as Bertrand Russell and Cassirer—he was not effusive about them.

In 1933 Husserl was forced out of his university position as a result of the Nazi racial laws against Jews. Heidegger's career in this political context, in turn, took off. When he was appointed rector to the university for a brief period, he publicly embraced the new political order, which was required of all university employees. Yet Heidegger never accepted a biological notion of race. He considered anti-Semitism "senseless" [*töricht*] and "abject" [*verwerflich*], continued to have close relationships with Jews, and mentored a great number of Jewish students.[24]

During these years, Husserl and Bergson agreed about the problems facing science, although they offered different solutions for it. In 1935, during a Vienna lecture, Husserl enumerated them. He blamed Einstein for part of the problem. Einstein's revolution, he argued, came at a high price. Science distanced itself from those aspects that had "meaning" for us, mainly our everyday sense of time flowing:

> Einstein's revolutionary innovations concern the formulae through which the idealized and naïvely objectified *physis* is dealt with. But how formulae in general, how mathematical objectification in general, receive meaning . . .—of this we learn nothing; and thus Einstein does not reform the space and time in which our vital life runs its course.[25]

A year later, in 1936, Husserl published the first two parts of one of his most influential texts about science, titled *The Crisis of European Sciences and Transcendental Phenomenology: An Introduction to Phenomenological Philosophy*. The work, which would be foundational for philosophers to come, provided a particular interpretation of Einstein's theory of relativity in relation to Michelson's experiment. Husserl knew well that "Einstein [used] Michelson experiments" to reach his conclusions. Yet he also knew that he used them in a particularly delimited way, one focused narrowly on "scales of measurement, coincidences established, etc."[26] Provocatively, he asked, why did Einstein stopped his investigations there? One could envision researchers taking more aspects of the experiment into consideration: "There is no doubt that everything that enters here—the persons, the apparatus, the room of the institute, etc.—can itself become a subject of investigation in the usual sense of objective inquiry, that of the positive science."[27] But there were good reasons why these additional topics should not and did not matter to scientists. "Einstein," he explained, "could make no use whatever of the theoretical psychological-psychophysical construction of the objective being of Mr. Michelson." How did the "Mr. Michelson" who actually built and performed observations (and who had distinctly personal opinions about Einstein's work) disappear from view, replaced by a seemingly generic and impartial observer who could ostensibly be just anyone? The ideal "Michelson" of the "Michelson experiments" was as distant from the actual "Mr. Michelson" as the virtuous Dr. Jekyll was from the vicious Mr. Hyde. The difference between them was key to setting boundaries between valid scientific experiments and nonscientific investigations. Who set these boundaries? They arose from "pre-scientific" "presuppositions" that were "common to all" arising from "the world of experience." Those were the very "premises" of scientific knowledge that Husserl offered to investigate.[28] Part of his philosophical project would be concerned with describing them in order to better understand them.

PARMENIDES AND HERACLITUS

A new generation of philosophers during these years agreed on a certain *diagnostic* for Europe's problems, one connected to the "crisis" brought about by science. Many thinkers would side with Husserl's phenomenology as a proposed *solution*. Not Heidegger. He distinguished himself from his teacher by offering a different interpretation of Einstein's and Bergson's contributions. In the winter semester 1942–1943 in Freiburg, Heidegger gave a couple of lectures on Parmenides of Elea and Heraclitus in which he referred explicitly to Bergson and implicitly to Einstein. The comparison was not entirely original, though it would prove increasingly fruitful. The philosopher of science Karl Popper, among others, compared Einstein to Parmenides. Heraclitus, in contrast, was seen as defending the view that reality was always ever changing. Russell, Reichenbach, and others compared Bergson to Heraclitus.[29]

How did Einstein's views compare to those of Parmenides? In the surviving lines of the ancient poem *On Nature*, Parmenides described the universe, in a section titled "The Way to Truth," as basically static and unchanging. Einstein and the mathematician Hermann Minkowski were often seen as offering a similar description and conceiving the universe as a block. "If all motional phenomena are looked at from this point of view," explained a writer on relativity, "they become timeless phenomena in four-dimensional space. The whole history of a physical system is laid out as a changeless whole."[30]

During the Freiburg lectures, Heidegger lamented that "clock time" and "lived time," the two traditional terms for understanding it, were heavily value-laden: the first associated with science and all its benefits, and the second with "the irrational" and "lived experience." But how could one go beyond these dichotomies and redistribute the heavy associations that came with them?

As a solution, Heidegger started by focusing on "everydayness" as a territory where these categories could not be differentiated. Our "everyday" was defined neither wholly by clock time nor by lived time—but rather by a combination of the two. In that territory, differences between past and future, left and right, and up and down mattered substantially. It made an important difference if we could "reach for [some thing], grasp it, look at it," or if we could not. Heidegger sought to give back the texture, meaning, and importance of things being left, right, above, below, before, after, far, or near. This texture was lost and obfuscated if these differences were thought about in terms of measured, objectively defined terms. For example: "When, for instance, a man wears a pair of spectacles which are so close to him distantially that they are 'sitting on his nose,' they are environmentally more remote from him than the picture on the opposite wall." Heidegger insisted that to properly think about the world we needed to consider the technologies that affected these distances—not only eyeglasses, but also the radio, the telephone, and a regular street.[31] The more "inconspicuous" these elements were in their effects on our environment, the more relevant they were for him philosophically. Precisely because of its "inconspicuousness," the "street" itself appeared more "remote than the acquaintance whom one encounters 'on the street.'"[32]

Heidegger's point was *not* that there were subjective understandings of distances, including temporal ones, that should be contrasted with objective ones, although he agreed that "one is inclined to pass off such estimates and interpretations . . . as 'subjective.'" Rather, a focus on these so-called subjective assessments as much more than merely "subjective" was what "'uncovers the 'Reality' of the world at its most Real."[33] At its core, Heidegger wanted to investigate experiences before they could be divided neatly into objective or subjective categories. His analysis of time included elements that had up to then been mostly ignored by philosophers, such as why certain people seem to never have enough time.[34]

During his last summer at Marburg, Heidegger referred explicitly to Einstein and Bergson as offering two distinct views on the nature of time and compared their work to that of Parmenides and Heraclitus. He paid compliments to Bergson: "Recently Bergson tried to conceive the concept of time more originally. He made it more clear than any previous philosopher that time is interwoven with consciousness." But he was not totally convinced: "But the essential thing remained unresolved in Bergson, without even becoming a problem."[35] Bergson had offered some of "the most intense analyses of time that we possess," he granted. After adding that Bergson's work could be "re-examined and improved," he accused the philosopher of "blocking" a solution. He boldly embarked on the problem of removing this obstacle.[36]

In the following lecture course during the winter semester (1929–1930), published as *The Fundamental Concepts of Metaphysics: World, Finitude, Solitude*, Heidegger continued his attempt to tackle aspects of time in a way that was not limited by the dual perspective of "clock" versus "lived" time. He tried to understand the characteristics of time that were traditionally understood subjectively, such as boredom, in a new way. When bored, "one feels timeless, one feels removed from the flow of time," but this sensation should not be understood in simple psychological terms.[37] Boredom and anxiety arose in particular situations and technological settings, such as when "waiting for a train," and—more important—they represented essential characteristics of existence, if understood more broadly.

Heidegger's winter lectures, delivered during a period of heavy setbacks for the German troops (1942–1943), communicated a new sense of urgency. Centering on Parmenides of Elea and Heraclitus, two topics that could seem extremely distant from his present-day concerns, they were directly tied to the much more recent confrontation between Bergson and Einstein. In those lectures, Heidegger pointed out that *Being and Time* was written as an alternative to both, explaining that it was urgent to find a way out of their impasse, which he saw as intimately tied to "the destiny of the West."[38] For him it meant an investigation into "the essential ground of ratio and of all thinking and saying." This analysis led Heidegger to think about technology in a new way and to continue his essay with a discussion of handwriting, the printing press, the typewriter, and electrification: "Insight into the 'metaphysical' essence of technology is for us historically necessary if the essence of Western historical man is to be saved."[39]

An essential part of Heidegger's philosophy consisted in focusing on what lay *between* tool and machine and *between* man and technology. Heidegger warned about the spread of new media technologies, which he called "an 'intermediate' thing, between tool and a machine." These things, such as the typewriter and printing press, were characteristically modern: "It is no accident that the invention of the printing press coincides with the inception of the modern period." These things concealed themselves "in the midst of [their] very obtrusiveness" but "transformed the relation of Being to his essence."[40] One could no longer "pretend as if 'technology' and 'man' were two 'masses.'" These two elements were complexly intertwined, and their interrelation was the reason, as well, why "the much discussed question of whether technology makes man its slave or whether man will be able to be the master of technology is already a superficial question." Heidegger expressed the need to "ponder the 'concrete' . . . and to remove the concealment thrust upon things by mere use and consumption." He asked his students to think about the typewriter. Anticipating that they might take his statements as "a digression," he underscored their central importance: "Λήθη [Lethe] and the typewriter—this is indeed not a digression." Also anticipating students that would "justifiably be asked what in the world that [the typewriter] has to do with Parmenides" he answered: "It has everything to do with it."[41]

By searching for a different "essential ground" and focusing on "intermediate" things, that were neither machines nor tools, Heidegger sought to bypass the impasse between "clock time" and "lived time," between the "rational" realm of science and the "irrational" ground of experience. Heidegger clearly identified the problem he saw around him and expanded his philosophy in a radically new direction in his attempts to find a solution.

This early analysis led Heidegger to think more carefully about paper in his later years. After the war, Heidegger used the example of paper to illustrate how modern technology and man could no longer be differentiated: "[The forester] is made subordinate to the orderability of cellulose, which for its part is challenged forth by the need for paper, which is then delivered to newspapers and illustrated magazines." Newspapers and magazines "set public opinion to swallowing what is printed, so that a set configuration of opinion becomes available on demand."[42] Even a man who could be seen to be immune to technological developments—the forester—was already part of a much larger system—print culture—from which he could not be separated. Even if the forester himself did not read—even if he was illiterate—he was directly affected by the printed opinion swallowed by others.

During the Enlightenment, "public opinion" emerged as a distinct category representing the voice of the people. Its emergence was tightly connected to new public spaces (such as coffeehouses) and new publication practices (such as the daily press). It could be a powerful force for keeping leaders from adopting unpopular authoritarian policies and could promote democratic causes. Could the very category of the "public" be understood independently of technology? No. The "public" could never, according to Heidegger, be the source of Enlightenment.[43] Why? Because a public composed of persons who were not separable from "technology" could never have sovereignty over it.

"THE MOST LOVELY AND PERTINENT OBJECTION": CRITICAL HISTORY AND THE FRANKFURT SCHOOL

As young men, Walter Benjamin and Heidegger sat together in the same classroom where they learned about Bergson's philosophy.[44] During those years, Benjamin zealously read what his classmate wrote. His assessment of Heidegger's work was, for the most part, critical. He disparaged it as penned for the benefit of "Catholic Germany."[45] But Benjamin was also very intrigued, especially about Heidegger's views on science. He wrote to a close friend asking him if Heidegger was right about his views on Einstein, expressing his suspicion that he must not be.[46] After that, Benjamin started to develop a different approach to history and philosophy than what Heidegger had to offer. This work culminated in his *Theses on the Philosophy of History*, a text exploring the relation between time and history in a new way.

Benjamin explained how, to combat an overly simplistic view of modern progress in society—one that had ended with the rise of fascist forces that he was fleeing—he first had to combat the concept of homogenous, empty time. He linked the modern concept of "progress," which he considered fraudulent and dangerous, to that particular conception of time: "The concept of the historical progress of mankind cannot be sundered from the concept of its progression through a homogenous, empty time." What is more, he called on intellectuals to question their views of time in order to better understand historical developments.[47] Benjamin was also critical of philosophies that wanted to turn homogenous time on its head and place "true" experience in its stead. He was unconvinced by the answer that came from the philosophy of life movement, within which he saw Bergson's work as "towering over this literature."[48]

For a growing number of theorists, understanding the limits and possibilities of historical development entailed an investigation into the relation between history and time. How could one move past the impasse of homogenous time and experienced time? Benjamin, along with many other intellectuals during this period, struggled with these two alternatives. He attempted to find a solution in which science was neither the solution nor the impediment for understanding time.[49] Benjamin died while crossing the border into Spain trying to flee from the Nazis—before he could tell us more.

In the 1930s, with the rise of Nazism in Germany, seemingly abstract discussions about the nature of time became ever more concrete and immediately relevant. Benjamin's mentor, Max Horkheimer forcefully criticized Bergson's focus on the fluctuating, continuous, and indivisible aspects of the world. For him, Bergson's focus on continuity and movement as essential characteristics of the fabric of reality belied the facts on the ground: a divided world. "The claim that reality is essentially indivisible contradicts the fact distinguishing history, at least in its form until now, that humanity is divided into the happy and the unhappy, the ruling and the ruled, the healthy and the sick," he explained.[50]

For the rest of his life, Horkheimer would struggle to find an alternative to the two dominant intellectual schools of his time: the neo-Kantian school (represented by Hermann Cohen and his students) and the *Lebensphilosophie*, or "philosophy of life" movement, which was tightly associated with Bergson. As an alternative to both options, Horkheimer proposed Critical Theory.[51]

Horkheimer started the Critical Theory project while he was the director of the Institut für Sozialforschung in Frankfurt (simply known as the Frankfurt School) and the chair of social philosophy at the University of Frankfurt. Like other Jews, he was eventually forced out of his positions and into emigration. Bergson reached out to Horkheimer. Although they did not agree on a solution that would bridge our sense of flowing time with static descriptions of it, both men agreed on the problem. In the United States, Horkheimer told his listeners that he had obtained "Bergson's personal acknowledgement that although he could not agree with me, he felt that it was the most lovely and pertinent objection which he had yet encountered."[52]

CHAPTER 12

Logical Positivism

Sitting in front of him in his classroom in Berlin, Einstein found two men recently released from active duty. They were both still eager to lead a battle, albeit of an intellectual sort. They would join the fight against Bergson and against Henri Poincaré's conventionalist approach to science. Einstein got to know them well. In years to come, he would teach them more than just physics. One of them, Hans Reichenbach, would become one of the most prominent defenders of logical positivism, arguably the most dominant philosophy of science movement in the twentieth century, sometimes referred to as logical empiricism. He found an ally and friend in Rudolf Carnap, an active and prominent member of the Vienna Circle, an association of intellectuals based in Austria that served as logical positivism's launching pad. Both men defended Einstein and attacked Bergson. Reichenbach, due to his Jewish background, and Carnap, because of his leftist and pacifist sympathies, would be forced into exile after the Nazis gained power. Both completed their careers in California. In influential works, and as founders of a journal with the immodest yet simple title *Knowledge* (*Erkenntnis* in the original German), they explained to a growing number of readers why science was so exceptional.

Although unnamed in many of their publications, Bergson was a clear target from the start to the end of their careers. Bergson was synonymous with a dangerous new enemy: metaphysics. When Carnap defined the term "metaphysics" in his influential *The Logical Structure of the World* (1928), he attributed to Bergson the view that it dealt with the "area of the nonrational," and he profoundly disagreed with how Bergson thought about its role in the development of science.[1] Later in life, he cited Bergson not just as an authority about what metaphysics was, but rather as a metaphysician. When one of his most influential articles was republished after the war, Carnap added a telling footnote. The article, initially published in 1932, had not referred to Bergson by name. Its target at the time was simply referred to as metaphysics. But safely in exile in UCLA, Carnap explained that one of its main targets had been Bergson, who appeared alongside Fichte, Schelling, Hegel, and Heidegger. He was now free to reveal the identities of his archenemies. Carnap went on to explain how he still found no use for "the field of alleged knowledge . . . which transcends the realm of empirically founded, inductive science" and which he associated with the "systems" proposed by these authors.[2] At stake in their fight against metaphysics was the status of their positivist conception of science in the modern world.

Reichenbach's distaste for Bergson also appeared in print late in his life: first in one of his most influential books, titled *The Rise of Scientific Philosophy* (1951), and then later after his death in *The Direction of Time* (published posthumously in 1956).

Logical positivism was a heterogeneous movement with many facets. Yet one of its defining principles maintained that science emerged from a strictly sensorial basis and was then built upward with clear logical principles. Most logical positivists were driven by two goals: that of building knowledge from a firm empirical foundation and that of distancing it from dangerous metaphysics. Part of the movement was shaped by a particular hatred for the work of one man—Bergson—and an intense appreciation of the work of another: Einstein. Some of its chief representatives, such as Reichenbach and Carnap, were extremely close to Einstein, owing much of the success of their careers to him. They strongly advocated limiting the role of philosophy when it came to science. "All the philosopher can do is to analyze the results of science, to construe their meaning and stake out their validity," explained Reichenbach.[3]

After being released from the army, Reichenbach attended Einstein's lectures in Berlin. He was hooked for the rest of his life. He dedicated his first book to the physicist, and in the years that followed, Reichenbach and Einstein would become close friends. Reichenbach met Carnap in 1923 and recognized him immediately as an ally and friend. Carnap first studied Einstein's work for his dissertation, *Der Raum* (completed in 1921, published in 1922). His aim throughout the text was to clarify and eliminate the preexisting tensions between mathematicians, philosophers, and physicists. In it, Carnap introduced a division of labor in which each (mathematicians, philosophers, physicists) would be limited to studying distinct notions of space (formal, intuitive, and physical, respectively). Conflict pertaining to who had authority on these topics, he argued, could be eliminated if readers properly understood that Einstein was a physicist and his description of space was physical. "All parties were correct and could have been easily reconciled if clarity had prevailed concerning the three different meanings of space."[4] If they adhered to these divisions, Carnap argued, most problems would be solved.

Carnap tried hard to show how physics did not need to engage with these other disciplines and could fare better if purified from them. He wrote *The Logical Structure of the World* to show how scientific knowledge could be traced back to simple sense data. By using clear, uncontested data as starting points, attaching them to numerical values, and combining them with clear mathematical rules, Carnap believed he could put science on firm, uncontestable footing.

LIMITING THE ROLE OF PHILOSOPHY

According to Reichenbach, Einstein had completely revolutionized the relation between physics and philosophy: "There is no separate entrance to truth for philosophers: the path of the philosopher is indicated by that of the scientist." Einstein, Reichenbach would repeat for the rest of his life, had finally pushed traditional philosophy aside: "It [modern science] has refused to recognize the authority of the philosopher who claims to know truth from intuition, from insight into a world of ideas or into the nature of reason or the principles of being, or from whatever super-empirical source."[5]

Einstein and Reichenbach would not always agree. In fact, Reichenbach's views about the firm foundations of science would eventually be much stronger than even Einstein's. But Reichenbach's early contributions to the physicist's work were essential in many respects. He was particularly successful in developing arguments against some of Einstein's colleagues—prominent mathematicians among them—who remained unconvinced about some of the physicist's claims.[6]

Was Einstein's "ideal clock"—the light clock based on light's constant velocity—simply a logical artifice? In 1917 the mathematician David Hilbert noted that Einstein's particular way of defining an ideal clock was insufficiently convincing as a physical concept, insisting that at a later point it "must appear as a consequence of the general theory."[7] Many others did not see a need for adopting the "ideal clock" proposed by Einstein. In light of these criticisms, some of Einstein's friends and advocates agreed that this concept made perfect sense, but only for temporary, pragmatic reasons pertaining to time measurement and coordination. Einstein himself initially espoused these practical justifications as well. In a letter to a student of the mathematician Hermann Weyl, he insisted on the theory's "didactic" benefits.[8] But neither Weyl nor his student were convinced that justifications based on didactic or practical benefits were enough for producing a theory of the universe, saying that "the task remains . . . to derive this fact."[9]

Reichenbach's contributions proved essential for facing these early criticisms. Einstein had first tried a different strategy. He responded boldly to Weyl's student, who had asked for clarification on exactly this matter. Einstein defended himself by placing the burden of proof on his critics. He was right in his assumption because to be wrong, someone would need to show him that "nature is different."[10] Others would have to disprove the pertinence of the "ideal clock" for physics. While the student was effectively silenced with this curt response, his teacher was not. Weyl continued to protest that Einstein's conclusions should appear as a "consequence of the developed theory."[11] The effects described by the theory *should* be a consequence of a much broader physical theory; many remained unconvinced that they indeed were. Soon thereafter Einstein had no choice but to admit that "the fact the measuring-rods and clock have to be introduced separately, instead of being constructed as solution" is a "logical weakness of the theory in the state today."[12]

Reichenbach continued to work hard to show how Einstein's theory was not simply a useful convention relaying on an artificial "ideal clock." His criticism of Poincaré's conventionalist philosophy was directly tied to his defense of Einstein. Einstein's theory was not simply one out of other possible ways of describing the universe mathematically, he insisted. Its mathematical structure was not a convenient tool: it revealed the shape of the universe itself. The philosopher and physicist Moritz Schlick was not convinced by Reichenbach and immediately wrote to Einstein to tell him: "Reichenbach does not seem to me to have done justice to Poincaré's theory of conventions." The problem, he continued, was that Reichenbach considered as fundamental and a priori certain aspects of measurement that could equally be considered conventional: "What he calls *a priori* correspondence principles and rightly distinguishes from empirical correspondence principles seems to me to be completely identical to Poincaré's 'conventions' and not to have any meaning beyond that."[13]

Einstein developed his most thought-out response to these criticisms in a lecture titled *Geometry and Experience* (1921), which was later expanded for publication.[14] His solution (developed in conversation with Reichenbach, Schlick, Weyl, and others) was frequently remarked upon.[15] The lecture was designed to show how the non-Euclidean geometry used in Einstein's work was not merely a useful mathematical formulation but, rather, a model of the actual geometry of the universe itself: that it was "a question of physics proper which must be answered by experience, and not a question of a convention to be chosen on grounds of mere expediency."[16] He aimed to convince readers that these mathematical techniques were more than simply tools used by physicists; they were actual models for the universe itself.

Reichenbach defended and improved Einstein's arguments. Always attentive to what his professor said and wrote, he joined in the fight and developed the lecture's central ideas more fully. Einstein's "ideal clock," argued Reichenbach, was based on a concept given "by definition," namely the constancy of the speed of light.[17] But this "by definition" solution was neither arbitrary nor merely convenient—it was also a "fact." To answer those who wanted to know if the velocity of light was actually constant and not merely defined as such, he stressed that it was "an empirical fact" that measurements of time and length could be and were undertaken in the manner described by Einstein;[18] it was "experimentally well-confirmed."[19] A few years later, Reichenbach once again explained why he thought that Einstein was entirely justified in his definition of time, arguing that it was "a matter of fact that our world" was a place where measurements were undertaken in this way.[20] The choice of measuring system could potentially be viewed as conventional—this was Poincaré's point, which even Einstein himself had admitted was "*sub specie aeterni*" right.[21] But the actual reality of how people measured was not. Actual measurement practices explained why certain measuring systems reflected the world as it really was and others did not. With these arguments, Reichenbach showed how Einstein's light clock was more than just ideal. Einstein's clock (based on the velocity of light as constant) was more than a convenient choice; it was ideal *by definition* and *empirically* so, as well.

Reichenbach sharpened the criticisms of Poincaré that Einstein had first introduced in *Geometry and Experience*. He attacked the Frenchman's claims about the relation of mathematics to physics—that mathematics was a tool and not a reflection of how the world actually was. He chastised him for espousing the view that scientists could chose between different geometries (such as that between Euclidian and multidimensional ones). From then to the end of his career, he would consistently claim that only one of them (the one used by Einstein) described the actual "geometry of the physical world."[22]

Einstein benefited enormously from Reichenbach's support, zeal, and talent. He gratefully accepted Reichenbach's generous dedication of *The Theory of Relativity and A Priori Knowledge* to him and offered a few suggestions. He gave the young philosopher some of the same advice he had earlier given to Cassirer: that he should not view the role of concepts for shaping knowledge as that meaningful. "Concepts are simply empty when they stop being firmly linked to experiences," he explained. "They resemble upstarts who are ashamed of their origins and want to disown them," he concluded.[23] In two sentences, Einstein gave his student a lesson in how not to appear too eager to advance professionally, as well as pointing him in a philosophical direction that would stress the importance of sense experience in science over the role of theoretical concepts.

CONSTANT—BY DEFINITION

Reichenbach's views were consonant with how he encountered and thought of radio technology as a practitioner. Early in his career

Reichenbach had worked as an engineer, yet he would eventually shed his technical training, becoming first a professor of physics (with Einstein's support) and then a professor of philosophy after his exile from Germany. While he was perfectly versed in technical engineering language, Reichenbach consistently downplayed the connection of the theory of relativity to contemporary technologies. His views about why relativity theory did not arise in connection with new technological discoveries stemmed from how he understood the relation of science to technology more generally.

"Telegraphy is as old as mankind," explained Reichenbach in his radio manual. By framing it as atemporal, Reichenbach naturalized the technology: "The prehistoric man who raised his arm up to wave to his contemporaries telegraphed."[24] This was "true wireless," he concluded. The only difference between the gestures of prehistoric man and contemporary telegraphy were "the thousands of years of scientific work that lay in between." Reichenbach highlighted the sole role of science in the thousands of years from prehistoric days to the present; he did not focus on the piecemeal practical development of technology. His descriptions elided the immediate war conditions that led Reichenbach to work with radio in the first place and the commercial interests of electromagnetic technologies (including those that led to the publication of the radio series with him as the editor).

Reichenbach's notion of scientific experiment—one of the most dominant models in the twentieth century—did not include a role for technology, let alone contemporary technologies, either scientific or commercial. In 1949 Reichenbach published an essay to appear alongside Einstein's autobiography. The essay summarized Reichenbach's views, which were dominant in Anglo-American circles and representative of logical empiricism. He explained how Einstein's work emerged from "an empiricism which recognizes only sense perception and the analytic principles of logic as sources of knowledge."[25] Intermediate material between sense perception and the final results of science were unaccounted for.

In America, Reichenbach set for himself the task of forming a new generation of philosophers of science. During these years, he described Bergson as a representative of the "philosophy of the nineteenth century," which was now superseded by the completely different approach he was forging. While the former was characterized by "persuasive solutions of systems that talk picture language and appeal to esthetic desires," honest work was now done by men "trained in the technique of the sciences, including mathematics, and who concentrated on philosophical analysis." This, according to Reichenbach, represented the "new generation" of the "professional philosopher of science." By virtue of the professional education of these new practitioners, philosophy would no longer remain a mere "by-product of scientific research" or a confused mass of metaphysical speculations.[26] In contrast to the outmoded discipline, the new profession he helped build as a professor of philosophy in California was one now cleansed of any aesthetic intention: "philosophy is not poetry . . . and picture language has no place in it."[27] Instead of thinking that science should be understood philosophically, Reichenbach believed that philosophy should be understood scientifically. Philosophy should become closer to science and leave behind its nineteenth-century antecedents, its connection to Bergson's system, and its recourse to "systems that talk picture language and appeal to esthetic desires." It should renounce an aesthetic appeal to become instead a truly "scientific philosophy."[28] Please keep it dry, gentlemen—one could almost hear him tell his students.

In 1951 Reichenbach published his most popular book, *The Rise of Scientific Philosophy*. Once again, poised to defend Einstein, he countered some of the objections to Einstein's work that considered its conclusions as emerging from contestable definitions. He again claimed that in Einstein's theory the constancy of light was given by definition.[29] Reichenbach argued in favor of simply *defining* the speed of light as a constant quality and then deriving all other important constants from it. But Reichenbach again added another piece to Einstein's argument: he explained that when Einstein said that "there can be no faster signal than light," he did not merely mean "that no faster signal is known to us." Rather, Einstein meant "the statement that light is the fastest signal is a law of nature." Was light the fastest signal (to date), or was the fastest signal light (now and forever)? The difference was important. The latter, answered Reichenbach.[30]

In the decades after the Second World War, logical positivism became firmly established in the United States because many of the school's members had obtained comfortable positions there, having been forced to emigrate from Europe. A rare challenge to logical positivism came from Harvard, where Alfred North Whitehead taught. One of Whitehead's students, the philosopher W.V.O. Quine, who had written one of his first papers at Harvard on Bergson, published the damaging "Two Dogmas of Empiricism" (1951), in which he directed arguments against Carnap and Reichenbach. It was a significant blow.[31]

CHAPTER 13

The Immediate Aftermath

AT THE HOME OF JEAN BECQUEREL, PARIS

"I am being splendidly regaled, as never before in my entire life," Einstein wrote to his wife on the Friday evening before he would encounter Bergson.[1] The physicist Jean Becquerel generously opened his home to him and invited about one hundred additional guests.[2] Discussions at "the library [of the Becquerel's], in *petit comité*," were "a supplement to the conferences at the Collège de France," explained a journalist.[3] "All who Paris could count of famous scientific personalities" were there, wrote another.[4] What was discussed in these private forums? How did these discussions affect the unraveling of the debate?

After inviting Einstein to his home and learning about the ensuing debate, Becquerel called on Bergson. Becquerel thought he had caught a mistake in the philosopher's interpretation of Einstein's work. Instead of confronting Bergson publicly, he decided, diplomatically, first to invite the philosopher over for a private meeting. The meeting did not go well. Becquerel felt he was not taken seriously by the philosopher, who apparently, did not "want to come around to the arguments" with which he was presented.[5] Bergson was not alone during his meeting with Becquerel. He was accompanied by Édouard Le Roy. Becquerel, as others, came to believe that "it was Le Roy who was the person responsible for Bergson's attitude."[6]

Volumes have been written about the reception of Einstein's work in France, but Bergson's role—generally dismissed as too egregiously off-topic for the history of physics—has been largely written out of these accounts.[7] However, records in the private correspondence of key scientists and clues in academic publications reveal that even the most prominent physicists in France and beyond seriously considered Einstein's work in terms of Bergson's arguments. The points raised by the philosopher remained central for decades to follow—for scientists.

A new cadre of scientists, such as Becquerel, joined Langevin in defending the physicist against the philosopher. When Bergson resisted the physicist's arguments, Becquerel retaliated by publishing a damaging article against the philosopher aimed at students. Becquerel would soon be joined in his fight against Bergson by André Metz, a practicing Catholic who had a brilliant military career and would emerge as a brave resistance fighter during the Second World War. In a series of back-and-forth exchanges, Bergson and Metz sharpened their arguments against each other to such an extent that agreement between them—and perhaps even mutual understanding—became impossible.

Within France, Einstein's invitation intensified preexisting tensions between philosophers and physicists of different schools and institutions. He was embraced by some members of the Collège de France (particularly by Paul Langevin, who invited him), greeted at the border by an astronomer from the Paris Observatory (Charles Nordmann met Einstein along with Langevin), courted by the Société française de philosophie (in which forum he debated with Bergson), admired by the Société astronomique de France (especially by its president, the prince Bonaparte), welcomed by the Société de chimie physique, and courted by the Société française de physique—for which Einstein had no time.[8]

The April 6, 1922 debate and its aftermath was hardly a confrontation of scientists against philosophers. Einstein was defended by scientists and philosophers of a particular bent, whereas Bergson was backed by a very different set of scientists and philosophers. Alliances and antagonisms between the different groups were subtle and complex. Many philosophers ended by siding with Einstein; many physicists, with Bergson.

"ERROR" OR "HALF ERROR"?

Jean Becquerel was the son of the eminent physicist Henri Becquerel, known as the discoverer of radioactivity. Jean was one of the first scientists in France to introduce classes on relativity at the École polytechnique and at the Muséum d'histoire naturelle, where he was a professor. He published two books on relativity in 1922, one of them designed for a general audience.

Soon after Einstein's Paris visit, he defended Einstein more forcefully than ever and attacked Bergson. When Becquerel read *Duration and Simultaneity*, he immediately thought Bergson had made an "error."

In an article published in the *Bulletin scientifique des étudiants de Paris* (March 1923), Becquerel made his disagreement with the philosopher public. After thoroughly criticizing Bergson's new book throughout the paper, he was slightly more diplomatic in a footnote. In a note, Becquerel explained that Bergson's error was really only a "half error," since it was an error mainly of "interpretation." But the main text of the article was unambiguously critical.[9]

Bergson made the disagreement with Becquerel public as well, although initially he had tried not to name his contradictor directly, explaining that at first he had decided to publish it "without a name, because otherwise, if one publishes the name, it immediately appears under the guise of a polemic." The second edition of *Duration and Simultaneity* (1923) included three new appendices written "to respond to Becquerel and Langevin, who have not accepted my critique of Einstein."[10] After unmasking his critics, he included the letter that Becquerel had sent to the philosopher. That letter and the three new appendices only intensified the debate between him and these physicists.

In his critique of Bergson, Becquerel repeated the usual story of the twins. Paul left Earth in a fast rocket ship, only to return and see that Pierre had aged much more than he. When they compared their clocks, they noticed that less time had passed according to Paul's. Who was right? "They are both right," insisted Becquerel. Why did Bergson fail to accept this? Becquerel explained in the preface to his book on relativity that for some people it was exceedingly difficult to rid themselves of "preconceived ideas" and of "ingrained habits," implying that Bergson was simply a stubborn old man.[11]

Bergson reprinted Becquerel's letter describing a scenario in which the clocks were "synchronized by light signals." In that example, Pierre, who stayed on Earth, would see his clock mark 8 hours passed while he would also see that Paul's would mark 4 hours. Becquerel illustrated how the slowing down of time of the traveling twin could be ascertained *every step of the journey*. This could be done in various ways. The twins could diligently compare their clocks by communicating with each other during the trip, exchanging light or radio signals along the way—so a comparison could be made even without having one of them return. Or, alternatively, the whole travel path could be strewn with clocks that showed Earth's time. The twins did not have to wait to compare their clocks once they met back on Earth. Paul could see how time was elapsing faster on Earth, and Pierre could see how Paul's clock was slowing down as soon as the trip started.

How did Bergson reply? He argued that nothing would prevent Paul from thinking that *his* time was real, whereas Pierre's was a mere "representation." And nothing would prevent Paul from thinking the same but exactly in reverse: that *his* time was real while Pierre's represented: "It is for this Paul simply represented and as a referent that 4 (represented) hours would have gone by, whereas for Pierre 8 (lived) hours would have passed. But Paul conscious, and therefore the referring one, would have lived 8 hours, because we need to apply everything that we have said about Pierre to him."[12] One of those times, Bergson, kept insisting, was a fiction. Determining when time was "represented" and when it was "real" in Einstein's theory was necessary to explore the difference between "representation" and "reality" more generally.

Becquerel added a further twist to the story, delineating what today is called the three clocks paradox. With this new example, he thought he had a clever and final argument against the philosopher. It aimed at solving their disagreements without bringing up the more complicated question of how one could bring the two clocks back together in order to compare their differences. A third space traveler could set his clock to the time of the departed one and bring it back to Earth. In this way, the twin on Earth could actually see that the traveling clock was behind his. In all of these examples the need for acceleration or change in direction was eliminated—the twin paradox could thus be explained solely within the framework of the special theory of relativity.

To Becquerel, this new illustration demonstrated that the time of Paul and that of Pierre should both be considered to be equally real. Why keep insisting that only one had to be chosen? "I am in complete disaccord with the eminent philosopher who affirms that Real Time is unique."[13] But Bergson was largely unfazed. Becquerel's example showed that the slowing down of times occurred even without a change in acceleration (and thus even in the realm of special relativity), but Bergson still believed the most important aspect of his argument still held. So what if Paul could see Pierre's clock traveling faster than his? So what if Pierre could see Paul's clock traveling slower than his? Whose time should be counted as real remained a question "that did not pertain to mathematical physics but, rather, belonged to philosophy."[14]

Bergson at first focused on differences in the twins' experience of effort and memory, but after criticisms were launched at him claiming that he was unnecessarily bringing in complications pertaining to living creatures, he focused on acceleration, a clear physical quantity. Differences in clock times arising from differences in acceleration proved that something in one of the twins' scenario was different from the other's. Their experiences of time were thus not *entirely* equal. Which of the twin's clock times was correct? For Einstein, both were equally correct. Back in 1915 he had argued that "our conception of time and space must be subjected to a fundamental revision," concluding that "the physical definition of time we are seeking is complete."[15] Bergson disagreed more than half a decade later. When the twins met back on Earth and compared their differing clocks, it was not immediately clear to him which of these times should be taken as valid. Although *physically* they could arguably have equal rights to both being valid, *philosophically* differences would remain between the two twins and the times shown by their respective clocks. Whose time would prevail on Earth would depend on how their disagreement was negotiated—psychologically, socially, politically, and *philosophically*.

"AT TIMES VIOLENT"

After he was unable to convince Bergson during their face-to-face meeting, Einstein, along with Becquerel, started promoting the work of an ambitious young writer. They endorsed a book on relativity by André Metz, an alumnus of the École polytechnique and a captain from the Rhine army positioned at Bonn.[16] Einstein explained how Metz's book "responded to a real need," "was completely exact," and contained the "refutation of the inexact assertion of other authors."[17] To observers at the time, it was clear that Metz and Einstein had a lot in common—even personally. "The personal philosophy of Einstein is similar to Metz's," explained one reader.[18]

As a soldier and scientist, Metz believed that confrontations, "at times violent," were "necessary conditions" for "making history." This maxim was true, he explained for "all domains"—including science. Describing his work opposing Bergson, Metz underlined how Einstein would only prevail if he fought hard and won. In a private letter to his friend the philosopher Émile Meyerson he described his intentions clearly:

> The triumph of people, ideas, or theories seems to me to have as a necessary precondition a fight, and a bitter struggle, sometimes violent. The names that remain in history are those of men who have fought, and who have fought in all areas . . . Corneille . . . Racine . . . Pasteur . . . Einstein himself, with his simple and benevolent disposition, owes his fame to the controversies his theory raises.[19]

Metz started his polemic by publishing damaging articles against Bergson in the *Revue de philosophie*.

In 1924 Metz wrote a direct response to Bergson's appended edition. Bergson responded once again to Metz in "Les Temps fictifs et le temps réel" (May 1924), in which he again tried to defend his philosophy.[20] He countered Metz's claim that he was professing a relativity theory that differed from Einstein's. All he was doing, he clearly insisted, was philosophy—*not physics*, and these two disciplines focused on distinct problems: "The role of philosophy is different. On a general level, it aims to distinguish the real from the symbolic."[21] Metz's claim that physicists had a "special competence" with respect to the questions of time and relativity was therefore inapplicable. And physicists, he added, were rarely philosophically competent: "One can be an eminent physicist and not be trained in the handling of philosophical ideas. . . . It is in vain that one argues here their special competence."[22] Bergson felt that questions of authority were being brought up gratuitously. Reacting against the growing authority of physicists, he concluded: "Besides, whether we are dealing here with physics or philosophy, the recourse to authority has no value."[23]

Metz was not alone in ignoring Bergson's insistence that he was doing philosophy—*not* physics. Einstein himself, although evidence suggests he knew better, took the same approach. In a private letter to Metz, he framed Bergson's mistake in terms of physics. "It is regrettable that Bergson should be so thoroughly mistaken, and his error *is really of a purely physical nature*, apart from any disagreement between philosophical schools," explained Einstein. He spelled out Bergson's "mistake" in detail: "Bergson forgets that the simultaneity . . . of two events that affect one and the same being is something absolute, independent of the system chosen."[24] By insisting on the "absolute" nature of relativity effects, Einstein tried to convince his followers that it was not necessary to think about relativity in terms of the *difference* between the two travelers. This strategy contradicted what he had written in his travel journal after reading *Duration and Simultaneity*, where he stated that Bergson "seems to really [*sachlich*] understand relativity theory and not to put himself in contradiction to it."[25] Why did both men now frame Bergson's contribution as mistaken in terms of physics?

Einstein wrote other letters explaining Bergson's mistake in this way. Miguel Masriera Rubio, a physicist from Barcelona, also received a letter in which Einstein exposed the philosopher's error: "In short, Bergson forgets that spacetime simultaneity has an absolute character according to the theory of relativity."[26]

Masriera Rubio and Metz did not hesitate to publish Einstein's responses. Anticipating that he may have violated Einstein's privacy by publishing the letter publicly, Metz explained to the physicist: "Clearly, the letter I am asking you for is meant to be inserted (unless you think otherwise) in the *Revue de philosophie*. It so happens that I have often made use of your letters (which have helped me to dissipate many errors), and I hope you don't see anything wrong with this."[27]

How did Bergson respond? The philosopher found it necessary to confront "only" one "new point" in Metz's litany of criticisms and corrections: "André Metz believes that automatic registering devices suffice, without the need of an observer to see what they mark," he explained.[28] Metz repeatedly claimed that Bergson was confused about relativity theory because he continuously tried to put living, flesh-and-blood observers in the theory to see how its results would change. In contrast to the philosopher, Metz believed that physicists did not need to talk about living people but were free to replace these by clocks. If the whole system was automated, Bergson agreed that the relativistic effect of time dilation would indeed take place. But the question of complete automation, and of completely eliminating human consciousness from the world, opened up a Pandora's box of additional philosophical problems. Bergson, in a footnote to his third appendix, considered a scenario in which observers were "nowhere."[29] But for Bergson, the complete elimination of human observers from the world was itself a philosophical riddle. In a world without consciousness, we might as well "say goodbye to the theory of relativity," he concluded.[30] And thus he continued to claim that the times described by Einstein were not all equally real. Some of them had to be, according to him, more "virtual," "fictional," and "represented" than the others.

In the end, Bergson simply gave up trying to convince Metz. Their mutual misunderstanding was simply insurmountable: "The meaning of my thoughts, as that of my book, has completely escaped him. There is nothing I can do."[31]

"COULD NOT BUT END IN FAILURE"

While repelled by Bergson's philosophy of science, Metz was increasingly attracted to an alternative school led by an old disciple of Bergson who had turned to Einstein after the debate: Émile Meyerson.[32] Bergson had initially doted on Meyerson, but the younger philosopher soon changed the course of his philosophy; he dedicated more and more time to thinking about the physicist than about the philosopher.[33] "Meyerson's peak of success as a philosopher of science came with *La Déduction relativiste*, a work endorsed by Einstein and Langevin and understood as having refuted Bergson's philosophical reading of relativity in his *Duration and Simultaneity*."[34] Einstein was elated to have a young French philosopher side with him and wrote a glowing review of Meyerson's book.[35] Meyerson was equally elated to have Einstein on his side. "Nothing else, in my career as philosopher, has made me more proud than the favorable judgment which you have given me," he wrote to the physicist.[36] With Meyerson, Einstein gained some allies from the philosophers' camp—men who were willing to free themselves from Bergson's hold.

After taking clear sides in the Einstein-Bergson polemic, Metz dedicated an entire book to the philosophy of Meyerson. Metz joined

Meyerson's cause, defending *that* philosophy with the same vigor with which he had defended Einstein against Bergson. "I have become his disciple and I work under his direction to spread his ideas," he explained to Einstein.[37] Meyerson would be eternally grateful. He included Metz in his last will, offering him 5000 francs from his inheritance.[38]

In January 1926 Einstein met with Metz and Meyerson in Paris, cementing his relationship to them.[39] Meyerson invited them to dine at his home, asking them to arrive after six in the evening.[40] Following the dinner, Metz asked Einstein if he could publish parts of their conversation, particularly those in which Einstein had praised Meyerson and in which Meyerson had praised Einstein. Einstein agreed, and Metz published the physicist's endorsement. Thanks to Metz's intermediary efforts, each were seen as—independently—legitimizing the other.[41]

Meyerson initially considered science as an attempt to substitute diversity with identity.[42] In *Identity and Reality* (1908), he had already begun a polemic against Bergson. He disagreed with how Bergson drew the boundary between science and other forms of knowledge, particularly by associating it solely with utilitarian concerns. "It is not right to say that science has action as its only aim, nor that it is only governed by the concern for the economy of action," Meyerson insisted.[43] He protested against the exultation of philosophical knowledge as allied with nobler, nonutilitarian concerns. Years later, in *Du cheminement de la pensée* (1931), he was even clearer in his disagreement with the eminent philosopher on exactly this same point: "Bergson intended to make thought a product of action . . . which could not but end in failure."[44] If science was solely associated with action, then philosophy would be in turn defined by inaction, lamented Meyerson.

Bergson disagreed with this characterization of his own work. He always insisted that philosophy, although not centrally concerned with the sphere of action characterizing means-ends utilitarian behavior, could lead to changes in that sphere. But many would repeat the same criticism that Meyerson had leveled against the philosopher. What is more, Becquerel's and Metz's ripostes to *Duration and Simultaneity*, its new appendices, and the follow-up publication "Les Temps fictifs et le temps réel" would seal Bergson's reputation of having been mistaken about relativity.

Metz continued his military career with evident success. When the political situation worsened after the Battle of France, Metz immigrated to London, joining Charles De Gaulle's Free France resistance movement with his two sons. "In 1940–44 all my things were kept by the Nazis and disappeared from my flat in Strasbourg. I succeeded to go to England (with my sons) to join the Free French Forces. . . . I never found again the letters received from Einstein (and from Eddington, Becquerel, Meyerson, etc. . . .)," he wrote years later. After he was able to contact Einstein's secretary, Helene Dukas, he asked her if any copies remained. Dukas did not find much. Some of Einstein's letters to Metz, especially those detailing "his opinion about Bergson," had been lost.[45]

When reminiscing about his life years later, Metz still showed some anger about how Bergson had treated him. He described Bergson's first response to him as "banal, since its tone was that of a professor who instructs . . . his student."[46] Clearly offended, he thought that it had been written in an inappropriate "tone for a polemic of this genre."[47] Another commentator similarly detected in their dialogue "the irritation and the barely dissimulated disdain that an illustrious philosopher must have felt vis-à-vis a young contradictor."[48] But the young contradictor had grown to be a hero—not in science and philosophy, but on the battlefield.[49] After all, he had learned from Einstein's example that those who "remain in history are those men who have fought" equally violently "in all areas."[50]

After the Second World War, during yet another animated discussion at the Société française de philosophie, Metz confidently argued that "there is nothing left" of *Duration and Simultaneity*.[51] A prominent biologist at the conference claimed just the opposite: "I believe, contrary to Monsieur Metz, that the critique of Einstein by Bergson is profound."[52] By then, a new generation of thinkers, many of whom (such as Gilles Deleuze) had lost close family members to yet another even more brutal war (but who had not been alive at the time of the debate or were too young to even remember it), were left to decide.

CHAPTER 14

An Imaginary Dialogue

KING'S COLLEGE, LONDON

During the summer of 1921, Einstein embarked on his first trip to the United States. Why not stop in England on his return voyage, deliver a lecture, dine at the elegant home of Lord Haldane, and meet Prime Minister Lloyd George?

On Wednesday, June 8, RMS *Celtic* carrying Einstein and his second wife, Elsa, docked in Liverpool. The following Monday, Einstein was scheduled to give a widely advertised lecture at King's College. Tickets were quickly snatched up, and eager students who were left outside had no other option than to peer in from the hallway. In Britain "the whole affair was well stage-managed."[1] The excitement was not confined to the community of scientists, let alone physicists. A British philosopher explained how Einstein's work "elicited such a concern among philosophers that it can adequately be compared to the revolution brought about by the introduction of the scalpel in medicine."[2]

Bergson was perhaps the only previous visitor to have caused such a stir at that same location. An observer that day "was reminded of a spectacle of a few years previously, when I had listened to Bergson speaking to a similar audience."[3] After the trip, Einstein expressed his admiration for "a formidable group of English scholars" who were "pacifists and refused to fight the war, e.g. Eddington, Russell."[4] These pacifists became Einstein's key allies.

When Arthur Eddington returned from a 1919 eclipse expedition carrying photographs showing the bending of light by the gravitational field of the Sun, Einstein was crowned as the man who forever changed our conception of the universe.[5] Eddington had avoided the draft by asking the government instead for permission to embark on an eclipse expedition to test relativity theory. At Burlington House, on November 6, 1919, he presented the results of two expeditions (one off the West Coast of Africa and the other one in Northern Brazil) in a combined session of the Royal Society and the Royal Astronomical Society. The next day, Einstein became world news. In global and international settings, the perceived merits and faults of their work intersected with key moments of world history, especially with the rise of fascism and World War II. It also intersected with key aspects of national histories.

Both Bergson and Einstein had a special relationship with England. Bergson's mother was of British and Irish descent. He had lived in London with his parents for a number of years before moving to France and becoming a French citizen. He learned to speak flawless English from his mother, and he would travel frequently to England. It was there that he met the Harvard professor William James, who would become one of his most important advocates in the English-speaking world.

THE PERIHELION OF MERCURY

During the presentation of his results, Eddington conveyed to the public how Einstein had placed a large bet and had convincingly won. The success of Einstein's theory was now "exceeding the advances associated with Copernicus, Newton and Darwin," he explained.[6] Eddington framed the eclipse expedition as a test of relativity theory's predictive powers. "The theoretical researches of Prof. Albert Einstein" were "strikingly confirmed" by the astronomical event.[7]

While the Michelson-Morley experiment is generally considered to have played an important role in the *special* theory of relativity, three other experimental results are often listed as proving the *general* theory (the perihelion of Mercury, the bending of light rays by the Sun, and the red-shift effect).[8] In the eyes of Einstein's supporters, the results of these demonstrated the virtues of Einstein's general theory. Were they (the "three classical tests" as they came to be known) not enough to blow Bergson's objections out of the water?

The history of Eddington's expedition and the example of the perihelion of Mercury have been widely used to showcase not only the power of Einstein's general theory of relativity but also the potential for science to predict previously unknown phenomena. In what sense did Einstein's theory predict a previously unknown effect?

Scientists and science enthusiasts had long known (decades before Einstein first started tackling the problem) that an advance in the perihelion of Mercury was "one of the most perplexing facts of astronomy."[9] The problem of the perihelion of Mercury started to preoccupy scientists intensely during the second half of the nineteenth century. A slight perturbation observed in the orbit of the planet could not be adequately explained with existing theories. According to Newtonian theory, Mercury's perihelion should advance by approximately 600″, but observations fell short of that number. The astronomer Simon Newcomb calculated the discrepancy between the calculated and observed values to be 43″. When Mercury's perihelion was calculated by using the best available numbers for the speed of light in combination with the most trustworthy formulas of celestial mechanics, the resulting values differed markedly from observed ones. Einstein's manuscripts show that he took careful notes of Newcomb's data.[10] Many scientists competed to explain this discrepancy, having already proposed various theories. Einstein found all of them "highly unsatisfactory." He considered his general theory of relativity as the "first realistic explanation for the motion of the perihelion of Mercury."[11]

How did the eclipse expedition confirm Einstein's theory? By showing how gravity bent light. As evidence, astronomers used photographs of stars taken during a solar eclipse. The difference between the calculated and observed values for the perihelion of Mercury could thus be explained by showing how the same factors that caused a curvature in the spacetime geometry around Mercury (and the strange advance in its perihelion) would *also* cause the bending of light around the Sun. Yet Einstein's theory was not the only one that considered light being bent by gravitational forces.

The idea that gravity bent light had also been considered for a long time—most notably during the eighteenth century—by eminent scientists. Since light was often described as "corpuscular" and made up of tiny particles (as per Isaac Newton), it was natural for scientists to think that gravity would affect these particles too. After Newton, Laplace speculated that if gravity was strong enough, light would could be forever trapped by a gravity source.[12] Einstein's own explanation, however, was unique and highly original—it was based on the idea that gravitational forces curved space and time, affecting the path of light—but it was hardly the only available one.

By 1916 Einstein sought to reinforce the credentials of his general theory by stressing how it explained certain things that the alternatives proposed by his colleagues did not. The strongest of these claims centered on Mercury. Einstein's theory explained the motions of that planet beautifully. Yet critics claimed that Einstein's theory did not *explain*—let alone predict—the perihelion but was actually only a theory that matched its value of 43″ beautifully.[13] Besides, even the calculation of the value of 43″ by Newcomb could itself be the subject of numerous errors. "Another astronomer" was apt to find "different values than those of Newcomb."[14] Einstein first described his own work to his colleagues in this way. "I derive quantitatively . . . the perihelion motion of Mercury," he wrote to David Hilbert, who was developing a competing solution, and did not claim then that his theory "predicted" this value.[15]

An alternative and highly popular explanation for the discrepancy between the observed and calculated values for the advance of Mercury's perihelion posited the existence of a planet named Vulcan, whose gravitational field ostensibly pulled the planet toward it. Scientists searched for it, and some claimed to have seen it (although accounts of its sighting remained suspect), but even those scientists who remained unconvinced by the Vulcan hypothesis knew they could posit invisible interplanetary masses spread in space affecting Mercury in the same way. Newcomb thought the difference could also be due to a flattening of the Sun, which changed the direction of its gravitational pull toward Mercury. Einstein knew of these alternative explanations but found them unconvincing.

ARTHUR EDDINGTON: BEFORE AND AFTER THE DEBATE

Eddington had personal and political reasons to support Einstein and become one of his most aggressive "bulldogs." In addition to being a Quaker, a determined pacifist, and a conscientious objector during the war, he had been chief assistant to the Astronomer Royal at Greenwich (since January 1906) and had struggled with the problem of finding a standard of measurement. Einstein offered him an elegant solution. Why not take it?

After catapulting Einstein to fame with his presentation of the expedition's results, Eddington continued to write and lecture widely about relativity, but soon his admiration of Einstein's work would start to be tempered in light of Bergson's philosophy. Bergson attentively read Eddington's *Space, Time and Gravitation* (1920), in which he found evidence for a particular weakness in Einstein's theory. He cited Eddington's description of Einstein's universe, where "events do not happen; they are just there, and we come across them."[16] Eddington would later agree with the philosopher's points of contention against the physicist, emphasizing again how in Einstein's work the "past and the future lie spread out before us as in a map," where our sense of time passing was nothing more than a mere illusory "fancy."[17] Eddington and Bergson considered that Einstein's spatial, cartographic characterization of time, in which our sense of flow and becoming were merely "illusory," was a glaring deficiency of his work.

In *The Nature of the Physical World* (1928), Eddington published a hypothetical dialogue "between the Astronomer Royal, and, let us say, Prof. Bergson, on the nature of time."[18] He generously acknowledged Bergson's "authority on the subject." Eddington surmised that if these two men entered into a discussion "there would then probably have been a keen disagreement, and I rather think that the philosopher would have had the best of the verbal argument." But winning the verbal argument did not amount to much. Since the "Astronomer Royal is entrusted with the duty of finding out time for our everyday use," Bergson would have had no choice but to use that time: "After showing that the Astronomer Royal's idea of time was quite non-sensical, Prof. Bergson would probably end the discussion by looking at his watch and rushing off to catch a train which was starting by the Astronomer Royal's time."

Throughout the text, he stereotyped Bergson's concept of time as "time lived," "private time," "time as estimated by consciousness," or "time of experience." The practical realities of time telling and distribution showed the relative meaningfulness of the astronomer's notion and the meaninglessness of the philosopher's: "Whatever may be time *de jure*, the Astronomer Royal's time is time *de facto*," he concluded.[19]

While this initial dialogue was largely dismissive of the philosopher's contributions, Eddington would soon make it clear that he was not inviting his readers to side with either of these simplistic positions. On the contrary, he clearly explained that he firmly opposed one of the main tenets of the general theory: that of considering time and space in similar terms. Instead of siding entirely with the imaginary Astronomer Royal, who could be seen as defending the position of most physicists, he took the place of a naive "beginner" who "is inclined to say: 'That is impossible. I feel it in my bones that time and space must be of entirely different nature. They cannot possibly be mixed up.'"[20] Eddington had tempered his enthusiasm and started to reevaluate the merits of Einstein's contributions.

Eddington's resolution of the "imaginary dialogue" was a strong statement critical of the direction science was taking during those years. Science had been very successful, Eddington admitted: "By long history of experiment and theory the results of physical investigation have been woven into a scheme which has on the whole proved wonderfully successful." But this success had come at a loss. Science was "only imperfectly representative of the time familiar to our consciousness."[21] Victory had come at a price. For this reason, certain scientific truths could easily lend themselves to ridicule. Statistics, for example, was often invoked to explain strange things that did happen and strange ones that never did. "If an army of monkeys were strumming on typewriters, they *might* write all the books in the British Museum," he quipped.[22] The very idea was so laughable that it showed a deep "imperfection" in science, one that became particularly visible when scientists dealt with real time-dependent events.

"Your protest in the name of common sense against a mixing of time and space is a feeling which I desire to encourage," he told his readers. "Time and space ought to be separated," he continued. Why? Because "the current representation of the enduring world as a three-dimensional space leaping from instant to instant through time is an unsuccessful attempt to separate them." Eddington was motivated to "resurrect the almost forgotten time of consciousness and find that it has a gratifying importance in the absolute scheme of Nature."[23] He felt entitled to craft and promote an intensely mystical view of science and nature.

Einstein later in life confessed to his friend Besso what he came to think about Eddington in a private letter. He called him a "man full of ideas, but deprived of critical spirit" and compared him to a "prima ballerina who does not fully believe herself in her elegant leaps."[24] But in the initial fight against Bergson, Eddington had proved himself a powerful asset, only to be abandoned later.

CHAPTER 15

"Full-Blooded" Time

"It was my good fortune," wrote the mathematician and philosopher Alfred North Whitehead, "to be present at the meeting of the Royal Society in London when the Astronomer Royal for England announced that the photographic plates of the famous eclipse . . . had verified the prediction of Einstein that rays of light are bent as they pass in the neighborhood of the sun."[1] The First World War had just ended, and Whitehead was immediately invited to write one of the first articles about relativity theory in the *Times Educational Supplement*. He remembered how that day seemed like "an intense Greek drama" with "a dramatic quality in the very staging."[2]

The mathematician was also present a few years later, when Einstein visited England again on his way back from America. Whitehead was among those in the audience at King's College. That evening he would have his first opportunity to talk to Einstein personally.[3] The physicist "roused the general public to such a pitch of enthusiasm" that his visit would remain unforgettable, reported Whitehead's student Herbert Dingle.[4]

Einstein had friends in England, but he also had strong critics.[5] Whitehead and Dingle would each develop sophisticated objections to Einstein's work, turning to Bergson for inspiration. Eddington was viewed as defending Einstein, and Whitehead as siding with Bergson. "The scientific world was split into two camps: Einstein and his lieutenant Eddington, Bergson, and his lieutenant Whitehead," explained one observer.[6] Although Bergson protested that "Whitehead should not be considered my lieutenant," he nonetheless agreed that they had arrived at very similar conclusions.[7]

By then, Whitehead had impressive accomplishments under his belt, counting among his titles those of dean of the Faculty of Science, Senator, chairman of the Academic Council, and professor at Imperial College. After dining with Einstein at Lord Haldane's house, their host "escorted [them] to his study and left them there alone. Saying they must have so much to say to each other."[8] Whatever the two men said to each other that night—we do not know what it was—developed into a growing rift between them and their understanding of time, history, and science.

Whitehead was an admirer *and* critic of Einstein's theory of relativity. He frequently expressed his admiration for Einstein as an individual, insisting that his criticisms did "not in any way represent my attitude towards him."[9] Whitehead's *The Principle of Relativity with Applications to Physical Science* (1922), which had been published when Whitehead was still a professor of mathematics at Imperial College, offered a meticulous reinterpretation of Einstein's theory. The book accepted all known empirical results, but not all of their philosophical import. A thoughtful reader of Whitehead explained how the philosopher considered Einstein's theory "adequate from the scientific standpoint and equally inadequate from the epistemological standpoint."[10]

Whitehead boldly offered a slight modification of Einstein's general relativity formula. While he left "for experiment to decide" about the usefulness of the modified formula, his attempt has been, for the most part, discounted as untenable. Yet this modification did not encompass the totality of his critique. It and his largely failed attempts to improve the theory hurt his standing among many of his peers, leading him to lose important allies. Whitehead knew well that "perhaps" his modified formula would not "survive further tests of other delicate observations," but even "in this event we are not at the end of our resources."[11]

For Whitehead and his supporters, the viability of the new formula was a minor component of a broader and much more significant philosophical project. The "conflict with Einstein's theory" could easily "cause one hastily to dismiss Whitehead's philosophy of science," explained an important philosopher.[12] But there was much more to his work. Whitehead would be soon be called to Harvard as a professor of philosophy.

ALFRED NORTH WHITEHEAD

What did Einstein think of Whitehead? Not much. A volume dedicated to Whitehead's thought published in 1941 reported that Einstein once confessed: "I simply do not understand Whitehead."[13] What did Bergson think of Whitehead? The admiration between Bergson and Whitehead was thorough and mutual. Whitehead's *The Concept of Nature* (1920) considered his work to be "in full accord with Bergson." In *Duration and Simultaneity*, Bergson judged Whitehead's *The Concept of Nature* to be "an admirable book" and "one of the most profound ever written."[14] Whitehead's admiration for Bergson's philosophy continued throughout his life. The editors of the volume for the *Library of Living Philosophers* dedicated to Whitehead's thought it was completely "appropriate" to include a facsimile of a letter by Bergson in the volume "since Whitehead has always acknowledged his indebtedness to this great French philosophical contemporary."[15]

In *Process and Reality*, which was based on lectures delivered in 1927–1928, Whitehead was even clearer about his mission and his engagement with the French philosopher. He wanted to take all of Bergson's philosophy except those aspects of it associated with anti-intellectualism: "I am also greatly indebted to Bergson, William James and John Dewey. One of my preoccupations has been to rescue their type of thought from the charge of anti-intellectualism, which rightly or wrongly has been associated with it."[16]

Whitehead's fascination with Bergson was sparked by his involvement with the Aristotelian Society, a forum that helped him turn to philosophy from mathematics. "My philosophic writings started in London, at the latter end of the war. The London Aristotelian Society was a pleasant center of discussion, and close friendships were formed," he recounted in his autobiographical notes.[17] In this context, he befriended H. Wildon Carr, an important popularizer of Bergson in the English-speaking world who also wrote extensively about the theory of relativity.

Whitehead, Wildon Carr, and Lord Haldane became influential supporters of Bergson's philosophy through the Aristotelian Society and beyond. Aristotelian philosophy, traditionally backed by the Catholic Church (and understood through St. Thomas Aquinas's interpretation), resonated positively in Anglican circles during this time. Whitehead came from an Anglican family with a father, brother, and uncles who were vicars. He was further exposed to Catholicism through marriage and thought of converting but desisted. He became intensely religious after the Great War, yet chose not to adhere to a particular church.

Lord Haldane, who had secured Einstein's invitation to King's College, was also a member of the Aristotelian Society. Haldane knew Bergson well—both men corresponded and discussed politics and philosophy avidly. One of his books, *The Reign of Relativity* (1921), discussed the work of both Einstein and Bergson. Haldane considered Whitehead's work on relativity as superior to Einstein's own: "Professor Whitehead seems to me to have brought out [the character of space and time] in his treatment of relativity more thoroughly than Einstein or even Minkowski himself has done."[18] Whitehead's understanding of relativity, Haldane continued, "is more thorough in the logical treatment of relativity than anything that I have so far become acquainted with in the works either of Einstein himself or his disciples in Germany."[19]

Lord Haldane and Whitehead worked closely. When Haldane wrote the highly successful *The Reign of Relativity*, he drew heavily on Whitehead's work and received direct help from him with the mathematical parts.[20] Bergson's influence was also clear in Haldane's book, not only through Whitehead, but also directly. After reading it, Bergson found the book "suggestive" and "inspiring," thanking Haldane for "having given me the honor of citing me extensively and examining my views closely."[21]

At Cambridge University, Whitehead quickly recognized a talented and aristocratic young student, Bertrand Russell. Whitehead became Russell's dissertation advisor as well as friend and collaborator. He coauthored with his student the landmark three-volume *Principia Mathematica* (1910, 1912, 1913), a text considered monumental in many respects. One philosopher explained how he found the difference between Whitehead and Russell to be slight at first. Their work showed a "family resemblance . . . with individual differences such as are found in all families except in case of identical twins."[22] But Whitehead eventually broke with his junior collaborator, and the differences between the two men sharpened.[23]

The particular divisive issue between Whitehead and Russell was directly connected to both men's view of Bergson. In a private letter to

Karl Popper written late in his life, Russell explained that he did not accept Whitehead's philosophy because of how it had been influenced by Bergson: "I never studied Whitehead's philosophical work at all thoroughly and I made a point of not saying anything publicly in criticism of it. What I did know of his philosophical work displeased me, partly because of what I thought unnecessary obscurity, and partly because of the trail of Bergson."[24]

BERTRAND RUSSELL

Einstein admired Eddington for refusing to fight during World War I and liked Bertrand Russell, who had been imprisoned for refusing to join the army, for similar reasons. As a sign of support for Russell, Einstein wrote the foreword to the German version of *Political Ideals* (published in 1922). A few years later Russell published one of the most important books designed to explain relativity to a general audience, *The ABC of Relativity* (1925).

Russell was one of the main advocates of analytical philosophy, a discipline that took natural science as a pinnacle of knowledge based on its foundation on sensations followed by uncontestable logical principles. He became such a persistent and resolute critic of Bergson that, decades later, scholars could still easily recognize in his work the "Russellian clichés about the alleged irrationalism and antiscientism of Bergson."[25] Bergson was well aware of Russell's antagonism, blaming it on a petty personal incident between them: he had once criticized Russell "in public" for attempting to give a materialist interpretation to Platonic ideas. Russell's subsequent attacks were the simple comebacks of a bruised ego, argued Bergson.[26]

Russell admired Einstein with the same fervor with which he hated Bergson. Russell and Einstein corresponded and worked together on many political fronts well into the Cold War era, when they authored the famous Russell-Einstein Manifesto (1955), warning of the dangers of nuclear weapons.

Bergson was not the first French philosopher whom Russell targeted. He started his earliest work by taking a stand against other key French philosophers who stressed the conventional or constructed nature of knowledge at the expense of its absolute character. *An Essay on the Foundations of Geometry* (1897), based on his Cambridge dissertation, attacked Poincaré.[27] Poincaré responded to Russell, but the philosophical bent of the two men would remain forever distinct.[28]

In 1912 Russell wrote one of his first polemical articles against the philosopher, titled "The Philosophy of Bergson."[29] Throughout it, Russell faulted Bergson for being "difficult," "not always easy to follow," and mostly just obscure.[30] Russell compared the Frenchman's philosophy, quite unflatteringly, to new advertisements of a popular instant beef-bouillon mix that not only tasted great but allegedly cured mostly everything from burns to pneumonia: he accused him of being a great "advertiser of himself and his theories," who "like the advertisers of Oxo," relied "upon picturesque and varied statement."[31] He intensely disliked Bergson's style, which he analyzed minutely: "The number of similes for life to be found in his works exceeds the number in any poet known to me."[32] His philosophy rested on "a mere play of words."[33] It was "an imaginative epic, to be judged on esthetic rather than on intellectual grounds."[34] It was, in short, a mere "poetic effort":

> Shakespeare says life's but a walking shadow, Shelley says it is like a dome of many-colored glass, Bergson says it is a shell which bursts into parts that are again shells. If you like Bergson's image better, it is just as legitimate.[35]

After writing this article, Russell delivered another lecture against Bergson, hosted by the Cambridge Heretics Society.[36] He restated some of the same points he had previously made: "One of the bad effects of an anti-intellectual philosophy, such as that of Bergson, is that it thrives upon errors and confusions of the intellect," representing the "bankruptcy of the intellect and triumph of intuition" that it allegedly advocated.[37] During those years, the differences between Bergson and Russell were considered so vast that they represented "two of the leading tendencies in modern philosophic thought."[38] Russell continued his tirade against Bergson in lectures delivered in Boston during the spring of 1914, later printed as *Our Knowledge of the External World*. For the rest of his life, Russell described Bergson as anti-intellectual, claiming, in a biting phrase—that, for Bergson, "intellect is the misfortune of man."[39]

Bergson and Russell disagreed about the foundations of mathematical logic. They disagreed about how to solve one of the most ancient intellectual puzzles, Zeno's paradox, often illustrated by the famous fable of the tortoise and Achilles. The puzzle pertained to the relation between continuous and discrete movement and touched on the question of infinity, a concept that was at once a mathematical, physical, metaphysical, and theological. Wildon Carr, the secretary of the Aristotelian Society, a defender of Bergson, and one of the most important writers who brought Bergson to English-speaking audiences, criticized Russell's attempt to base modern mathematics on "a paradoxical definition of continuity." He pointed out the inadequacies of Russell's critique of Bergson and of his philosophy of mathematics. Russell had argued "that continuity is infinite divisibility," and this definition, many agreed, could be used to solve Zeno's paradox. Carr readily admitted that certain things were infinitely divisible, but they forgot Bergson's insistence that one thing was *not* divisible: real "movement."[40] Carr concluded that Russell's definition of continuity "removes one paradox only to leave me with a greater" one.[41]

Asked to weigh in on the discussion, Bergson accepted Carr's response to Russell wholeheartedly, writing to the editor of the *Cambridge Magazine*: "I find that the response made already by Wildon Carr is excellent."[42] For decades to come, Carr would pit Bergson against Einstein.[43] He considered that "Einstein is a philosopher, despite himself" and compared him to the famous doctor described by Molière whose circular arguments explained sleep-inducing substances in terms of their dormitive qualities.[44]

Another important difference between Russell and Bergson centered on the spatial qualities of time. Bergson had violently fought against treating time as a kind of space, insisting that they were entirely different. Russell, on the contrary, took them as so similar that he introduced the concept of "time corpuscles" as an exact counterpart to "space corpuscles":

> In saying this I am only urging the same kind of division in time as we are accustomed to acknowledge in the case of space. A body which fills a cubic foot will be admitted to consist of many smaller bodies, each occupying only a very tiny volume; similarly a thing which persists for

an hour is to be regarded as composed of many things of less duration. A true theory of matter requires a division of things into time-corpuscles as well as into space-corpuscles.[45]

In the post–World War II years, Russell made one of the most damaging criticisms that Bergson would ever face, associating his philosophy with that of Nazi-occupied France. The bestseller *A History of Modern Philosophy* claimed that Bergson's philosophy "harmonized easily with the movement which culminated in Vichy."[46]

Could there be any truth in Russell's accusation? Bergson openly warned against the "formidable wave of anti-Semitism which will fall on this world"—even as the Vichy government offered to exempt him from the obligations imposed on other Jews.[47] When others fled, he chose "to remain among those who tomorrow will be the persecuted ones."[48] In protest to the government, he resigned from all of his official positions. He actively tried to help colleagues who were being forced from their jobs, such as Jean Wahl, through his contacts with important officials of the Vichy government, especially Jacques Chevalier, a friend who sympathized with Bergson's efforts and helped. Few would know the exact details of Bergson's opposition to the Nazis. Russell's book, with the damaging allegations of Bergson's politics, sold like hot bread—the Englishman would derive an important part of his income for the rest of his life from its royalties.

"HAVE A CARE, HERE IS SOMETHING THAT MATTERS!"

What aspects of Bergson's work inspired Whitehead? The Englishman shared with Bergson a desire to think of time as something different from space. But how did the work of the two differ? In contrast to the French philosopher, Whitehead would use the phrase "passage of nature" to describe what Bergson sometimes referred to as time. Whitehead's use of a different term stemmed from his desire to prevent this philosophical notion from being confused with "the measureable time of science and of civilized life generally."[49]

What did Whitehead dislike about Einstein's approach? His answer was simple. "The concept of the passage of time has been lost," he lamented.[50] To save it, Whitehead rejected Einstein's cavalier dismissal of psychological time in favor of physical time: "It follows from my refusal to bifurcate nature into individual experience and external cause that we must reject the distinction between psychological time which is personal and impersonal time as it is in nature."[51] While Einstein referred to single "world lines" to refer to paths traveled by objects across a single space-time continuum, Whitehead wanted to keep space as different from time, distinguishing a "spatial" route from a "temporal" route. He also wanted to distinguish salient temporal events that stood out among others, using the term "historical route" to signal these differences. The theory was much more complicated, but the benefit of his approach, he argued, was that it would allow scientists to think of certain events as more "significant" than others. These would retain the "full-bloodedness of a moment of time."[52] His hope was that certain aspects of our notions of historical time would thus be part of scientists' understanding of physical time. Additionally, Whitehead did not just take the results of measurements at face value and build a scientific theory from them, he also strove to study why certain "events" led us to take measurements that were meaningful for us. He cared about why we measured time and space more than about the results of the measurements themselves. "We ask, why this pathetic trust in the yard-measure and the clock?" and answered clearly.[53] The reason why we were driven to measure, he argued, was because we cared about how particular "events" were related to each other: "The yard-measure is merely a device for making evident obscure relations between those events in which it appears."[54]

From July 13 to 16, 1923, three groups met together: the Aristotelian Society, the Mind Association, and the Scots Philosophical Club.[55] During the meeting Whitehead considered the travelers in the twin paradox as connected to each other via the transmission of time signals. Whitehead conceived relativity theory in terms of the actual comparison of differing times. He thought of the twin paradox in terms of twins who could be connected to each other through communications technology. The traveling twin could "count the days on Earth by means of a signal transmitted from Greenwich each day at noon," or he could be "not attending to the Earth at all. He takes his clock with him."[56] Both clocks would show different times. Whitehead agreed that neither of these times could be simply dismissed or that one of them could be considered "fictional" or "represented," as Bergson had claimed. But he also argued against Einstein's interpretation of them as equally real and equivalent to time-in-general.[57]

Whitehead, as Bergson would do in the appendices to *Duration and Simultaneity*, placed a strong emphasis on the seemingly minor differences between the twins. He argued that the *meaning* of the transmitted times had changed for each: "the lapse of clock time is a lapse of time according to the traveler's meaning, and the meaning differs from that for Earth time." Both clocks were "congruent." They "ran truly," but Whitehead urged readers not to forget that "the meanings for time were different in the two cases."[58] The experience of each twin was different. Scientists should not forget that something caused "a diversity of history which produces the discordance of chronology."[59] Changes in clock times could only arise in connection to changes in "the life history of any body," be it of a traveling of a twin or of a molecule.[60]

For the rest of his life, Whitehead explicitly fought against "the bifurcation of nature" into dualistic camps (that of physics and psychology, matter and mind, and others), inviting us instead to look at the links between these seemingly opposing concepts. Whitehead's answer to the impasse consisted in crafting a philosophy that denied a distinction between nature and experience: "Nature is thus a totality including individual experiences, so that we must reject the distinction between nature as it really is and experiences of it which are purely psychological. Our experiences of the apparent world are nature itself."[61] The philosopher refused to accept the dichotomies that came to characterize modernity: the local and the distant, the psychological and the physical, but also others, such as the organic and the mechanical. At times, he used the term "organic mechanism" to identify his philosophical viewpoint. What is more, he considered the reasons marking such divisions differently, championing new directions for research, where "science is taking on a new aspect which is neither purely physical, nor purely biological. It is becoming the study of organisms. Biology is the study of the larger organisms; whereas physics is the study of the smaller organisms."[62] The philosopher did not want to separate human experience from physics: "We should either admit dualism, at least as a provisional doctrine, or we should point out the identical elements connecting human experience with physical science."[63] Fulfilling the option he proposed was much harder: it required going back to the drawing room to construct a physics that did not set itself up against lived experience.

Whitehead fought against thinking of the world in terms of "bare events." He instead proposed to think of an event that mattered as "a drop of experience." Instead of focusing on matters of fact, he placed emphasis on facts that mattered:

> Our enjoyment of actuality is a realization of worth, good or bad. It is a value-experience. Its basic expression is—Have a care, here is something that matters! Yes—that is the best phrase—the primary glimmering of consciousness reveals, something that matters. . . . The dim meaning of fact—or actuality—is intrinsic importance for itself, for the others, and for the whole.[64]

While Einstein had difficulty understanding Whitehead, many philosophers understood Whitehead more than they understood Einstein. One philosopher explained how they felt intimidated by relativity theory, claiming they "had all become defeatists, and drew into our own shells, where we might hope to withstand the assaults of the mystical giant Abracadabra, who could make the less appear the greater length." Luckily for them, Whitehead drafted his own version: "Such was the general situation when along came Mr. Whitehead, speaking a language that for the most part we could understand and employing equations that after considerable brushing up we could follow." For them it was only after Whitehead started working on it that they learned that "there might be some sense in relativity after all."[65] But the sense that they got from these expositions was radically transformed, and patently Bergsonian.

HERBERT DINGLE AND THE HISTORY OF SCIENCE

How could scientists deal with "a diversity of history which produces the discordance of chronology" that led to the paradoxes of relativity theory?[66] Wasn't this task better left to historians? Herbert Dingle, one of the most important historians of science in the post–World War II era, would grapple with this question, and with Bergson's and Whitehead's objections, throughout his life. After the war, he became professor and head of History and Philosophy of Science at University College London, founded the *British Journal for the Philosophy of Science*, and became one of the founders of the British Society for the Philosophy of Science. He was active in the International Union for the History of Science (where he held the title of vice president), wrote more than twenty articles on the topic and various books, and praised the initiative of Harvard's president James Conant to include classes on the history of science in the general undergraduate curriculum.[67]

As a young man, Dingle had studied physics and astronomy. He enrolled in Whitehead's three-year course at Imperial College and "went to all of Whitehead's lectures that he could."[68] When he combined what he learned from Whitehead with his experience listening to Einstein, his defense of Bergson and his critical stance toward the physicist was forever sealed. Whitehead encouraged him to write a popular version of the theory, which Dingle titled *Relativity to All*. He was a strong advocate of relativity theory, but he would soon find an essential inconsistency in it.

Dingle was able to attend Einstein's lecture at King's College with a press pass, which he obtained because he had published an article on Einstein's theory in one of London's evening newspapers.[69] His initial campaign to investigate the respective merits of Einstein and Bergson soon became much more—a full-blown controversy at the center of broader debates about social planning and the value of a scientific education over one based on the arts and humanities. It was not easy to defend Bergson during these years. Just for trying, Dingle gained the reputation in the scientific world of being a stubborn crackpot who refused to accept Einstein's accomplishments.

In Dingle's writings, the Einstein-Bergson debate intersected with key themes of British liberalism and its controversies. His work in the history and philosophy of science refused to reduce science and technology to a by-product of social factors and relations. He wanted a space for thinking about science that was not limited to the narrow Marxist view that was then gaining ground.

In the 1950s Dingle came across a footnote in the book *The Foreseeable Future*. It contained an extremely short—just a few sentences long—reference to the twin paradox. A "returning astronaut" explained the author, "would, in fact, find that time had gone more rapidly on the Earth than on his spaceship."[70] The book had been written by the scientist and Nobel laureate Sir George Thomson, the only son of the discoverer of the electron, J. J. Thomson. That a scientist from the respectable Thomson lineage believed this statement was true represented, for Dingle, everything that was wrong with the world at midcentury. It certainly did not help that the rest of the book was a long triumphal narrative of science and technology that extolled, over and above other recent discoveries, the benefits of atomic power. Dingle set out on a mission to correct these views.

"A revival of interest in Bergson's philosophy would be salutary," insisted Dingle.[71] His most forceful defense of Bergson and attack of Einstein appeared in 1965, as an introduction to a new English translation of *Duration and Simultaneity*, and continued for the rest of his life.[72] As a Quaker who had started a career in science, he criticized the preponderant role of scientists in modern society. They were "the most insidious and most dangerous of all usurpers."[73] He insisted that certain aspects of the theory of relativity were conventional and the result of current technological limitations rather than universal. As many others before him, such as Whitehead, he did not believe that the speed of light was an unsurpassable velocity that would remain so for all of eternity: "There is no reason at all for doubting that material velocities exceeding that of light are possible and may well be attained before long."[74] When these velocities would be discovered, presaged Dingle, Einstein's sins would come back to haunt him.

The dialogue between Dingle and his contradictors was "parried in a communication breakdown" filled with mutual "incomprehension." Although most of his peers claimed to have proven Dingle wrong on a quite basic aspect of relativity theory, others, including Dingle himself, continued to claim that his main objection to the theory was simply not up to falsification by scientists. Consensus between him and his interlocutors could not be reached: "Its end was an impasse marked by mere recognition of where the unyielding disagreement lay, or sometimes even a lack of ability or willingness to comprehend the disagreement."[75]

Dingle advocated that science should become more philosophical: we should aim to have a *philosophical science* rather than a *scientific philosophy*. According to Dingle, science was, in fact, nothing more than philosophy, but a particularly successful one at that: "Indeed, the outstanding fact about modern science is that it is not a practical undertaking at all . . . but the most abstract of all things—a philosophy."[76] While frequently those trained in philosophy aspired to gain scientific knowledge, Dingle—who was trained in science—aspired to be more of a philosopher, by way of Bergson.

Dingle drew a direct connection between Einstein's theory and the philosophical movement, logical positivism, that allegedly sustained it. He was highly critical of "ideas of the logical positivist type that originated in relativity theory." His work sought to supersede them with a philosophy inspired, instead, by Bergson.[77] By the end of his life, Dingle had few allies. He, however, greatly admired the work of the physicist Percy Bridgman at Harvard, whose criticisms of relativity theory had inspired him deeply and whose work also echoed with Bergson's critique.[78]

Writing in 1965, Dingle still criticized Einstein for having acted in such a rash way. Einstein had "arbitrarily postulat[ed] a certain method of timing distant events" and *then* had claimed to have discovered a "fundamental fact of nature." Dingle refused to grant such a fundamental role to effects connected to these timing procedures. The paradoxes of relativity were "a fortuitous consequence" of various "physical effects."[79] Einstein's method was not, as the physicist claimed, "the one and only natural way of timing distant events." It was entirely conventional: "the time of a distant event can be chosen as we wish."[80]

The historian of science looked closely at the exact timeline of Einstein's work. He noticed that the physicist had *first* defined a particular procedure of synchronizing clocks at a distance, which *then* resulted in constant measurements for the speed of light, not the other way around:

> What Einstein succeeded in doing was to define a procedure for timing that event so that the observers, on applying it, did in fact time the event differently and in such a way that they both arrived at the same velocity of light.[81]

Dingle reminded readers how in 1921, during his lectures at Princeton, Einstein had described the process through which he conceived of the theory of relativity *in the actual order* that it had taken place. If seen in this way, the use of a constant value for the speed of light when timing distant events appeared once again as a *choice* rather than as an absolute *necessity*. Einstein explained his theory to Princeton students in this way:

> The theory of relativity is often criticized for giving, without justification, a central theoretical role to the propagation of light, in that it founds the concept of time upon the law of propagation of light. The situation, however, is somewhat as follows. In order to give a physical significance to the concept of time, processes of some kind are required which enable relations to be established between different places. It is immaterial what kind of processes one chooses for such a definition of time. It is advantageous, however, for the theory, to choose only those processes concerning which we know something certain. This holds for the propagation of light *in vacuo* in a higher degree than for any other process which could be considered.[82]

Speaking to a throng of eager listeners, he was forthright about his theory's weakness. If the "central theoretical role" given to the propagation of light in his understanding of "the concept of time" appeared "without justification," why adopt his theory? He proceeded to justify himself: "The situation, however, is somewhat as follows." He felt entitled to treat light in such a special way "in order to give a physical significance to the concept of time." Since a material reference for time had to be chosen, it was "advantageous . . . to choose only those processes concerning which we know something certain." Light, at the moment, was the best available solution because "this holds for the propagation of light *in vacuo* in a higher degree than for any other process which could be considered."[83]

After his debate with Bergson, Einstein developed a bolder response. He worked much harder to prove why clocks based on light defined as a constant were "ideal" for more than these practical "advantageous" reasons—they were also ideal for foundational ones.

FROM "TWO CULTURES" TO THE CONTINENTAL-ANALYTIC DIVIDE

Was Dingle simply being pedantic by reminding his readers of the actual time line of events in which Einstein first treated the light clock as a convenient device for determining time and later as an ideal clock that revealed a fundamental property of time? Dingle had deeper motives for concern. His fear resided in what he saw as a growing rift between two fundamental ways of obtaining knowledge in the modern world—where one, scientific knowledge, was increasingly gaining ground only by seeming to lay outside of history. In consequence, it appeared increasingly divorced from everyday human affairs, interests, and needs.

C. P. Snow, a contemporary of Dingle, lamented the separation of sciences and the humanities into "two cultures" in his influential 1959 lectures. During those years, Dingle raised similar alarms. He warned of the potential dangers that a "hostility to the arts" and an excessively practical (Marxist) view of science could have on the world. He directly targeted J. D. Bernal's *The Social Function of Science* (1939). "It is one of the paradoxes of modern life that while . . . it is dominated by science, very few people have any clear idea what science is," he explained.[84] Dingle connected the sociology of science, which he associated with Robert K. Merton in the United States and with Bernal in Britain, to logical positivism. Through the British Society for the History of Science, he sought a different route for understanding science, a way that was not based on scientific materialism, logical positivism, or reductive sociology.

With rare exceptions coming from the history of science, science and philosophy grew further and further apart from each other. Within philosophy, practitioners increasingly differentiated analytical philosophy (often grouped together with logical positivism) from Continental philosophy. Because of these divisions, argued Dingle, scientists rarely considered their work, its import and implications for society, in a broad sense. The scientist, he lamented, "understands what he is doing about as well as a centipede understands how he walks."[85]

CHAPTER 16

The Previous Spring

CHICAGO, USA

On May 3, 1921, Einstein traveled to Chicago to deliver three lectures on the theory of relativity.[1] Over dinner at the Quadrangle Club, he met the American philosopher George Herbert Mead, one of the founders of American pragmatism. Mead had studied Bergson's work intently decades earlier. After learning about Einstein's work, he started asking how the physicist's work fit—or did not fit—with the philosopher's.[2]

Bergson had close connections to the United States. In 1913 he delivered a famous lecture in New York's City College credited for causing one of the city's worst traffic jams.[3] During the Great War, he embarked on a mission to try to convince President Wilson to enter into the war against Germany. Intellectually, he was closest to William James and John Dewey. Two of his closest supporters in England, Wildon Carr and Whitehead, would later relocate to America (Carr to the University of Southern California and Whitehead to Harvard in 1924) and bring with them many of Bergson's lessons.

After dining with Einstein in Chicago, Mead felt "discouraged," leaving that evening with a feeling that he had not fully understood the importance of the theory. To a friend, he confessed: "I have recovered somewhat from a discouraged feeling I brought away from the Einstein dinner—that I had not grasped the import of Relativity."[4] In the years that followed, Mead dedicated himself to understanding the theory of relativity, just as he continued to study Bergson apace.[5]

American pragmatism, with its desire for civil, democratic, and practical solutions to pertinent debates, would soon be engulfed by the debate across the Atlantic. Mead, who was a student of John Dewey, Josiah Royce, and William James, struggled to understand how relativity fit with Bergson's work, crafting novel answers to some of the same questions that preoccupied his illustrious mentors. Concerned with how to mediate the interests of private individuals in a collective society, he asked how Bergson's view of time could be made to fit with Einstein's.

Einstein's trip to the United States, a fundraising initiative for Zionist causes, was controversial for many reasons. Not only was it radical to host a member of the enemy German nation, Zionism was also tightly associated with British imperial ambitions. After hearing about the physicist's travel plans to "fraternize with Englishmen and their friends," the scientist Fritz Haber tied to convince his friend to change his plans: "Do you want to sweep aside so much bloodshed and suffering by German Jews with your conduct? And what do you gain by going on this journey now, instead of delaying it by a year, and waiting for more peaceful relations among the nations?"[6] The furor over his visit was such that even President Harding met with him. American journalists were fascinated. In a year, Einstein would visit another enemy nation. The French press would get their chance.

When Einstein arrived in Chicago, Mead was posed to ask how relativity fit with Bergson's philosophy. Mead had first studied Bergson's work during the early decades of the twentieth century, writing a thorough review of *Creative Evolution*, and returned to his work years later, during the summer of 1920.[7] Mead's final engagement with relativity theory, which included a close examination of the Michelson-Morley experiment, was not published until after his death.[8] Acknowledging that there is "no position eternally pegged down that could not from another watchtower be seen to move," he reached an original conclusion: "Our grasp of the innermost structure of things is experimental."[9] Even at this unfinished stage, his solution to the Bergson-Einstein impasse was framed in terms characteristic of American pragmatism, in which reality could be considered as essentially perspectival, intersubjective, and dynamic at the same time as universal, because of the legitimacy of the experimental method.

PRAGMATISM

In a letter written in the summer of 1920, Mead explained that Bergson had "the same problem before him that confronts pragmatism—that of bringing the immediate experience which has in the past been relegated to the field of psychology into that of the reality which science assumes without question."[10] Bergson had failed to solve this problem, which now Mead considered a task left up to himself. Pragmatism had its goals clearly delineated because Bergson's answer to this question proved unsatisfactory. From Bergson's error, Mead drew an important lesson: "Don't make your philosophy out of the temporary defeats of science," but rather pay to science its due respect.[11]

Einstein's visit to Chicago only increased Mead's curiosity about the potential conflict between the physicist and the philosopher. Mead associated relativity with the belief in a universal, predetermined reality, and Bergson with its near opposite—a belief in "the emergence of the novel." Intellectuals, he argued were torn into two opposite directions: "In one direction we move toward the scientific ideal of the world at an instant, while in the other we moved toward the heightened temporal intuition of a Bergsonian picture of the world."[12] Mead considered it imperative that philosophy reconcile these two points of view:

> It is the task of the philosophy of today to bring into congruence with each other this universality of determination which is the text of modern science, and the emergence of the novel which belongs not only to the experience of human social organisms, but is found also in a nature which science and the philosophy that has followed it have separated from human nature.[13]

Mead denied the division of labor in which science dealt with nature (both human and nonhuman) and human experience was chalked up as an epiphenomenon worthy only of secondary attention. The experience of time was one of his central examples, proving that the physical investigation of time should not have to be done at the expense of our experience of time.

In the summer of 1927 Mead taught a course on Bergson's philosophy.[14] Later, during the Carus Lectures delivered in Berkeley in 1930, Mead disagreed intensely with the block, or four-dimensional, view of the universe, in which the past, the present, and the future were already laid out, predetermined, and in which our sense of time was a mere illusion:

> This view of reality as an infinite scroll unrolling in snatches before our intermittent vision receives another variant in the picture of reality as a four-dimensional continuum of space-time, of events and intervals, forever determined by its own geometry . . . whose present character is a function of our minds and not of any section of the ordered events in the universe.[15]

Mead's attempt at a complete reconciliation of Bergson's philosophy with relativity was left unfinished. "At the time of my last conversation with him, in the week before his death," explained his colleague the philosopher Arthur E. Murphy, "he was at work on Bergson's *Duration and Simultaneity* in its relation to his own account of relativity."[16]

HARVARD

On September 14, 1926, Mead delivered an important lecture at the Sixth International Congress of Philosophy at Harvard University. The section was titled "Physics and Metaphysics, with Special Reference to the Problem of Time." Presenting along with him was Whitehead.[17] "There are metaphysical abysms between us," wrote Mead to his daughter-in-law after the event, "but what are they between gentlemen."[18] Mead had turned to Eddington's and Whitehead's work to deepen his understanding of Einstein's. Whitehead's philosophy, in particular, then took him back to a study of Bergson.

Whitehead's work on relativity, argued Mead, had an essential advantage over Einstein's in that it accepted the reality of change and movement, instead of claiming that it was simply an illusion. Whitehead's project, he argued, was "undertaken to preserve motion and change within a relativistic universe."[19] Yet at one point, Whitehead's philosophy also failed to convince Mead. Whitehead conceived of "events" as "eternal objects," rendering them unnecessarily abstract. Mead "on the contrary" chose to "recognize what becomes as the event which in its relation to other events gives structure to time."[20] This simple recognition led him to a solution in which he "viewed himself as siding with Bergson in retaining the centrality of duration or process as the crucial element," and in which he corrected Whitehead's philosophy by finding that "universals arise in social behavior and entail neither an intellectual distortion of reality [Einstein's block universe] nor a realm of eternal objects [Whitehead's events]."[21] Scientific observations had to be confirmed "in the mouths of two witnesses at least,"—and this essential characteristic revealed the social, communicative nature of science.[22]

Whitehead had recently moved to Harvard to teach philosophy. It was a significant transition from his previous career in mathematics. For several decades already, Harvard had been a key place for housing important advocates of Bergson, such as William James, and was distinctly cold toward Einstein. During his 1921 trip to the United States, Einstein was invited as a visitor but not to lecture. To complicate matters further, Harvard's president A. Lawrence Lowell spoke to him in French, whereas in Princeton he had been greeted in German.[23]

William James, one of Bergson's most ardent admirers, had taught the philosopher and historian Arthur O. Lovejoy, who had written an important essay criticizing Einstein and defending Bergson's approach.[24] In it, Bergson appeared as Lovejoy's key ally against Einstein.[25] After a thorough investigation of Einstein's theory, Lovejoy concluded by transcribing, without translating, Bergson's conclusion: "We arrive, in the end, by a different road, at the conclusion already expressed by Bergson: *nous prétendons que le temps unique subsiste dans l'hypothèse d'Einstein à l'état pure; il reste ce qu'il a toujours été pour le sens commun.*"[26] Part of Lovejoy's argument rested on the importance that should be given to the disagreement between the two persons observing simultaneous events: "The fact that two observers disagree concerning the simultaneity of a pair of distant events . . . is no more significant than would be a disagreement between two illiterate persons over the question of whether a whale is a fish." One could mean by "'fish' any free-swimming animal that lives in the water," while another one could take it to be "gill-breathing and cold-blooded animals."[27] Lovejoy agreed that even if under special relativity two persons disagreed about their assessments of simultaneity and measurements of time, he refused to conclude from this that both were right.[28]

At Harvard, Whitehead would become the advisor of W.V.O. Quine, a philosopher largely remembered for dealing the first significant challenge to the logical empiricist tradition often associated with Einstein's work. One of Quine's first assignments—his final paper for his course Contemporary Philosophy—centered squarely on Bergson.[29] From that moment onward, Quine, like Lovejoy, was hooked on French philosophy.

OPERATIONALISM: PERCY BRIDGMAN

While Whitehead was in the philosophy department teaching courses on Bergson, Percy Bridgman (who would win the Nobel Prize in physics in 1946) was in the physics department. Bridgman wrote extensively about the philosophy of science, developing a perspective known as operationalism.[30] Although he did not name the French philosopher directly, at least one reader recognized certain of his arguments as distinctly Bergsonian. Bridgman's strategy, surmised William Marias Malisoff, a philosopher of science who worked on armament technology (and who allegedly spied for the KGB), probably resulted from the obvious reason that in America "the comparison to Bergson may not flatter." Malisoff was convinced that he had discovered that Bergson was the main source of Bridgman's privileged information.[31] Certain similarities were obvious. Whereas Bergson used the words "virtual" and "phantasm" to describe one of the travelers in relativity theory, Bridgman, in his own analysis of relativity, described them as "vicarious" and "ghostly."[32]

Bridgman presented his critique of Einstein's theory of relativity in *The Nature of Physical Theory* (1936), an elaboration of three lectures given earlier that year at Princeton University. Almost a decade and a half later, his "indictment" of Einstein was even stronger. The physicist argued that Einstein's expansion of his 1905 essay into the general theory was so "uncritical" that it "conceals the possibility of disaster."[33]

Bridgman adored Einstein's early work, particularly because of how it was attentive to the actual operations of measurement, because it was modest about drawing any broader implications, and because it was simple and procedural. Einstein's methodology in 1905, he claimed, was perfectly consonant with the "operational point of view." His later work was not. "Einstein did not carry over into his general relativity theory the lessons and insight which he himself has taught us in his special theory," Bridgman lamented.[34]

In evaluating Einstein's work, Bridgman disagreed with the physicist's refusal to accept a single frame of reference as unique. Bridgman, in contrast to Einstein, believed that one frame of reference was different from most others. Because "our measuring operations" generally took place in one frame, this frame had a special validity. Accepting the singular importance of one frame of reference over others hardly implied a return to a Newtonian universe, because, Bridgman argued, the preferred "starting-point" could be "different for each individual observer." But in denying the importance of a unique frame of reference, Einstein did great harm. He led science down a path where it stood in opposition to "the obvious structure of experience."[35]

The question of which traveler's reality should be accepted remained pertinent during these years. Bridgman introduced a "third ghostly observer" as required by the revised "three clocks paradox," arguing that it did not eliminate these questions. (Debates about how to best reach agreement about matters of time would continue into the 1950s.) The two famous observers of the twin paradox still had to come to an agreement about their differences, he explained. They needed to be "similar" in some sense and they had to be able to "communicate" with each other.[36]

When Bridgman reevaluated Einstein's work in 1949, he was immediately led to think about the meaning of communication. He understood that Einstein's interpretation of time dilation only held "if the two observers are allowed to communicate with each other."[37] But more important, he argued that communication between them needed to be *meaningful*. The whole theory, he argued, was based on "the assumption that the two observers are able to communicate meanings to each other."[38] Bridgman pointed out thorny issues pertaining to the transfer of meaning in theories of communication. For meaningful communication between them to be possible "a certain similarity between the observers themselves is necessary."[39] In conclusion, he insisted that the philosophical problem of reaching agreement about matters of time had not disappeared.

"Even the inhabitants of this planet with different cultural backgrounds do not find it always easy to communicate," stated Bridgman.[40]

Why would it be different for the observers in the famous twin paradox? For Bridgman, debates about the twin paradox could not be solved simply by reference to technical procedures for time coordination. More complex issues pertaining to "communication," "sameness," and "agreement" remained at the center. In his view, the challenge of determining how some "other" being—even if their otherness was as slight as could be in the case of a twin—experienced time had not been solved. In the coming decades, a number of scientists and philosophers followed Bridgman in thinking about these questions in terms of the *communicative* structures that allowed for the transfer of meaning.

CHAPTER 17

The Church

INSTITUT CATHOLIQUE, PARIS

How did the Catholic Church react to the Einstein-Bergson debate? The Church's response to Bergson and Einstein was varied and changed through time. Both thinkers had strong views about the institutionalized religions of their era and of their relation to other forms of spiritualism and mysticism. While initially, during the first decades of the century, Bergson was considered a dangerous enemy to Catholicism, some people later saw him as an ally. The Church itself was divided about the merits of the philosopher: was he an enemy or a source of inspiration? In his last will, Bergson asked for his burial to be officiated by a Catholic priest, fearing that his fellow Jews "will probably hound my legacy after my death."[1] He was right. When he died, during the Vichy period, a priest officiated the ceremony. A moving obituary written by a friend and minister to the government was transmitted through the radio.[2] In disregard of official German policy, occupied France publicly honored the death of a Jewish philosopher—without mentioning his "non-Aryan" heritage.[3]

The relation of Einstein toward religion and spirituality is no less complicated than Bergson's. Although he was brought up in a secular home, Einstein eventually became deeply spiritual. Volumes have been written about his religious beliefs. Many more have been penned about the relation of the Catholic Church to science. In light of the conflict between Einstein and Bergson, a new part of this history becomes apparent.

During the first decades of the twentieth century, the Catholic Church benefited from a backlash against certain anticlerical writers of the previous century, such as Hippolyte Taine and Ernest Renan. A growing number of readers now thought that these authors had placed too much hope in science. Numerous scientists and intellectuals around 1900 argued that science was "bankrupt" and in "crisis." In France, "the crisis of science" was a common diagnostic for society's ills.

In the context of the "crisis of science" movement, Bergson's philosophy appeared to many as a salutary alternative, one that did not follow a blind allegiance to science as the solution to all ills (moral as well as physical) but that was not limited to following official Catholic philosophy either, which was still largely dominated by the Thomist revival of Aristotle.[4] Thomist clergy at first responded most violently against Bergson, although the Jesuits and the *Revue du clergé français* were in general less virulent.[5] The figure of Galileo appeared as a key historical reference, as a new generation of thinkers considered science to be the new Inquisition of their era.

JACQUES MARITAIN

A talented Catholic intellectual jumped to Bergson's defense after the philosopher's debate with Einstein: Jacques Maritain. Maritain referred to the "discussions of April 1922" that took place "at the Société de philosophie, following the eloquent contribution of Bergson."[6] His defense was surprising, because—before then—he had been one of Bergson's most hardened critics.

The Church was larger than Jacques Maritain, but if there is a single man who could represent Catholic philosophy of science from the first decades of the century up to the Second Vatican Council, that would be he. He was author of the standard textbook for philosophy used in Catholic schools and seminaries, *Elements de philosophie*. Since its appearance in the 1920s, it has remained a standard text. His influence was such that he was involved in drafting the preamble to the Constitution of the Fourth French Republic (1946), for which he corresponded regularly with Charles de Gaulle. He also helped craft the United Nations Declaration of Human Rights (1948) and the Canadian Charter of Rights and Freedoms.

Maritain is known for having coined the word "scientism." His "Reason and Modern Science" (1910) article, which appeared in the *Revue de Philosophie*, argued that science was gaining an unwonted influence in the world, larger than that of religion but also—more radically—larger than that of reason itself.

Maritain came from a Protestant family, and his wife, Raïssa, from a Jewish one. He converted to Catholicism in 1906, and throughout most of his life he worked at Catholic institutions. He was a friend and mentor to Pope Paul VI, and his work left a mark on the pope's encyclicals. Paul VI placed in his hands a copy of the "Message to Men of Thought and of Science" at the close of Vatican II. He also influenced John Paul II, criticizing some of the reforms of the Second Vatican Council. When Raïssa died in 1960, he chose retirement at the monastery of the Little Brothers of Jesus in Toulouse, where he remained until his death.

Before Jacques and Raïssa turned to Catholicism, they had turned to Bergson. When they met at the Sorbonne they promised to commit suicide together unless someone saved them from the arid positivism dominating the intellectual life around them. Their savior was Bergson. But the enchantment would not last long. By 1911 Maritain had broken with him.[7] In 1913 Maritain condemned Bergsonian philosophy for being "radically incompatible with Christian philosophy," and particularly incompatible with "the philosophy of Saint Thomas." He claimed that "it led, inevitable, to modernism," referring to the controversial reform movement within the Church that would become Pius X's bête noire.[8] In 1914, in no small part because of Maritain's influence, Bergson was placed on the Index of Prohibited Books.

Einstein's debate with Bergson brought Maritain, at the time one of Bergson's fiercest critics, to reevaluate the contributions of the philosopher. It was a shock to see two men who were "profoundly dissimilar being in accord" right after the debate.[9] After his confrontation with Einstein, some of Bergson's old enemies became new friends. Maritain's and Bergson's rapprochement was particularly bewildering since previously "these two men assuredly represent[ed] in France the two opposing poles of contemporary philosophy."[10] Why did the debate produce such a marked shift in Maritain's assessment of philosophy, science, Bergson, and Einstein?

In 1921, before *Duration and Simultaneity* appeared, Maritain published a book, *Théonas ou les Entretiens d'un sage et de deux philosophes sur diverses matières inégalement actuelles*, which dealt with Théonas (bishop of Alexandria during the years 283–301), Einstein, his paradoxes, and the mathematization of time. He employed arguments "very similar to those that Bergson would use."[11] The book argued that Einstein confused reality with measurement and that the physicist simply dealt with mathematical time, whereas common sense and philosophers were concerned with real time.

Einstein, argued Maritain, had revealed himself to be a good physicist but a bad philosopher: "Like most modern scientists, Einstein seems to have studied only very superficially metaphysical and critical problems," he claimed. "How desirable would it be for [Einstein's] philosophy to have been good!" Maritain diagnosed Einstein's philosophical bent: transcendental idealism of the worst Kantian type. And the most lamentable thing was that Einstein did not even suspect that he suffered from these gross "metaphysical prejudices."[12]

After the Bergson-Einstein debate, Maritain referred to Einstein's "naïve" and "monstrous" theory of relativity as "the new physical dogma." It was a "parade" that resulted in "poor intellectual buffoonery" and "grand metaphysical misery."[13] Worst still, Einstein's inadequate philosophical reflections seemed to "confirm the frequently made claim that modern science is less a kind of real knowledge [*connaissance proprement dite*], than a sort of art and fabricated logic." Einstein showed himself to be a "virtuoso at the great keyboards of signs" but terrible as a "contemplator of being."[14]

The theory, allegedly, depended "on the most false of metaphysics." Maritain's critique was different from the one frequently launched by the anti-relativity campaign of Germany. It was neither explicitly motivated by anti-Semitism, nor did it object to it as a scientific theory. Maritain had absolutely no qualms against "the value of his [Einstein's] scientific theories, as pure physical-mathematical theories." On numerous occasions he insisted "that it should be well understood that in the present debate the physical-mathematical theory of Einstein is not put into question (that would be the task of specialists to investigate its validity)." He was questioning only "the philosophy of nature and the metaphysics of Einstein."[15] His qualms lay with "interpretative hypothesis that attach them to the real, and all the conceptual material that clothes them."[16] Einstein was, in the end, a "metaphysician (despite himself perhaps)" who wanted to "introduce us surreptitiously into the most false of metaphysics."[17] He should be admired as a "pure physicist," but we should also hold a "complete aversion" for him as "pseudo-metaphysician."[18]

Michelson's null experimental results were correct, but the replication of the experiment in 1922 at a mountain-top height of 1,800 meters showed contrary results. Maritain paid close attention to these ongoing investigations. He concluded that the results of the experiment did not "impose [Einstein's theory] as a necessary conclusion." Its success was only achieved with the "help of a wide niche of presupposed hypotheses."[19] Einstein's main mistake, according to Maritain, was to have a restricted notion of reality—as *that which could potentially be measured.*

Maritain accused Einstein of putting the cart before the horse. Einstein defined equality as a comparison of two measuring rods, but Maritain argued that in order to start comparing measuring rods (in the first place) a person already had to have an *intuitive* knowledge of the concept of equality. The same argument applied to simultaneity. Einstein defined it by the example of "the pointing of the small hand of my clock to 7 and the arrival of the train," but the philosopher argued that this procedure could not be used unless that person had a prior commonsense notion of simultaneity. The pernicious consequence of following Einstein's method was that, after all the mathematical manipulations were completed, scientists would have an impression that the result "represented through his eyes *time*, 'real time.'" But that quantity was "a new and totally different concept."[20] This confusion led to a second negative consequence. It made the commonsense concept of time appear as "fraudulent." What outraged Maritain was similar to what infuriated Bergson. Their call to dutiful action as a corrective motivated their work: "It is the office of the philosopher to denounce such encroachments." Physicists should not feel they had the right to "revise common notions of space, time and simultaneity, the elucidation of which belongs to a superior science that completely escapes their competency."[21] Einsteinian views were "systematically poisoning" the public, as many were led to "accept the absurd and to lose all confidence in common sense." The result was alarming: an "amputation of the intellectual faculty itself."[22]

Simultaneity, when measured in a certain way, was relative. Was it always so? Maritain granted to Einstein that "simultaneity *thus measured*" was relative, but not simultaneity in general.[23] He cited Bergson's insistence that the popular notion of simultaneity was not only different from the scientist's, but also valuable. He complained about the "marvelous presumptuousness of scientists," who believe they had the right to revise everyday notions of space and time.[24]

Maritain, who would be one of the authors of the Universal Declaration of Human Rights after World War II, framed the fight against Einstein and the defense of Bergson in terms of *human rights*. He argued for "the right" to defend what were perhaps vulgar and imprecise, but nonetheless valuable, commonsensical notions. Yes, he agreed that these imperfect notions were corrected by scientists, but this should not lead us to their complete abandonment. He restated his central claim again and again: Einstein's theory of relativity, although invaluable for physics, should not lead us to invalidate traditional concepts of time and space.

"Metaphysical larva"—those were strong words, but they were not directed against Einstein. Their target was the German philosopher Immanuel Kant. Kant's philosophy was "the metaphysical larva that not only the relativists profess, but a great number of modern scientists."[25] Many historians of science and philosophy have compared Einstein's empirical notion of reality to that of the physicist and philosopher Ernst Mach, but Maritain traced it instead to Kant. Maritain considered it ethically wrong to place so much emphasis on the human senses, to define reality as it appeared through them, and to devalue higher forms of intelligence and common sense. Einstein's work, he lamented, smacked of "embarrassing residues of Kantianism."[26] The disturbing word "parasite" followed that of "larva" in Maritain's descriptions of the metaphysical underpinning of Einstein's relativity. Einstein's science came along with a "pseudo-philosophy that was its parasite."[27]

Maritain stressed how measurements referred only *indirectly* to the thing measured. Speed-of-light measurements were an indirect indication of the true nature of light. To fully know objects, these needed to be considered as "beings" under much more "diverse aspects," including "accidents, quantity, quality, etc." But even their "sensory measure" taken in its entirety could only reveal a limited aspect of them.[28] Because of how the physicist confused time with measured time, he explained, "Einsteinian time" remained in large part "fictional or imaginary."[29]

"Trains, revolving sidewalks, railway tracks, kilometric boundaries made of clocks, spaceship travelers, observers smoking cigars and holding mirrors" populated popular accounts of relativity.[30] Maritain was as irritated by these examples as he was by its grandiose conclusions. He accused Einstein's popularizer Arthur Eddington of being fatally prone to bouts of exaggeration, ridiculing his claim that he who travels at the speed of light would "possess immortality and eternal youth." Maritain quipped instead: he would be "dead" because he would be "flat, which is very uncomfortable, even if time has stopped."[31] Maritain saw Einstein's science as part of a fashionable new trend, similar to new fads in poetry, music, and entertainment. "These comical objects," he explained, "occupy in the current Philosophy of nature the same place as the jazz-band and the cocktail in the new Poetry (which was new yesterday, since Poetry's clock ticks even faster than that of Physics.)"[32] He sharpened his criticism in a book with the telling title of *Antimodern*, in which the jazz-bar cocktail appeared as despicable as the non-Euclidean geometries of Lobachevsky and Riemann. The book called on physicists to "renounce the ambition of handing to us reality in the raw."[33] He urged them to admit the hypothetical and limited nature of their claims, as he denounced the arrogance implicit in the famous phrase attributed to Newton: *Hypothesis non fingo!*

Maritain revised his initial views about Einstein and relativity in important ways after the April 6, 1922, discussion. After corresponding with various individuals (including Metz himself) and reading hot-off-the-press publications, he recanted some claims in the third appendix to the second edition of his book. He was even more explicit in his acknowledgment that there was no contradiction in the theory's "internal logic" and that it had a complete "indemnity from logical sin."[34] But even after these qualifications and slight factual corrections, he maintained that his main philosophical point still applied perfectly to it.

Maritain agreed with the general point made by Bergson, that since physicists were not talking about the time of everybody else, it would be best for them to "give to the concept of simultaneity used by physicists another name."[35] The problem, according to Bergson and Maritain, would be easily solved in this way. Otherwise, conflict would clearly ensue. "What is the advantage that you have, as a physicist, to use the word simultaneity in these two senses," both asked?[36] In addition, they agreed that the intuitive concept of simultaneity should have preference over the measured concept.

Maritain initially criticized Bergson for placing too much emphasis on the value of nonhuman perspectives. According to Maritain, if Bergson wanted to include the perspective of different forms of consciousness into philosophy he "would himself have to become materially vegetable or mineral."[37] But Einstein proved to be worse, according to Maritain, by going in the opposite direction. One of the reasons why he combated Einstein was because he thought the physicist's emphasis on sensory impressions was one that conceived of men *as* simply animals: "It consisted, to say it truthfully, in a *brutalization* of the human sciences." In Einstein's theory of time, he argued, humans were considered as "purely sensing beings," reduced to noting basic sense impressions, where a man was no more sophisticated than a "thinking brute."[38] Maritain argued for a more "intelligent" and less "sensorial" definition of time, where humans were considered positively (built in the image of God) rather than brutish or animalistic.

While the Kulturkampf of the 1920s pitted Maritain firmly against Einstein, the post–World War II period brought remorse. The Catholic philosopher then worked hard to help Holocaust survivors and tried, albeit unsuccessfully, to convince the Pope to speak about the evils of anti-Semitism and the Holocaust. But by then it was too late. The crime of the century had already taken place—a crime infinitely more severe than that of the Church against Galileo.

GALILEO'S LEGACY

A few months before the Germans invaded Poland, Einstein read an inspiring biography, *The Life of Galileo*. He felt personally touched. When witnessing the conflict between Einstein and Bergson, both men and many of their interlocutors were reminded of Galileo's trial. The name of Galileo came up over and over again in discussions about the merits and faults of Einstein's work.

In the seventeenth century, the Roman Inquisition convicted Galileo for defending the radical ideas of the astronomer Nicolaus Copernicus.

Copernicus had argued against Ptolemy's theory, insisting that the Earth was not at the center of the universe but that it revolved around the Sun, and Galileo zealously defended his ideas. Galileo's trial hinged on evidence that Copernicus's theories conflicted with key biblical passages. But more important, Galileo was wholly condemned for promoting an exaggerated view of the power of science (particularly his own understanding of it) compared to other forms of knowledge. The Holy See mobilized its powers to curb the excesses of a single man. Centuries later, the fate of Einstein's theory of relativity (and its critics) was still evaluated in direct comparison to Galileo's work (and his inquisitors). What role did Bergson and Einstein play in this portrayal?

For nearly half a millennium, the figure of Galileo remained a powerful narrative representing the power of science over other kinds of tyranny. In the 1930s, the famous playwright Bertolt Brecht penned one of his most famous plays, *The Life of Galileo*. Einstein read the script and immediately wrote a letter to the author telling him "how much joy you gave me with your Galileo." Brecht, according to Einstein, had perfectly captured "the attitude of pre-Galilean science toward experiment." But the power of the play's narrative, derived most of all, from "its strong links to the political problems of today."[39] A sensational theatrical production premiered a few years later in Zurich.

Einstein associated himself strongly with Galileo. In 1918 he responded to critics of relativity by parodying Galileo's *Dialogue Concerning the Two Chief World Systems*, precisely the text that had brought the astronomer into conflict with the Catholic Church. In 1953 he wrote the foreword to the republication of this same famous text. In it, the physicist claimed that even today "we are by no means so far removed from such a situation" as the one that confronted Galileo when he searched for unbiased truth in the face of the Holy Inquisition's attempt to quash it.

Galileo's critics gained important allies more than two hundred years later, around 1900. The physicist and philosopher Pierre Duhem, who was an important influence on Bergson and many of his disciples, started publishing revisionist articles in the *Annales de philosophie chrétienne*. This work culminated in a book titled *Sauver les phénomènes* (1908), arguing for the pertinence of Galileo's prosecutors. Bergson found profound affinities between his philosophy and that of Duhem, who was by then the most ardent Catholic defender of Galileo's critics.

When Bergson argued that he and Henri Poincaré had "arrived at similar conclusions," he added that "the physicist Duhem has preceded" both of them "in this critical project."[40] Maritain was similarly full of praise for Duhem. Einstein's theory of relativity, he argued, "represented in particular a considerable intellectual regression in comparison to the conception of physics advanced by Duhem—a grandiose conception of a superior reasonable soul, and that remains the wisest."[41]

Bergson's debate with Einstein added a new wrinkle to the old story of Galileo and his inquisitors. Einstein knew the work of Duhem well; his friend Friedrich Adler had introduced it to him. Yet Einstein identified himself strongly with Galileo—not with his critics or with their twentieth-century apologists.[42]

After working on Galileo, Duhem challenged the concept of crucial experiment. Was a crucial experiment a test used by scientists to decide in favor of one hypothesis over another? Did it guarantee that scientific knowledge rested firmly on experimental proof? Duhem questioned the very possibility of experimental verification by emphasizing how scientists could never sharply distinguish their results from the theoretical presuppositions involved in obtaining their observations in the first place. He stressed how all observations with instruments, even an apparently simple glance through a magnifying glass, were tainted by our theoretical suppositions of that instrument and how it worked. For this reason, even simple measurements, just as much as the results of more complicated experiments, arose from particular theoretical outlooks. Could something as apparently simple as determining if the Earth rotated around its axis—or if it was, on the contrary, fixed while everything else rotated around it—remain unproved experimentally? Was it a matter to be determined by evidence-based science aided by instrumental innovations (such as the telescope) to increase the range of our observations, or was it mostly dependent on theoretical presuppositions, conceptual paradigms, and worldviews?

Bergson's student, Édouard Le Roy, who was responsible for prodding Bergson to speak during the debate, adopted a position similar to Duhem's. He claimed that scientific instruments were "materialized theories" that left a mark on scientific results.[43] Theoretical presuppositions, he argued, affected the outcomes produced with them. Many others since then, such as Alexandre Koyré and Gaston Bachelard, would continue to highlight their role in shaping scientific knowledge. Bachelard drew heavily from Le Roy's work when he claimed that "an instrument, in modern science is truly a reified theorem."[44]

Inspired by this work, the contributions of Galileo were revaluated by key historians of science well into the twentieth century. Thomas Kuhn, an avid reader of Koyré and Bachelard, author of *The Structure of Scientific Revolutions* (1962), and famous for coining the term "paradigm shift," argued that the change from a Ptolemaic to a Copernican worldview was mostly conceptual. "Available observational tests . . . provided no basis for a choice between them [Ptolemy and Copernicus]," he claimed.[45]

COPERNICUS'S LEGACY

Was the Catholic Church's response to the Einstein-Bergson debate to be as controversial as its earlier response to Galileo? Who was right: Copernicus or Ptolemy? What was more important for the development of science: experiment or theory? These questions were disquieting and continued to constitute the terms of various debates in the history and philosophy of science to the end of the twentieth century.

Relativity was often described as another Copernican Revolution. These comparisons started even before Einstein wrote his groundbreaking paper on the subject. Poincaré claimed that the revolution brought about by this new science could be as important as the one that "befell the system of Ptolemy by the intervention of Copernicus."[46] In Germany, Max Planck considered the two revolutions as strikingly similar.[47] Langevin also used the comparison in many of his writings. The theory was "more than a discovery"; it was, rather, "a change of point of view comparable only to that introduced by Copernicus when he put the Earth in its place in the system of the world."[48]

Comparisons with Copernicus during this time cut two ways. Had Copernicus indeed proved anything, or had he offered merely a different worldview, albeit better (in some ways) than the previous one? Certain scientists and philosophers around the turn of the century argued that the Copernican Revolution was merely a different, more efficient in some respects, formulation than Ptolemy's. Le Roy started this argument in “Science and Philosophy,” which was published in the *Revue de métaphysique et de morale* in 1899. He developed this idea further in a paper he delivered at the International Congress of Philosophy in Paris in 1900, where Bergson and Poincaré were present. He even ventured to claim that the Catholic Church had done nothing wrong in condemning Galileo.[49]

In his address to the congress, Poincaré framed Copernicus’s revolution as nothing more than a more convenient formulation than the preceding ones. His comments set off a scandal. Was his interpretation of these historical events a sign that he sided with Ptolemy over Copernicus, or even worse, with the Inquisition over Galileo? Was he denying the inevitable progress of science? “Every reactionary journal in France had asked me to prove that the Earth goes around the Sun,” he later recalled.[50] Poincaré responded to these accusations in “La Terre tourne-t-elle?” Opening with the provocative title “The Earth, Does It Turn?” his reply was written as a letter to the scientist and popularizer of science Camille Flammarion. Poincaré explained that he was in no way defending the Inquisitors who condemned Galileo, but Poincaré nonetheless continued to argue for a different interpretation of the Galileo affair. “‘The Earth turns round,’ and ‘it is more convenient to suppose that the Earth turns round,’ ” he explained, “have one and the same meaning.”[51] To differentiate his position from that of previous critics of Galileo, he insisted on the strengths of the “conventional.” Conventions were not artificial—they had a tight connection to reality. Poincaré sided strongly with Galileo, but for other reasons than most did: “The truth, for which Galileo suffered, remains the truth, although it does not have quite the sense as it popularly does and its true sense is much more subtle, profound and rich.”[52] Truth based on the strength of convenience, he argued, was much richer than when it was not based on it.

THE NEW ABSOLUTE

By the end of 1900, Duhem, Le Roy, and Poincaré had all offered another way of interpreting Galileo’s contributions, and they each had provided new models for understanding the Copernican Revolution and the progress of science. Maritain followed their work carefully. So did Bergson.

Bergson claimed that Einstein's so-called relativistic conclusion ended by confirming our belief in one single and universal concept of time: "The theses of Einstein do not seem to contradict, but they in fact confirm the natural belief of men in a single and universal time."[53] By democratizing all systems and having none be privileged, Einstein was, underhandedly, introducing a new cult of the absolute: "The suppression of a privileged system is the essence itself of the theory of relativity. Therefore this theory, instead of eliminating the hypothesis of a unique time, confirms it and gives it a superior justification [*intelligibilité superieur*]."[54] The astronomer Charles Nordmann explained it in similar terms: "The premises of Einstein and the facts on which it rests lead us to the contrary to negate the relativity of time and to prove the existence of absolute time."[55] "All is relative and only that is absolute," he concluded, with marked irony.[56]

Alongside Bergson, numerous other commentators considered that Einstein had returned to the old theological concept of time, yet it was now clothed in ostensibly secular garb. If the time of Paul was relative, if that of Pierre was relative, if all observers were relative to each other in the same way, and so on, then it was theoretically possible to account for all positions to gain a total, complete, and absolute representation of time. The constancy of the speed of light—one of the pillars of relativity theory—was cited as a new form of absolutism: "Light is for Einstein the new and only absolute in a world where all else is relative."[57] Many interpreters who followed Bergson claimed that Einstein, along with the mathematician Minkowski, who had helped develop the mathematics for the theory, were actually setting science on even more absolutist foundations than before. "Minkowski showed how to recover the absolute," explained Eddington, "by searching more deeply."[58] Minkowski himself preferred to call relativity theory the "postulate of the absolute world," and Einstein at one point thought that the label "relativity" was indeed a misnomer. For Bergson, the return of these absolute and conservative notions in science represented a real threat to the advance of knowledge.

A number of faculty members at the prestigious Institut Catholique de Paris (Catholic University of Paris) started defending Bergson. In the decades that followed the debate, Bergson's stature among some Catholics grew considerably. He was completely reevaluated by key Catholic thinkers, who found in his philosophy an entryway into religious thought and life. Some also attacked Einstein, turning to a rapidly growing arsenal of arguments against him. Bergson's views on Einstein were defended by the Jesuit Father Auguste Valensin. The admiration was mutual. Bergson told a close friend how Valensin possessed a "subtle mind and taste."[59] He happily accepted the father's argument against Einstein and in favor of "real local times."[60] Father Augustin Sesmat at the Catholic University in Paris argued for "great prudence" in judging Einstein's theory. He did not want to accept that all frames of reference should be equal and would continue to argue that "even in Relativity theory there are still naturally privileged systems, whose choice is not at all arbitrary."[61]

In the 1940s, Bergson's work was positively reevaluated by Father Antonin Sertillanges, who published his conversations with the philosopher and penned a number of short pamphlets defending Bergson as an important inspiration for Catholics all over the world.[62] Bergson had "opened the road to access" Catholicism, because he brought "minds closer to metaphysics." His genius resided in fighting against two enemies. First, by "repudiating the 'religion of science,'" so prevalent during those years, and second by combating empty faithless intellectualism. Bergson placed himself valiantly "as Samson between the two columns of the temple." His philosophy, argued the Dominican thinker, could be used to "make room for an architecture sister to cathedrals."[63] When Bergson died, Sertillanges explained how his death "was a loss for the whole universe," but it was a particular loss "for Catholicism."[64]

Bergson, explained a recent scholar, attempted in the twentieth century what the Catholic Church did in the seventeenth: "Cardinal Bellarmine had urged Galileo to speak 'hypothetically and not absolutely.' Similarly, Henri Bergson proposed to Einstein that he eliminate the paradoxical appearance of his theory."[65] But despite attempts to associate Bergson with the Catholic Church and its role during Galileo's inquisition, Bergson presented his critique of Einstein as completely the opposite. Bergson would not side with Galileo's critics as others (such as Duhem) had done, nor would he back a theological notion of time. His position would be most similar to Poincaré's, in that he gave an important place to convention in science. By arguing that science had conventional aspects to it, he was not arguing that it was artificial, false, or unreal. Conventions were real: "This is the reality, if one agrees to call representative of the real any convention adopted for expressing mathematically physical facts."[66] This permitted Bergson to make an even more radical point against Einstein: by ignoring the conventional aspects of science, the present day continuators of Galileo's scientific program (such as Einstein) were now as dogmatic as Galileo's inquisitors had once been. He framed Einstein's revolution as completely undergirded by an old conception, a single and universal notion of time, which had clear theological roots. In this interpretation, Einstein appeared closer to the men who had been driven by theological concerns than to other scientists (such as Poincaré), who by acknowledging the role played by conventions in our world, actually came closer to seizing the "really real." Conventional reality was a slightly but significantly different kind of reality from the "really real." "If you define reality by a mathematical convention, you end up with a conventional reality," explained Bergson. This conventional reality was different from the "really real," which he defined as "that which is perceived or can be."[67]

A new generation of Catholic writers admired Bergson for professing a new nondogmatic view of knowledge, neither dogmatically religious nor dogmatically secular—and for having a modesty that neither Galileo's inquisitors or Einstein had. The pious poet Charles Peguy, venerated for espousing a new form of religiosity in a secular century, admired Bergson primarily for these reasons: "A great philosophy is not that which passes final judgments, which takes a seat in final truth. It is that which introduces uneasiness, which opens the door to commotion."[68]

CHAPTER 18

The End of Universal Time

BERN, SWITZERLAND

Édouard Guillaume participated in the second meeting at the Collège de France on April 5, 1922, the day before Einstein and Bergson met. Guillaume had been Einstein's office mate as a lowly second-class patent clerk in Bern, but now the two former friends were estranged. He was the younger cousin of the Nobel Prize winner Charles Édouard Guillaume, who was an expert on standards (including time standards) and head of the Bureau international des poids et mesures.

Einstein had moved on to bigger and better things. Rumors circulated that he would soon be receiving the Nobel Prize. Guillaume had stayed behind at the patent office in Bern. When Guillaume learned that his old colleague was going to Paris to lecture, he bought a train ticket, packed his bags, and jumped on the train to meet him there. Einstein's ex-colleague was prepared to "demolish" the physicist.[1] For the rest of his life, Guillaume insisted that he could easily turn Einstein's interpretation upside down to prove exactly the opposite of what Einstein had found—that time was *not* multiple and relative but, rather, single and universal. All that Guillaume had to do was to show scientists how they could use only one variable t, instead of the many variables, such as t_1 and t_2, that characterized Einstein's work. His method, he bragged, matched all the observed results associated with relativity.

Only "one day before [their meeting] the majority of newspapers inserted a note . . . claiming that this physicist [Guillaume] had found in Einstein's theory gross errors in his calculations and that he, coram populo, was going to demonstrate." Overeager newspapers announced that "these errors should naturally produce the complete crash of Einstein's synthesis, the total bankruptcy of the law of science."[2] But when Guillaume arrived, he found the public set against him in what was an "agitated" meeting. The press referred to him as "an intolerant spirit" and, perhaps most damagingly, an "impolite" one. After one "impetuous" interruption launched by someone in the audience, Guillaume simply stood up in ire and left the room.[3] For the rest of his life he continued working on a rival theory that would prove the existence of universal, absolute time.

One newspaper described the confrontation that day as one in which "Guillaume was out of the ring in two rounds." The arguments between the two men could be described in simple terms. Guillaume claimed that what Einstein described as a sphere, "seemed to him to be an ellipse." Einstein, "with a big grin," replied: "Perhaps, it could, in a particular system, be an ellipse." But he was hardly conceding. "When it is an ellipse, nobody cares," he concluded.[4]

What was at stake in the Guillaume-Einstein confrontation and how did it fit with the famous affair that was to take place the next day? Bergson started to follow Guillaume's work closely; Guillaume, in turn, followed the work of the philosopher and corresponded with him. After Bergson closely inspected the premises underlying Guillaume's work, he decided not to side with it in its entirety. A few months before Einstein and Bergson met, Bergson confided to a friend that his own argument "comes close to the conclusions of Guillaume."[5] Bergson referred positively to Guillaume in *Duration and Simultaneity*, but he also explained how their projects were hardly identical; they differed in very meaningful ways. Bergson accepted Guillaume's points but insisted that "there is more" to the critique offered by Einstein's former colleague. He did not want to return to the concept of absolute time, including Guillaume's. Throughout these discussions, three distinct notions—Guillaume's "universal time," Einstein's "relative time," and Bergson's "duration"—became ever more differentiated.

Although their relationship would end in acrimony, Einstein and Guillaume had been friendly in the early years. Guillaume translated Einstein's first popular exposition of relativity in 1910, helping Einstein reach a Francophone audience for the first time.[6] Problems between the office mates started brewing around 1917, after years of apparently amicable comradeship. After reading one of Guillaume's first critical papers (which had been sent to the Swiss Society), Einstein insisted that his colleague's interpretation "was impossible." But Einstein offered not to say it publicly, on the condition that Guillaume drop his case: "I shall not return to the matter publicly if you do not force me to by emphasizing it constantly."[7] But Guillaume wanted and welcomed a public confrontation, replying to Einstein a few days later "it would be my pleasure to have your refutation appear in the *Archives*. For I believe that it is worthwhile."[8]

ONE *T* FOR TIME

In Paris, during Einstein's lecture, Guillaume insisted in using one *t* for time. When Langevin asked him "What is that *t*?" Guillaume answered, simply, that it was time. What time? That of the station or that of the train? asked Langevin. "Everybody's time," replied Guillaume.[9] The practice of using a single variable, *t*, for time had been standard practice for scientists since the early nineteenth century. Bergson referred to it negatively in *Creative Evolution* (1907), "pointing out that the abstract time *t* attributed by science to a material object . . . remains the same, whatever be the nature of the intervals between the correspondences."[10] The single *t* was commonly associated with the Newtonian concept of an absolute and universal time, which had been the dominant notion—until Lorentz derived the equations central to relativity theory that used t_1 and t_2.

For Newton, time was God-given and as absolute and perfect as the deity himself. A century later, a new "secularized" interpretation of time gained prominence. The philosopher Immanuel Kant argued that time, alongside space, could not be studied directly; rather, that they were two a priori concepts shaping our involvement with the world. Although in Kant's view they were no longer a feature of the God-given universe itself, time nonetheless remained a universal and single concept. Einstein's special relativity work dispensed with these prior notions since it was based on the variables of t_1 and t_2, which could be expanded in an infinite series represented by t_n. It needed neither God nor consciousness to sustain it; it could be described perfectly by simple recourse to clocks.

Guillaume believed that a single *t* could be reconciled with all observed experimental facts; it could be an averaged and comprehensive value that would stand for all the other $t_1, t_2, t_3 \ldots t_n$. Einstein repeatedly claimed that this *t* was a concept that he did not "understand" because he could not put his finger on it.

Although thinkers who believed in the traditional notion of absolute time (such as Guillaume) frequently protested against Einstein and sided with Bergson, Bergson's own intentions were never to defend or attack this notion. Bergson's contribution resided instead in criticizing scientists' tendency to "ignore the cardinal difference between *concrete* time, along which a real system develops, and that *abstract* time which enters into our speculation on artificial systems."[11] When they used *t* for time, they were substituting the real universe for an artificial one. He had developed these thoughts in *Creative Evolution*. Scientists and mathematicians, he explained, often used equations for time to predict future states quite successfully. But they should not forget that they were modeling the future and thus working with a concept that was different from the actual universe, which disappeared in the process of abstraction:

> When the mathematician calculates the future state of a system at the end of time *t*, there is nothing to prevent him from supposing that the universe vanishes from this moment till that, and suddenly reappears.[12]

Guillaume was unconcerned with the complex interplay between the abstract and the concrete. He deeply admired the philosopher's work, ignoring important differences in their distinct approaches. When *Duration and Simultaneity* appeared, he referred to the book as "a magisterial study" that would shed light on the "passionate debate" surrounding Einstein's visit.[13] During the summer of 1922, Guillaume completed the draft of an article dedicated completely to studying the philosopher's work.[14] He immediately sent his article to Bergson, who responded with a letter thanking him for it and politely commenting how Guillaume had "placed" Bergson's philosophy "on a different ground" from which it truly belonged.[15] Guillaume's intention in his article was to prove that Bergson's views on time were relevant for physics, and not only for psychology, whereas Bergson responded that he never had considered the import of his work to be merely psychological.

UNIVERSAL TIME, SECULARIZED

Before he turned to Bergson for inspiration, Guillaume turned to Lorentz. Years earlier Lorentz had claimed that "the evaluation of these concepts (relativity, time) belong mainly within epistemology, and the verdict can also be left to this field."[16] Guillaume was emboldened by this statement. In a letter, he relayed the verdict by such a preeminent authority back to Einstein. He cited this passage verbatim, making sure that the physicist was aware of it, and he also expanded Lorentz's conclusion. In contrast to Lorentz, Guillaume did not believe that the debate could be solved solely through an epistemological discussion: "I am convinced, though, that epistemology *alone* will never be in a position to throw much light on the question." He wanted to prove it: "You know, of course, what the Greek did to show that motion is possible: He walked!"[17] So while Lorentz claimed that alternative interpretations could be valid for different reasons, Guillaume worked hard to prove one, and only one, of them.

Guillaume tried to do exactly the opposite of what Einstein had done—a job completely "reciprocal" to Einstein's. Einstein had postulated the relativity of time; Guillaume, on the other hand, explored what would happen "when 'universal' time is simply *postulated*." While Einstein had eliminated the strange contraction that appeared in Lorentz's interpretation through *relative* simultaneity, "if *absolute* simultaneity is introduced, the contraction must disappear" as well. Guillaume was initially modest in his intentions. Rather than seeking a new theory of the universe, he aimed to "bring clarity to all these issues."[18] But Einstein did not take his ex-colleague's provocations lightly. "That parameter *t* [used by Guillaume] just does not exist," he replied.[19] "Upon calm consideration, you will yourself become convinced of the nonexistence of a *t* to which the role of universal time could be ascribed," he concluded.[20]

Clearly exasperated with his ex–office mate, Einstein advised Guillaume to talk to Langevin. Guillaume dutifully continued his correspondence with Langevin, arguing the same case and receiving similar responses.[21]

But these negative responses did not prevent him from publishing numerous articles and a book.[22]

In the summer of 1920, Einstein's friend, mentor, and collaborator, the mathematician Marcel Grossman, was brought into the fray when Einstein and Guillaume started to communicate with Grossman as the intermediary.[23] When Grossmann asked Einstein what he thought of Guillaume's recent work, Einstein responded by referring to his former colleague as "this man" and describing him as distinctly stupid. Guillaume's piece was as "stupid [*blöde*], as everything else that this man writes about relativity."[24]

Guillaume wrote to Grossmann complaining that "he would want Einstein to pay more attention to his work, which is held in high esteem by important people in France."[25] When he asked Grossmann about Einstein's impression of his work, which he could no longer get directly, Grossmann was more diplomatic, softening Einstein's words. Instead of telling Guillaume—in all honesty—that Einstein had called his work "stupid," he told him that Einstein claimed it was "non-sense [*Unsinn*]."[26]

By the fall of 1920, Grossman warned Einstein that a "cult" was developing around Guillaume and urged him to respond publicly. Three days later Einstein referred to the "Guillaumiade," clearly frustrated with his ex-colleague. He instructed Grossmann to send a notice to the *Archives* saying he could not understand Guillaume because "no clear chain of reasoning underlies Guillaume's explications." Einstein concluded that "the statement is hard, but I can find no other way; this nonsense has gone *too* far!"[27]

Grossmann did not follow Einstein's instructions, withholding the note and publicly saying instead that Guillaume's theory was tautological and did not carry any physical or mathematical importance.[28] Einstein had refuted Guillaume, he told the press.[29] This prompted Guillaume to write directly to Einstein, telling him what Grossmann had told him and how it had "hurt him to the highest degree." He added that "it would be fruitful if you could substantiate your position sometime. . . . That is why I must ask you, dear Einstein, for a reply that I can publicize." He pleaded: "This you really cannot refuse your old office mate."[30] Einstein did refuse but tried to make amends of some sort, saying that "if I wrote to Grossmann that it was nonsense, this must be understood by reference to me, or better yet, to the present state of my brain, nonsense is what one calls whatever one cannot grasp."[31] Since Einstein refused to take him seriously, Guillaume decided to respond publicly.[32]

Einstein's position was consistent throughout his correspondence with Guillaume. Einstein could not find a concrete meaning for "universal time." "Universal time" could not be *actually* measured, and since it could not be measured, Einstein was not going to consider it. For this reason, he could not "understand" it as a relevant concept for physics. For Guillaume, in contrast, the proof that universal time was real resided in our "intuition" of it. Guillaume wanted to debate Einstein publicly, but Einstein refused again, saying he was "too busy."[33]

Guillaume had important allies. The writer Lucien Fabre, one of the early supporters of relativity in France, wrote to Einstein, finding in Guillaume's work, "which you already know . . . one of the most serious and original attempts to give to your brilliant theory a new, supplementary basis that is capable of harmonizing with it, adding to its intellectual seduction a new element of certainty."[34] To Einstein's disgust, Fabre published a book, prefaced by Einstein, in which his work appeared next to Guillaume's.[35] Paul Dupont, ingénieur des Ponts-et-Chaussées, a man described as "verif[ing] the interpretation given by Bergson to Einstein's theory," defended Guillaume.[36] Like Guillaume, Dupont wanted to go further than Bergson, whom he criticized because "his conclusion is a simple negation of the absolute theory of relativity."[37]

Einstein felt that Guillaume's crusade was motivated by ill will and by political differences. He explained to Guillaume that his theory "has nothing to do with the Manifesto of the 93."[38] The comparison referred to a text signed by important German intellectuals defending the actions of Germany during the Great War. Perhaps his ex-colleague's animosity was motivated by anti-Germany sentiments? Guillaume did not accept these accusations, nor did he acquiesce. He published an article in the *Revue général des sciences* (January 15, 1922) titled, "Is There an Error in the Calculations of Einstein?"

As time passed, Guillaume became more and more attached to Bergson's way of thinking: "The conclusions of Mr. Bergson's study are of essential importance for physicists."[39] Like Poincaré, Lorentz, and Bergson, Guillaume accepted a difference between accelerated time and unaccelerated time, to which he referred in the standard scientific notation as t and t'. "Consider the difference $t - t'$. It is not equal to zero. What does it signify?" for Guillaume the answer was the same as Bergson's, that t' was "fictional" and "unreal." It was "a time without duration, where events do not occur, and things do not subsist, and where beings do not age."[40] Alongside Bergson, Guillaume cited Poincaré every chance he had, peppering his articles with "said Poincaré" and "with Poincaré." He even edited *La Mécanique nouvelle*, a book detailing Poincaré's contributions to relativity, writing the introduction for him.[41]

Guillaume followed closely the red-shift experiments that were being conducted and awaited the results, but he did not believe that they would be decisive in proving Einstein's theory. He would grant Einstein the credit for having "divined a correct formula," but not much more.[42] "In our opinion, the point of view defended by Bergson would not be weakened," he concluded.

Guillaume liked to point out that "in practice, all temporal measurements refer to spatial ones." To a question of distance, such as how far Paris is from Berlin, you could answer with time, such as a day and a half by train. This frequent practice of interchanging of time and space standards permeated even common discourses. What was its significance for physicists? Poincaré had noted the possibility of redefining length in terms of the speed of light. Two lengths could be considered equal if each were defined as "the distance covered by light in equal periods of time."[43] If the velocity of light was constant—scientists and engineers generally considered it so—the paradoxes of relativity would clearly follow. What, then, was Einstein's particular contribution? Guillaume asked.

Bergson and Guillaume agreed that scientists could consider the speed of light as a constant velocity in order to *make* it into a measurement standard. They could use it as a new way of measuring time and space since they clearly needed better standards. The deficiencies of the sidereal clock, especially after the deceleration of the Earth's rotation became widely acknowledged, were widely known. Bergson agreed with Guillaume's insistence that relativity theory consisted in "making a clock out of the propagation of light instead of the rotation of the Earth."[44] Guillaume positively cited Bergson's call for the "adoption of a new clock . . . the optical clock—that is to say the propagation of light."[45] But he, like Bergson, insisted that espousing a system of time measurements based on the constancy of the speed of light should not be confused with time itself.

Part of Guillaume's critique was informed by current problems in metrology. Guillaume was the younger cousin of Charles-Édouard Guillaume, who had also studied at the Zurich polytechnic and who, as director of the Bureau international des poids et mesures that had sponsored Michelson's work, was one of the world's most respected experts on metrology.[46] Charles-Édouard received the Nobel Prize in 1920, two years before Einstein, for his work in this area. A few days before Einstein's lectures, Charles-Édouard gave a talk to the Association générale des étudiants in which he noted current problems and challenges facing science.[47]

Yet Guillaume's criticisms differed in important ways from the problems faced by metrologists, such as those detailed by his older cousin. While metrologists tried hard to forge ideal standards for practical use, Guillaume began by assuming that universal time existed. He then sought to reconcile all the effects of relativity with that concept.

GOD'S TIME

What was at stake? "Before God, a thousand years are as one day," went a common saying. Common understandings of God often attributed to him immense powers in his perception of time. In Christian theology, an eternity was defined as a mere instant for God.[48] Angels followed next in this ability.[49] Humans usually came in third.

The astronomer Charles Nordmann succinctly explained the deity's power when it came to time: "For an infinitely perfect being equal to a God, past sensations would be as actual as the present ones and time would not exist."[50] Other "pure spirits," such as angels, were not as powerful as God in their ability to seize time, occupying an intermediate position between God and humans. For them "a more or less long chunk of our time" could seem like an instant. Humans were worst off, since for them, instants were not divisible and since the latter did not correspond to meaningful lengths of time, let alone eternities. For us imperfect souls, instants were "an indivisible limit between the time before and the time after."[51]

A popular mid-nineteenth century science writer clearly explained the particular power of an omniscient consciousness in grasping time: "Omniscience, with respect to the past, becomes identical and one and the same thing with actual Omnipresence with regard to space. For, if we imagine the eye of God present at every point of space, the whole course of the history of the world appears to Him immediately at once."[52] In this view, events did not take time to reach God: "For, since God is not spatially separated from things, it seems no definite lapse of time can occur either between his prehension of them or theirs of him. There can be no transmission with the velocity of light from an event to the divine observer."[53] How did Einstein's theory of relativity account for the existence or inexistence, or even *the mere possibility of existence*, of such an observer?

Since its inception, the idea of universal time was tightly coupled with that of God. Newton famously described time in both theological and scientific terms. For him, and for many thinkers after him, absolute time was defined by recourse to an absolute observer—an omniscient consciousness—which he attributed directly to God. The "sensorium of God" guaranteed the existence of absolute time: Because he "endures forever and is everywhere present; and, by existing always and everywhere, he constitutes duration and space."[54] Newton's association of absolute time and an omniscient consciousness was widely shared.

During the French Revolution, Newton's view of time and its connection to God was questioned and secularized. The French scientist Pierre-Simon Laplace, in his famous *Mécanique céleste* (1799), provided an alternative model of the universe and of temporal development—one that no longer needed a place for God and his rather large sensorium. By midcentury, Newton's science was ridiculed for having depended so centrally on God. The French scientist François Arago lambasted Newton, who "believed that a powerful hand should intervene every once in a while to fix the disorder."[55]

But even when direct references to God were absent, allusions to godlike perspectives continued to appear well into the twentieth century. Even Laplace, who was rumored to have answered "Sir, I have no need of that hypothesis" when Napoleon asked him if God had to intervene to keep the universe stable, could not help but talk about "an intellect who at any given moment knew all of the forces that animate nature and the mutual positions of the beings that compose it." Laplace speculated that "if this intellect were vast enough to submit the data to analysis," then "for such an intellect nothing could be uncertain and the future just like the past would be present before its eyes."[56]

Scientists who came after Laplace frequently debated if this superior intellect actually existed or not, taking sides for or against. Yet even those who denied its existence (and who followed Arago in mocking Newton for relying on it) often continued to describe a similar concept, even if they referred to it only as hypothetical.[57] "Even atheists themselves," according to Poincaré, "place themselves in the place of an omnipresent God when they discuss time."[58]

Léon Brunschvicg, who asked Einstein a difficult question on that April evening in 1922, considered the existence of a *surobservateur* as "capable of coordinating in a unique representation the different systems that the groups of observers make of the universe according to the different circumstances of their observations." For him, this was precisely the fiction "that even Newton could not sustain other than by appeal to the image, perhaps contradictory, of an entity contemporaneous with all times, of a 'super' observer." This entity, according to Brunschvicg, did not exist: "There is no 'super' observer," he argued.[59] But others were not as sure.

The physicist Hendrik Lorentz also invoked a "universal spirit" when he disagreed with Einstein. Although he did not argue for the existence of this entity, he did argue that humans were most likely similar to it. Lorentz was ready to concede that imperfect spirits like us, at this moment in time, could not determine a difference in kind between t_1 and t_2, but he argued that a "universal spirit" could. And he could be sure of another thing: that "surely we are not so vastly different" from this "universal spirit."[60] Striving to place himself in its perspective, Lorentz felt justified in his search for something that would help scientists find a better reference for time. Lorentz, working within a Christian view in which humans were made in the image of God, argued that we were built like this universal spirit. For this reason, we should be allowed to *think in the same way it would*.

Bergson was sometimes, although erroneously, considered to defend a concept of absolute time based on these theological considerations: "As . . . Bergson [has] said, the inability of human beings . . . to determine a unique cosmic present or simultaneity need not prevent God, who knows things directly, from experiencing such a present," explained one of his commentators.[61] Bergson did invoke the examples of "a superhuman consciousness coextensive with the totality of all things" and "a superman with a giant's vision" in his confrontation with Einstein, but his purpose was not to prove its existence or take its side.[62] It was only to prove that the scientist's perspective was only one of many others that needed to be considered.

"AN OLD QUESTION THAT CONTINUES TO BOTHER ME"

After decades of silence and yet another war, Guillaume wrote a letter to Einstein, hoping to "end with you an old question that continues to bother me."[63] It was 1948, and the letter was intimate and friendly. Einstein's ex-colleague reminisced about the "wonderful hours that I passed with you at Bollwerk and Aegertenstrasse." He referred to the café just around the corner from the patent office where they worked together and to the physicist's home, which he had frequently visited. He recounted how he had finally retired from his last job at the insurance company La Neuchâteloise. Einstein was by then ensconced at Princeton.

The physicist was not amused by the return of the past. He responded to the author's "familiar tune" in a familiar way, repeating the same answer he had always given once again: that he did not understand what universal time was, since it was not something that could be measured with clocks.

Guillaume was flattered simply by receiving a response from Einstein. It "produced in him a great joy." He sent another letter on the very day of the anniversary of their contentious meeting, on April 5. Einstein did not respond. Guillaume wrote another letter that summer. No response. Guillaume finally stopped writing.

The old concept of universal time, so passionately defended by Guillaume on April 5, 1922, could not be revived. After debating Guillaume, Einstein had completely eliminated it from the competition. Only Bergson's remained to be dealt with—the next day.

CHAPTER 19

Quantum Mechanics

SOCIÉTÉ FRANÇAISE DE PHILOSOPHIE, PARIS

History repeats itself twice, "the first time as tragedy, the second time as farce," wrote Karl Marx.[1] History seemed to repeat itself at the Société française de philosophie. In 1929 Einstein visited the venerable institution for the second time. Many of the participants who had been there in 1922, most notably Langevin and Le Roy, were there again. Bergson was notably absent—yet he was defended by his old advocates as if he were an invisible ghost in the room. During that trip, Einstein paid the philosopher a visit to his home, accompanied by the poet Paul Valéry.[2]

"A new fashion," noted Einstein "has arisen in physics."[3] Seven years later, the discussion at the Société no longer centered on the theory of relativity, but on a "new physics" soon to be known worldwide as quantum mechanics. Bergson's philosophy started to be strongly associated with this "new physics." Quantum theory was creating "a revolution of much greater philosophical import than the one brought about by relativity theory," explained one of Bergson's allies.[4] A reviewer of Bergson's final book *La Pensée et le mouvant* (1934), saw in it the "close harmony between Bergsonism and the metaphysics of some of the modern physicists."[5] How would this new revolution affect the stature of Bergson vis-à-vis Einstein?

Einstein would never accept some of the central tenets of quantum mechanics. In a famous showdown at the Solvay Conference of 1930, he locked horns with one of its main representatives, the Danish physicist Niels Bohr. During this meeting, it appeared to many of the attendees that it was Einstein who had lost. Some of those who followed these new debates saw in them, and in the triumph of quantum mechanics, the final vindication of Bergson's lifework.

One of the first instances in which Bergson's philosophy was seen to bear directly on quantum mechanics pertained to the dual nature of light. Was light wavelike or particulate? The question had haunted physics for centuries. Newton had famously considered it particulate, but in the nineteenth century, a dramatic scientific demonstration proved to be one of the most damaging reversals for Newton. In 1818 Agustin Fresnel placed an opaque screen in front of a beam of light and saw the light at the other end increase, not decrease, in intensity. One possible explanation for this unexpected result was that light waves had traveled around the disk. When they tried to model the experiment using wave equations, the results matched observations perfectly. Later, in famous experiments in which light was sent through a current of water and compared to when it traveled freely, many became even more convinced of its wavelike nature. In France, the benefit of overcoming Newton's theory had clear nationalistic overtones: French research overturned the results of an untouchable, and very British, scientist. But the case was not entirely closed. Einstein, in his research on the photoelectric effect, had shown once again the usefulness of treating light in terms of particles, which he now called photons.

New research on X-rays led scientists not only to consider light as waves once again, but also, even more strangely, to consider it as both wavelike and particulate. Its dual nature was soon considered as a vindication of Bergson's philosophy.[6] "After light was reduced by the theory of emission to the movement of corpuscles . . . now it appears to us, especially after following the recent research on X rays, as born out of continuity and discontinuity," explained Jacques Chevalier, a scientist and friend of Bergson, who would become minister of education during Vichy.[7] This duality, according to Chevalier, was comparable to the one "that Bergson speaks of, which would be as a whole a superior synthesis of absolute continuity and absolute discontinuity, or more exactly, a reality of which these two concepts are only partial expressions."[8] In years to come, the wave-particle duality of light would mark the new science of quantum mechanics. Light should either be wavelike or particulate, Einstein would insist, but not both at once.

In his work on X-rays, the physicist Maurice de Broglie showed that the rays seemed to behave both as waves and as particles. His famous brother, Louis, would repeatedly associate these new insights from physics with Bergson's philosophy. De Broglie paired up with Chevalier, who was speaking in various forums about "science as swinging pendulum-wise from an insistence on continuity to an insistence on discreetness, these two being in fact complementary and inseparable aspects of one physical reality."[9] De Broglie and Chevalier would soon be joined in their efforts by Le Roy. Together they published a volume titled *Continu et discontinu* (1929).[10] In it, de Broglie's chapter focused on the "crisis of quanta" affecting contemporary physics, explaining that it "has undermined the edifice of our knowledge in accentuating the antinomies between the continuous and the discontinuous."[11]

Some of the authors of *Continu et discontinu* came from a different background from many of the scientists that became Einstein's closest supporters. The aristocratic background of the de Broglie brothers contrasted starkly with that of other physicists of their generation, especially those with leftist sympathies such as Langevin, Perrin, and Marie Curie, who were Einstein's friends.[12] Chevalier was a devout Catholic who came from an elite military family.[13] He was the godchild of General Philippe Pétain, who would lead France during the Vichy occupation. As a government official Chevalier fought against the principle of *laïcité* and tried to reintroduce religious catechism in elementary schools.

Quantum mechanics resulted from the work of many individuals, Erwin Schrödinger, Max Planck, Werner Heisenberg, Niels Bohr, and Louis de Broglie being the most well known. Einstein himself was a reluctant contributor to the field. His work on the photoelectric effect, which dealt with light as particulate, was seen as an important contribution. The "group of studies, for which in particular Einstein has received the Nobel Prize, falls within the domain of the quantum theory founded by Planck in 1900," explained the presenter of the prize after citing Bergson's objections as the reason why relativity should not be included in it.[14] Yet Einstein never accepted the new physics entirely, being especially allergic to its espousal of indeterminism, action at a distance, and "complementarity"—a label used to explain how physical reality could be understood simultaneously in contradictory terms (such as waves and particles).

Quantum mechanics broke with relativity theory and classical physics in various respects. First, it revolutionized the concept of measurement by claiming that the act of measurement itself changed the experimental system. Second, it introduced limits to what could be known with measurement. The measurement of a particle's position introduced a certain degree of uncertainty in measurements of momentum. These two principles had a third, essential consequence. Quantum mechanics forced scientists to reevaluate the idea of physical causality, reintroducing into the universe an essential indeterministic quality. This last revolution was never accepted by Einstein, who used the famous line "God does not play dice with the Universe" against it. This line was cited by Heisenberg in *Physics and Beyond* and repeated by Bohr as well, as both men marked their differences with Einstein.

When some scientists working from the end of the 1920s to the end of the 1930s argued for an essential indeterministic aspect of the world, they were rapidly accused of being Bergsonian. One of the reasons why Bergson, after all, was interested in the topic of time came from his desire to emphasize those aspects of it that could lead to change. "And the more we descend from the motionless idea, wound on itself, to the words that unwind it, the more room is left for contingency and choice," he explained in *Creative Evolution*.[15] Readers across the globe were quick to point out the connections between quantum mechanics and Bergson: "In casting aside logic and the principles of science, are not these physicists approaching the position much lauded by Bergson . . . ?"[16] "Attacks on causal knowledge" were labeled as Bergsonian in important journals of philosophy.[17]

Bergson's work, which had long been considered a philosophical defense of indeterminism, was seen to bear directly on Heisenberg's principle of indeterminacy. The "uncertainty" principle claimed that the velocity of a particle could not be known at the same time as its position. Arthur Lovejoy understood it as "a logical puzzle as ancient as Zeno of Elea and as modern as Bergson." Bergson, an even stronger advocate argued, had offered "a remarkable anticipation of the principle of indeterminacy," beating Heisenberg by twenty years.[18]

By the 1930s, most of the educated public had heard that atoms seemed to be "misbehaving" in the microscopic realm. The poet Paul Valéry followed this research closely. "In the most intimate part of the atom, nothing seems as from the outside," he informed Bergson.[19] After he received a copy of Bergson's book *La Pensée et le mouvant*, he was particularly intrigued by a long footnote "on the subject of the *grande affaire* of Relativity." Referring to the recent advances in quantum mechanics that Einstein famously resisted, he asked if these "up-to-date microphysics" could be brought to bear on "some of your conceptions?"[20] A friend of Valéry remembers a conversation with Bergson in which the philosopher had responded to these comparisons by saying: "That which is funniest [*drôle*], is that physicists have come back to find liberty!"[21]

While some continued to criticize *Duration and Simultaneity*, others applauded the alleged "quantum mechanical" aspects of Bergson's philosophy. At the time of Bergson's death, the literary critic André Rousseaux claimed that it was *not* Bergson who had been wrong on science, it was science that had been wrong on Bergson. Bergson had been misjudged by the "science of his time," which was one that "was . . . marked by caducity."[22] New scientific discoveries associated with quantum mechanics proved that the "Bergsonian revolution will be doubled by a scientific revolution that, on its own, would have demanded the philosophical revolution that Bergson led, even if he had not done it."[23] Science had finally caught up with Bergson: "The chains that Bergson broke, and that one thought were constrained by science, are ones from which science itself, by the new discoveries, is now liberated."[24] The connection between the philosopher and one of the founders of the new discipline was so clear for Rousseaux that he titled his article "From Bergson to Louis de Broglie."[25]

QUANTUM MECHANICS: LOUIS DE BROGLIE

The Japanese-born physicist Satosi Watanabe, a student of Louis de Broglie who was sent by the imperial government to study first in France and later in Germany with Heisenberg, referenced Bergson repeatedly. De Broglie wrote the preface to his thesis on thermodynamics, comparing Watanabe's views explicitly with Bergson's.[26]

In the 1940s de Broglie cited Bergson's first published book, *Time and Free Will* (*Essai sur les données immédiates de la conscience*, 1889) celebrating its trenchant anti-determinism.[27] In another paper entirely dedicated to the question of the relation between quantum mechanics and Bergson's philosophy, aptly titled "Les Conceptions de la physique contemporaine et les idées de Bergson sur le temps et sur le mouvement," (Concepts of Contemporary Physics and Bergson's Ideas of Time and Motion) in the *Revue de métaphysique et de morale* (1941), the connection was sealed. De Broglie argued that "if Bergson could have studied quantum theory in detail . . . he could doubtless have repeated, as in *The Creative Mind* [La Pensée et le mouvant], that 'time is this very Hesitation or it is nothing.'"[28] *Physique et microphysique* (1947) drew on this earlier paper and continued to detail the numerous parallels between quantum mechanics and Bergson's philosophy.

Some supporters of quantum mechanics argued that it was a much more fundamental, relevant, and profound branch of knowledge than relativity had ever been. De Broglie downplayed the importance of relativity theory when compared to these new scientific developments: "The theory of relativity itself now appears to us as simply a macroscopic and statistical view of the phenomena: it describes things approximately and in bulk and does not descend profoundly enough into the detailed description of elementary processes." Although he described Bergson's *Duration of Simultaneity* as "the least estimable of his books" and one that "has justly been criticized," he nonetheless found that the philosopher's other work "consequently antedates by forty years the ideas of Niels Bohr and Werner Heisenberg on the physical interpretation of wave mechanics." In his view, Bergson's philosophy stood in opposition to relativity because Einstein's work simply extended classical conceptions of space and time, which only quantum mechanics truly questioned: "In truth, relativity physics seemed to be in flagrant opposition to Bergson's views, precisely because it pushed the specialization of time and the geometrization of space to their extreme limits, and for this reason it is the final development of classical physics."[29] Bergson approved of de Broglie's research, claiming in his last writings that his own conclusions were "the direct consequences of the theory of Louis de Broglie."[30]

Many other quantum physicists continued to draw on Bergson, especially after the Second World War. At the Maison Franco-Japonaise in

Tokyo, Watanabe delivered a talk exploring the connections between Bergson's work and quantum mechanics.[31] In his presentation, the physicist categorically claimed that "modern [quantum] physics has found the lost link between psychic time and physical time."[32]

LIVED TIME, CYBERNETICS, AND CHAOS THEORY

The physicist Olivier Costa de Beauregard, a pious Catholic and student of Louis de Broglie, claimed that Bergson's arguments about relativity were "absolutely erroneous" at the same time he stressed the relevance of other aspects of Bergson's philosophy for quantum mechanics.[33] For the rest of his life, Costa de Beauregard would return to "the illustrious philosopher of *duration*" to reintegrate into physics the time that Einstein lost: "The irreparable time that flees, the impossibility of remaking the past, and also of knowing the future." For him, these aspects of time were the most important ones: "These are the lessons of everyday experience, even more banal (because of their more vital importance) than the impossibility of monothermal transformation of heat into work."[34]

Quantum mechanics, cybernetics, information theory, and the concept of negentropy—alongside Bergson's philosophy—aided Costa de Beauregard in the arduous task of recovering our experiential sense of time and integrating it back into science.

In the 1960s, the French physicist paired up with a group of radical American scientists who would "save" physics from the "stagnation" it faced after the Second World War. Money to the discipline coming from national defense-oriented agencies started to dry up. The federal government had bankrolled the Manhattan Project and other wartime initiatives, benefiting both experimental and theoretical physics.[35] How should a new generation of physicists survive? Not only were traditional sources of funding no longer forthcoming, practitioners in these new fields faced strong opposition from conservative Cold War scientists. John Wheeler, a collaborator of Einstein's who worked on the atomic and hydrogen bomb projects, went on a "warpath" to "drive the pseudos out of the workshop of science," but despite these obstacles, they went on to develop quantum information theory.[36]

Quantum mechanics and information theory were not the only sciences that were frequently understood by reference to Bergson's philosophy. Across a growing number of disciplines, from psychology to cybernetics, scientists stressed the pertinence of Bergson's conclusions. Eugène Minkowski, a psychologist and friend of Bergson, decided to reverse the trend of thinking about time only in a quantitative manner and to write about how we experienced time in its flow. *Le Temps vécu* (1933) inaugurated a psychopathological study of a person's experience of time that drew heavily on Bergson. To deny differences between the past, present, and future, he argued, amounted to a "scientific barbarism." Our most intimate and relevant feelings, he noticed, were intimately connected to these categories. "Disappointment and regret" were for the past, and "desire and hope" for the future. By the late 1930s Bergson was thrilled to see the work of his friend "cited more and more" and the concept of "lived time" take flight.[37]

In America, the mathematician Norbert Wiener revived Bergsonian time in his famous book *Cybernetics: Control and Communication in the Animal and the Machine* (1947) and returned to the concept again in *The Human Use of Human Beings: Cybernetics and Society* (1950).[38] Wiener explained how the differences between Bergsonian time, with its focus on irreversibility, and Newtonian time, which was reversible, had once been used to mark the difference between the living and the mechanical. But, according to Wiener, the boundary between the living and the mechanical shifted due to "the great mechanization of the Second World War." It was now much more useful, he argued, to extend Bergsonian time to non-living systems. A specific set of nonliving mechanisms (servomechanisms or control mechanisms) seemed to "live" under this kind of time. These machines functioned in real time and adapted themselves to such situations via feedback loops. "Thus the modern automaton exists in the same sort of Bergsonian time as the living organism," he concluded.[39]

The vindication of Bergson's work by scientists strengthened in the late 1960s and continued for the rest of the century. *Duration and Simultaneity* appeared in English in 1965. A few years later (1969), Bergson's debate with Einstein was translated and republished. The editor, P.A.Y. Gunter, collected various essays by scientists and philosophers who drew from Bergson, ranging from de Broglie to Milič Čapek. His conclusion was clear: "The charge of anti-scientific intention which Bergson has been saddled is seen to be not merely misleading but radically false: Bergson's philosophy of intuition is the affirmation, not the negation of science."[40]

PART 3

THE THINGS

CHAPTER 20

Things

What do we find when we look even more carefully behind the scenes of the debate? We stumble upon certain things that drove "adversaries" into "absolutely opposite positions."[1] Einstein and Bergson played key roles in bequeathing us a world split by two irreconcilable theses, yet they themselves were led in particular directions by a variety of elements that surrounded them: clocks, the telegraph, telephone and radio communications, cinematographic cameras and film, atoms and molecules. Einstein and Bergson disagreed about the meaning, use, and importance of all of these things. They were not only objects brought into their discussions; they were not merely illustrative examples; nor were they just tools used to communicate their arguments—they played a central role in the twentieth-century divisions often associated with Bergson and Einstein.

With a time-lapse focus on April 6, 1922, we can see how seemingly unimportant things played a prominent role in the unraveling of the larger conflict.[2] Einstein and Bergson recede into the background and no longer appear at the center of the debate they detonated. They no doubt made history, but they did not make it as they pleased.

Enough time has passed from the initial *cri de guerre* that we can safely imagine the path new recruits would take—traveling further and further apart from each other. But we can also take a step back and ask new questions. What fueled the conflict and how can we move beyond it? What cuts across the "absolutely opposite positions" represented by the two men?

Part 3 focuses on the debate by taking us beyond the men, asking instead what drove them to fall into such a stark impasse in the first place. It centers on elements that appeared as much in ordinary workplaces as in sophisticated philosophical discussions; on what was equally present on busy Monday mornings as on Sunday afternoons; on what changed how we thought, felt, and experienced the here and now; on things that lay ambiguously between animals and ghosts, meek (in that they are often unacknowledged) but powerful (in creating divisions). Things that were used all the time become particularly important—affecting time because they were parasitical on time. A common ground underlying otherwise uncompromising positions is made up of seemingly minor things.

BEYOND TECHNOLOGY

It was abundantly clear by 1922 that Einstein's relativity theory agreed with experimental results and that it was elegant and logically consistent. In decades to come, it would gain even more support. Our contemporary understanding of how certain modern technologies work, from radio to GPS, now depends on it.

But Einstein's theory was much more than a good explanation of existing experimental results. Other theories, most notably Lorentz's and Poincaré's, also matched known results. Why should Einstein's be accepted over the others? What was the relation of Einstein's work to technology? What was its relation to concrete observations? The physicist offered much more than a simple procedure for time coordination. But his theory could not be entirely disconnected from technology either—it had to maintain a real connection to cutting-edge, experimentally verifiable processes. A fine line separated the grandiose theoretical claims of the theory from its mundane practical applications and verifiable observations.

Einstein and Bergson held an extremely low opinion of technology—especially when compared to science. They were skeptical about its alleged benefits. "What comes to the mind of a sensible person when hearing the word technology?" asked Einstein. "Avarice, exploitation, social divisions amongst people, class hatred," he responded. Technology, in his view, could easily be considered the "wayward son of our era."[3]

Bergson was similarly pessimistic. In *The Two Sources of Morality and Religion* (1932), the philosopher noted how "for a long time it was taken for granted that industrialism and mechanization would bring happiness to mankind." But this belief had recently been proven wrong: "Today one is ready to lay to their door all the ills from which we suffer." "Never, it is said, was humanity more athirst for pleasure, luxury, wealth," he concluded.[4] When he was awarded the Nobel Prize for literature (in 1928 for the year 1927), he used the occasion to warn listeners about the dangers of technology:

> If the nineteenth century made tremendous progress in mechanical inventions, it too often assumed that these inventions, by the sheer accumulation of their material effects, would raise the moral level of mankind. Increasing experience has proved, on the contrary, that the technological development of a society does not automatically result in the moral perfection of the men living in it, and that an increase in the material means at the disposal of humanity may even present dangers unless it is accompanied by a corresponding spiritual effort.[5]

The criticisms of technology launched by Einstein and Bergson were hardly anomalous. They echoed those of many others such as Sigmund Freud's, who similarly took technology as one of the main culprits of, instead of a solution for, society's ills. In their view, technology was separate and inferior to science.

Einstein knew much more about technological innovations than did Bergson. He was well-versed in the operating principles of the electrical technologies that constituted his family's business; he had studied at a polytechnic university; and when working at the patent office, he had the benefit of reviewing the most cutting-edge inventions of his era. Historians have documented his keen interest in things ranging from clocks to refrigerators. But professionally, Einstein's career centered on *theoretical* physics, a discipline that in the first decades of the twentieth century emerged as distinct from experimental physics and even farther from engineering.

As much as they resisted the very idea of "technology," neither Bergson nor Einstein were immune from living through vast changes in the material culture of their era. Scientific laboratories were increasingly equipped with innovative instruments that would be widely employed in advanced industrial societies. These instruments were not simply mechanical, but also electromechanical, and they appeared alongside new forms of life, spectacle, and telecommunications that would characterize post-Fordist economies.

"EXPERIMENTALLY REALIZED"

Henri Piéron, a professional experimental psychologist, was one of the last to speak during the debate on April 6. He raised a crucial point shortly before Bergson delivered his closing remarks that day.[6] After sitting silently throughout the entire meeting, Piéron finally decided to bring to the attention of the audience an example in which this *verbal* argument seemed to have been realized *experimentally*. In scientific laboratories, he argued, there "existed cases where the confrontation is experimentally realized." While in front of him he saw two *men* reach different conclusions about the nature of time, in *laboratories* he had witnessed time itself emerging in two different guises: physical time and psychological time. In scientific laboratories, he argued, there "existed cases where the confrontation [between Einstein and Bergson] is experimentally realized."[7]

Piéron referred to laboratories of experimental psychology that housed new instruments for measuring how humans reacted to a variety of stimuli. The daily work of these laboratories consisted largely in investigating a subject's "sense of time, succession and simultaneity" by comparing it against clock time.[8] Laboratories of experimental psychology had expanded significantly in recent years, growing in importance due to their connection to the communications, transportation, advertisement, and entertainment industries. Initially, they had been furnished mainly with chronographs—stopwatches that measured the time between stimulus and response, but Piéron was interested in a broader set of "signaling instruments [*appareils-signaux*]."[9] With these new machines, investigations could increase in complexity, from trials initially focused on measuring reactions to simple stimuli such as pendulum beats and electric shocks, to experiments run with more complicated stimuli: words, longer texts, moving images, audio, and even film.

People usually agreed about when two events were simultaneous or not. Simply seeing two events occurring at the same time was a fine way of establishing simultaneity. "The arrival dead-head of two race-horses are simultaneous if we see them at the same time," explained an astronomer.[10] But in a fast-paced world, such determinations were insufficient. The meaning of "at the same time" was hardly clear. The discrepancy between clock time and human time appeared prominently when events occurred in very close succession. Time appeared most perplexing in those instances. Since the middle of the nineteenth century, scientists had realized that the determination of the simultaneity of two events was highly imprecise—even among talented observers. Because all "sensations separated by a very real interval will be falsely noted as simultaneous," determinations of simultaneity—even locally—varied widely.[11] Poincaré, like most scientists of his generation, was well aware of these complications. When reviewing the work of a colleague in 1886, he cautioned against the casual use of the phrase *à la fois* (at once). The colleague accepted Poincaré's criticism, revising his work to acknowledge that we often judge "sensations" as "successive or simultaneous," and that our judgment did not often correspond with "exterior" events: "in between the exterior phenomena and the sensation which reaches our consciousness, there is almost always a delay."[12]

Every day, argued Piéron, laboratories of experimental psychology displayed a schism between how humans and instruments measured time. They showed how a person's determination of "at the same time" often did not coincide with that of a much more precise instrument. Experiments undertaken with these instruments, Piéron continued, showed why Einstein and Bergson were really talking about two different aspects of time: one psychological and the other physical. "Therefore Bergsonian duration seems to me to remain a stranger to physical time in general and in particular to Einsteinian time," he concluded.[13]

The laboratories described by Piéron in 1922 would increasingly become sites for studying key aspects of industrial interwar and postwar environments. Initially centered on reaction-time experiments, they were soon connected to the transportation, entertainment, and telecommunications industries. New instruments were adapted to optimize the time necessary for reacting, reading, and writing. An increasing number of scientists became concerned with perfecting techniques for inscribing, recording, transmitting, and processing all sorts of events occurring in time. Speed was central to modern technological culture, and experimental psychologists studied how to survive and advance in that culture. The quest to attenuate or eliminate human errors in assessing and reacting to fast events motivated some of the most important changes in science and laboratory techniques, from the development of cinematography to "impersonal" micrometers, direct-reading instruments, and automatic inscription devices. As the differences between Einstein and Bergson worsened, an increasing number of fingers and eyes were set on keys and screens.[14]

That evening in Paris, Bergson was not content at all to declare a tie between himself and Einstein. He was unwilling to divide the booty into two camps: the error-prone Bergsonian duration and a precision-friendly Einsteinian time. Yes, he agreed that laboratory experiments unambiguously showed a difference between psychological and physical assessments of time. Yes, he also agreed that one of these could be associated with his concept of duration and the other one with the physicists' time. Yet Bergson warned listeners against believing in a hard-and-fast distinction between these two concepts. Any physical measurement of time, he argued, contained an irreducible psychological element. Psychological and physical concepts of time joined in key ways—no matter how hard scientists tried to separate them; no matter how hard scientists worked to set a clear hierarchy between them.

Bergson set up his first assignment in *Duration and Simultaneity* to be that of determining how Einstein's theory traveled from its mathematical structure to concrete reality and back. It was hard work: "To get at this, we went over Lorentz's formulae term by term, seeking the concrete reality, the perceived or perceptible thing, to which each term corresponded." But the effort paid off: "This examination gave us a quite unexpected result," he wrote in the preface to his book. It confirmed his own—not Einstein's—theory of time. If his exercise was followed carefully, readers could learn that they did not need to accept all of Einstein's conclusions. Bergson criticized Einstein for mixing up two things that were not the same: the abstract and the concrete. He accused the physicist of letting "two different conceptions of relativity, one abstract and the other full of imagery, one incomplete and the other finished," coexist uncritically.[15] His task was to separate these concepts and reintroduce clarity after the physicist had distorted them beyond recognition.

Both men were surrounded by a host of new innovations that did not adequately fall within the label of "technology"—especially not in the way that it was understood at the time, when it mainly referred to the large machines of the Industrial Revolution, such as engines, pumps, looms, and transportation machines. What were they? Many of these things were so novel at the time of the debate that a clear category to describe them did not exist.[16] One strange quality united them: they were used for the storage and transmission of events in time. Today, we may refer to them as telecommunications media, but neither of these terms were in use at the time of the debate between the two men. References to light-based transmitting, recording, and timing devices appeared over and over again in the discussion between Einstein and Bergson and their interlocutors.

WHY LIGHT?

Throughout Einstein's life, two related questions were repeatedly brought up in critiques of relativity. First, why define time exclusively by reference to clocks? Second, why consider clocks in terms of the behavior of light? Light, he answered, behaved differently from most other things. "The comparison of light with other 'stuff' is not permissible," Einstein told an audience who he was finding difficult to convince.[17] At the time, in 1911, they remained largely unconvinced. "He always comes to perceive the world around us by way of light signals," protested one of his listeners.[18]

In Paris, Langevin was facing similar questions from some of the same philosophers Einstein would encounter more than a decade later. Why change our common understanding of time and space in the universe just because of certain properties of light? Langevin responded by clarifying the different steps that led him to those conclusions. If one accepted a particular property of light (that its velocity was independent of the movement of its source), then "we are necessarily driven to the conclusions that I have developed for *optical* space and time, that is, as they are measured by light signals."[19] The next step consisted in convincing his listeners that these light-based measurement techniques were better than the other ones based on the old mechanics. "The other ways of measuring were infinitely coarser," he explained. If these old measurement techniques could be "brought to a degree precision comparable" to those of the optical ones, then old notions about the "form of space and time" would also change.[20] Were these advances in precision worth the trouble?

Even some of Einstein's strongest supporters, such as Eddington, continued to be surprised about the preponderant role played by light in relativity theory. Eddington confronted head-on the prickly accusation of the singular importance given by Einstein to light signals, answering the "objection" that "is sometimes raised to the extravagantly important part taken by light-signals and light-propagation in Einstein's discussion of space and time."[21] Eddington, like Einstein, considered the speed of light as a "universal" and "fundamental" constant. How did such a "fundamental velocity" relate to the actual velocity of light, something that could seem too tied to the limitations of current signaling technology. By "a lucky coincidence" both of them were nearly identical: "Luckily, there is a physical entity—light—that moves habitually at the same speed of the fundamental velocity."[22] Because of its fundamental nature, the speed of light typically "crops up in all kinds of problems whether light is concerned or not," explained Eddington.[23]

The year before he confronted Bergson, Einstein was asked at Princeton why he had defined time by reference to light. Acknowledging that "it is immaterial what kind of processes one chooses for such a definition of time," he told students that it was desirable to agree on one, and that light had certain distinct advantages over other means.[24] In Paris and afterward his argument defending the importance of light for determinations of time would grow stronger: the speed of light was a fundamental constant providing the necessary foundation for an "objective" definition of time.

Charles Nordmann followed the discussions in Paris closely. He had just published a widely read popular account of relativity, titled *Einstein et l'univers,* which sold some 45,000 copies and was translated into English. He had been one of Einstein's strongest supporters, but after hearing the discussions at the Collège de France, he was forced to add a new chapter, "Einstein à Paris," to later editions. He concluded his account with a note of caution: "There remains something infinitely troubling in the Einsteinian system." This was due to how it depended on a "certain conception about the propagation of light." Nordmann knew fully well that measurements of light's velocity showed that it was constant in all directions and independent of the motion of its source, but he also knew that this result depended on how scientists thought of time and length measurement units. He still considered that "all the science of Einstein, however coherent it may be, stands on a mystery, exactly like revealed religions."[25] After learning about Bergson's interventions, the astronomer became more and more interested in what the philosopher had to say. He would later write an entire book weighing in on the respective merits of each of the two men's perspectives, *The Tyranny of Time, Einstein or Bergson?*

In England, Whitehead was even more ambitious than Einstein. Bothered by how the new theory of the universe seemed too tightly connected to current signaling technologies, he developed his own version in a way that was no longer dependent on them. Contemporary technological limitations in signaling speed were the reason why the speed of light appeared as a constant in Einstein theory. That is why "we are driven to the *convention* that light, as our quickest system of signals, is moving with uniform velocity."[26] In his own development of relativity theory, Whitehead instead used a "critical velocity *c*," but one that was "defined without reference to the velocity of light."[27] If later experiments found transmission speeds faster than light, Einstein's theory would have to go, whereas Whitehead's could survive. One could merely use that newfound speed for the value of the critical velocity. Whitehead boasted that in his theory "light is given no privileged position."[28] Was it an improvement?

In Paris, a younger generation of historians and philosophers would reconcile aspects of relativity theory that many earlier commentators had considered to be contradictory. The controversy between Einstein and Bergson directly influenced a new generation of scientists and philosophers who started their careers sometime around the time of the April 6, 1922 debate.

Alexandre Koyré, a Russian émigré living in Paris who would become one of the most important historians of science of the century, stressed the dual role of light as "real" and "formal." He remarked on the "double role played by light in the theory of relativity": it was a "universal constant" and therefore "an ontological formal constituent of nature," but it was also a "real process of nature."[29] Bergson's mistake, he continued, arose from "not having noticed the *dual role* of light in the special theory of relativity." If only he had accepted it without objection, he would have admitted the "real character" of the changes described by the theory of relativity. Light was a thing too.

BACHELARD: "TALES OF PASSING TRAINS"

How are we to think of science, philosophy and the arts in relation to the things around us, including light? The year Einstein and Bergson met, Gaston Bachelard graduated with a degree in philosophy (*agrégé de philosophie*). He had previously been teaching physics and chemistry at the Collège de Bar-sur-Aube, a small medieval town in the northeast of France. There Einstein "awakened" him from a "dogmatic slumber" that—ironically—would lead him to accept "almost all" of Bergson's philosophy. Both of his thesis examiners for the title of Docteur ès lettres from the Sorbonne—Abel Rey and Léon Brunschvicg—were present during the debate and had been involved in the discussion in key ways. Would they urge students to continue to confront science? Or would they bend to its growing authority? If so, how would they mitigate the risk of having philosophy become increasingly irrelevant in comparison to it? Bergson had seen the problem early on and had warned his readers about it: "For wanting to prevent all conflict between science and philosophy, we have sacrificed philosophy without any appreciable gain to science."[30] Could one hope for a different outcome?

After graduating, Bachelard published a book on the value of relativity theory. In it he remarked on how current determinations of the speed of light, something so directly tied to contemporary experiments, emerged as a law of nature: "In effect, in relativity, the speed of light does not appear as a reality that was found through experiment, but rather as a reality affirmed by a law."[31] Bachelard appeared to have taken a side against Bergson and for Einstein. His *Essai sur la connaissance approchée* (1928), which was based on his doctoral dissertation, sought to rehabilitate the concept of quantity over that of quality—exactly the opposite task of Bergson's *Time and Free Will*. But his engagement with the work of both men was so thorough, and so profound, that his philosophy offered an alternative to both. Could one take the best of Einstein and the best of Bergson to understand the laws of the universe and how we have come to know them? Bachelard's initial "implacable" critique of Bergson gave way to a philosophy that took some parts of it and discarded others.[32] By the time *La Dialectique de la durée* appeared in 1936, the author described himself as an almost exact inversion of Bergson.[33]

Bachelard understood knowledge as much more than a utilitarian technique. He asked readers to accept the concrete in the abstract and the abstract in the concrete (arguing for the use of a single "abstract-concrete" concept). Like other intellectuals at the time, he was confounded by Einstein's use of thought experiments and by the proliferation of these beyond the confines of physics. "All the tales of passing trains which signal an observer standing in a station, of aviators who smoke cigars in lengthened or contracted periods of time—to what purpose are they?—or, more precisely, for whom are they designed?"[34] The answer was complicated. Bachelard did not think they were created for "those who have not understood" relativity nor were they conceived "for those who have understood" it already. Rather, they were necessary for enacting a total reconfiguration of a "space-time notion"—one that was not only limited to the realm of specialized science but that required a connection to general and even literary culture.

Bachelard responded to the Einstein-Bergson impasse by using all his insight and instinct against separating science from other areas of culture, by reincorporating in it the role of material culture, literature, and even poetry. For this reason, he would be remembered equally for his contributions to the philosophy of science, for his poetry, and for his theories of poetry. He not only asked us to think of science poetically, but to consider poetry scientifically, both approaches were equally important and complementary in many ways. Science had a poetic force and poetry an eerie connection to a truth.

Bachelard, who eventually became "the major emblematic figure" of French philosophy of science, started his career by considering Einstein and Bergson in a manner that significantly changed the common interpretations of their work.[35] In light of Bachelard's work, Bergson was described as the "last metaphysician"; Bachelard would be crowned as France's "first epistemologist."[36] "With Bergson there is a world that ends," noted a recent historian, who continued, asking "With Bachelard is there a world that begins?"[37]

CHAPTER 21

Clocks and Wristwatches

AT THE PARIS OBSERVATORY

It would be tempting to think that scientists could solve the debate on time by reference to what instruments measured—that cutting-age clocks and automatic recording devices could measure time and that scientists need not worry about it any longer; that they could finally conclude endless debates among philosophers and scientists.

In the sixteenth century, Charles V of Spain famously remarked that it was harder to govern clocks than it was to govern men. Bergson would probably have agreed. After all, he insisted that when scientists measured time, they removed from it what was most important, its flow and its relation to duration. Bergson's perspective on time measurement could not be more different from Einstein's. The philosopher was convinced about the importance of the unquantifiable aspects of time, whereas the physicist was equally convinced of the opposite.

"Before, it was the astronomer who surveyed the clock. . . . Now, it is the clock that frequently surveys the astronomer, and rectifies his results," explained the astronomer Charles Nordmann.[1] Could clocks demystify time? Could they keep philosophy, and messy human needs and concerns, at bay?

THE UNWINDING UNIVERSE

How was time actually determined at the time of the Einstein-Bergson debate? By the end of the nineteenth century, scientists' and philosophers' trust in clocks and the timekeeping and distribution networks connected to them had fallen to an unexpected low point—even as their use continued to increase. Although the time system in place worked well for practical purposes (except for the occasional train wreck due to clocking errors), almost nobody thought that a proper theory of time could be based on it. For most users, the time provided by service networks—the time that was used by the public at large—did not adequately reflect cosmological time and its passing. This method was practically and theoretically the best one scientists had. It worked, albeit imperfectly. Scientists who worked hard to establish it were as aware of its deficiencies just as much as the general public who used it and the philosophers who thought about it. A commentator on Einstein's visit claimed that the physicist had not succeeded in convincing his listeners of his new theory of time because of the simple fact that "under my eyes, at the time that I am writing these lines, the pneumatic clocks . . . proclaim respectively nine thirty-five, nine thirty, nine thirty-two."[2]

Since ancient times, time had been measured by using the stars. Time keeping methods based on the Earth's rotation against the fixed stars were commonly called sidereal clocks. If one considered the solar system as moving with a constant and stable velocity, then sidereal clocks based on it would be similarly constant and stable. The notion of the clockwork universe is usually traced back to Isaac Newton, who believed that the solar system, once wound up and set in motion by God, had simply continued ticking along with occasional help from the Almighty. During the Enlightenment, a clockwork-universe view remained prevalent. Clockwork was the perfect metaphor for a universe in which the future unfolded at a predictable and constant velocity. But the metaphor had its problems. On-the ground realities seemed quite different. Clocks ran fast and wound down. Why would the universe be any different?

By 1850 the idea of a stable clockwork universe was challenged on many fronts. The universe, just like a regular engine, appeared to be running out of steam. The repercussions were dramatic. They would lead—eventually—to the demise of the solar system and to the end of the world. Friction, even that of waves hitting the shore, acted as a brake on the Earth's rotation. Every time the Earth turned and every time waves splashed back and forth, the expenditure of the Earth's energy became irrecoverable.

"The tides must act as a brake on the Earth's rotation," wrote Bergson, who kept abreast of the most recent research. He came to the conclusion that the sidereal clock was simply too imperfect. Scientists needed a "new clock" and the "light clock" was the best alternative, he explained.[3] Bergson was hardly alone in acknowledging the deficiencies of previous timekeeping systems. Each year was getting shorter by 53 hundredths of a second. Careful observations of the moon's motion showed that sidereal days were getting longer. In a century they would be a quarter to a half an hour longer. Because of these effects, a perfect mechanical clock set in motion in 1800 would be 22 seconds ahead of a sidereal clock by 1900.[4] Astronomers also noticed that Encke's comet was orbiting the sun in slightly smaller and smaller ellipses, concluding that the changes in its orbit were due to the friction of a resisting medium that was slowing down the solar system. One astronomer recommended to "ban from our conversations" the saying "regular as the sun."[5]

These problems were precipitated by a sea change in how time was thought of across a wide range of disciplines. Natural historians, such as Charles Lyell, marshaled new geological evidence as proof that the universe was gradually evolving. The religious belief in an original flood followed by centuries of stability (a theory sometimes referred to as catastrophism) no longer appeared convincing to many. Following in the footsteps of these researchers, Charles Darwin famously found evidence of evolution not only in geology but, controversially, in living nature, including in humankind.

In physics and astronomy, a new temporal consciousness was linked to the science of thermodynamics and the law of entropy. Thermodynamics had important repercussions not only for ideas about time but also for practical timekeeping. Clocks wound down due to friction, losing some of the wonder that had previously surrounded them; describing God as a watchmaker no longer seemed like such a compliment.

Even the dimensions of the solar system were changing in step with the velocities of the orbits of the planets. The Earth-Sun distance, a unit that had been considered stable and fixed across time, was changing apace, slowly but surely. Scientists, engineers, philosophers—and even the public at large—were confronted with a more general, and more vexing, question: did natural standards even exist? The famous Michelson-Morley experiment was undertaken with such questions in mind. Would the Earth's orbital and rotational speed disturb time standards based on light waves?

TIME WITHIN

If changes in the sky above affected our units of time, why not try to go bottom-up? Could we not refer a standard unit to a portion of our bodies to measure the world and the universe? After all, colloquially, measures of length were based on body parts, such as the foot.

"Man is the measure of all things," claimed the Greek philosopher Protagoras in a widely cited phrase. During the Enlightenment, the philosopher John Locke related physical measures back to bodily standards in his *Essay Concerning Human Understanding* (1690). Our sense of physical time, he argued, arose from the temporality of our own bodily processes: "So that to me it seems, that the constant and regular succession of ideas in waking man, is, as it were, the measure and standard of all other successions." Locke's notion can be compared to Immanuel Kant's understanding of time and space as a priori concepts. In Kant's view, our sense of physical time and space arose from a certain primary proclivity of our minds to organize experience in these terms.

But the problems with basing a time standard on the body were many. Which body should be used? Why this one and not another? Would bodies change in proportion to other physical changes?

Individual differences marred time measurements. Values changed slightly depending on the person who obtained the measurement. Scientists proposed various techniques and improvements to mitigate these individual differences. In the seventeenth century, Ole Rømer, the Danish astronomer famous for his measurements of the speed of light, built an instrument to measure time by following the movement of the stars across the sky. Previously, observers had focused on the center of the Sun and tracked its movement, but scientists like Rømer quickly learned that different people estimated its center differently. Soon after this realization, astronomers started to measure time by reference to the stars instead of to the Sun. By finding a star that moved in the same direction as the Sun, but that could be easier to bisect, astronomers minimized some of these errors. By the late nineteenth century, astronomers at various observatories in Europe and beyond measured time in terms of the rotation of the Earth against the stars by bisecting a star using a mobile crosswire (called an "impersonal micrometer") that recorded its position electrically as it followed the star's movement. They then used the electrically recorded movement of the micrometer to automatically regulate clocks. But even then "a personal equation"—a term referring to these slight individual differences in an observer's assessment—influenced the results. Einstein and Bergson drew different conclusions from these slight discrepancies. While Einstein saw in them the reason for the untrustworthiness of psychological assessments of time, for Bergson they posed much more complicated questions about the relation of physics to psychology.[6]

THE BEST CLOCK IN THE WORLD: 1228 L

The international time network based in Paris depended on four clocks housed 27 meters below ground in the vaults of the observatory. They were caged in hermetic vaults to keep the pressure regular. Nobody could touch them. They were regulated electrically and connected to other clocks within the observatory. The best, most regular one of the four, was named 1228 L. "She" had ticked along since Christmas Eve of 1919 without apparent fatigue. Nobody touched "her" until three years later, when her electric contacts were carefully cleaned. Although the most predictable, astronomers found that on occasion even she "changes sometimes, and without an apparent reason, brusquely from 3 to 6 hundredths of a second."[7] The four clocks were set on time with an impersonal micrometer that followed the rotation of Earth vis-à-vis the stars. 1228 L seemed to keep time so well—even better than the selfsame astronomer who took it from the stars—that it introduced a strange power reversal between man and machine.

Not all scientists agreed with the system. Even within Paris, competing systems were plainly evident. In 1922 the clock at the Paris Observatory and those regulating the city disagreed by 9 minutes and 21 seconds. While the official city clocks marked Greenwich time, the Paris Observatory remained firmly against the time sent by their competition across the channel. If the astronomers, who historically had valiantly fought against the well-endowed Royal Greenwich Observatory, gave in, this would represent a "scientific Waterloo" and a sore national "abdication" on the part of France.[8] But the Paris Observatory and the city were not the only ones who disagreed. Railroad time was sent from the city of Rouen, where the directorship of the main railway system was located. Thus, in Paris, the time inside a railway station differed from that outside by 5 minutes.

"THE OLD SONG WITH THE OLD MOTIF"

In his 1905 relativity paper, Einstein described a typical way of understanding time: "If, for example, I say that 'the train arrives here at 7 o'clock,' that means, more or less, 'the pointing of the small hand of my clock to 7 and the arrival of the train are simultaneous events.'"[9] Coordinated clocks and watches were everywhere when Einstein wrote these lines. In this simple way, Einstein defined the nature of simultaneity and time simply and succinctly for the scientist. This definition may seem uncontroversial enough, but in 1922, it was anything but that.

Why did Einstein understand time in such a basic, procedural way? Part of the answer could be found on Einstein's own left wrist. The physicist had been using the same silver watch for more than a decade—he trusted the little instrument like he probably trusted no one else. Einstein had obtained this watch around the time he was fourteen years old and wore it "for a good 28 years." In the fall of 1921, when it was "still running as excellently as on the first day," he decided to pass it on to his eldest son, Hans Albert, who had just turned seventeen.[10] Einstein must have soon thereafter bought a new wristwatch. A photograph taken during his trip to Japan after the Paris meeting captured him with a watch on his left wrist. The instrument would have probably remained hidden under his sleeve had Einstein not extended his arm in order to get a drink. A news photographer released his fast shutter just at this very photo-opportune moment.

Wristwatches started to become common in the 1920s, when the public at large adopted the fashion of soldiers who had found the pocket version impractical and cumbersome during the First World War. In contrast to pocket watches, they could be consulted more frequently and rapidly than when they were kept inside coats and pockets. They were usually place on the left hand, so that they could be easily wound up with the right hand without the need to remove them. "The very determining of time," noted Heidegger, "should claim as little time as possible."[11] Wristwatches met this goal.

By the time Einstein and Bergson debated, two different ways of noting time, one human and subjective and the other one clocklike and objective, were widely noted. Subjective understandings of time were frequently related to meaningful moments, places, and events, which were increasingly described in literary and poetic ways. With the spread of pocket and wristwatches, these differences became even more frequently noted. In 1922 Kafka, in his private diary, described an "inner clock" and an "outer one," which did not agree:

> It is impossible to sleep, impossible to wake, impossible to bear life or, more precisely, the successiveness of life. The clocks don't agree. The inner one rushes along in a devilish or demonic—in any case, inhuman—way while the outer one goes, falteringly, its accustomed pace.[12]

The proliferation of clocklike technologies during the twentieth century exacerbated the differences between these two senses of time.

Throughout his life Einstein noted these two ways of experiencing time. He considered his external life to be regulated somewhat like a clock. To Pauline Winteler, a family friend with whom Einstein boarded while he finished secondary school, he explained: "There is very little that is of interest in my external life: in fact, the latter is so philistine that people could use it for setting their watches—except that their watches would be somewhat late in the morning." This clocklike regularity, however, characterized only his "external life."[13]

Einstein's personal correspondence up to the time of his death included detailed statements about how he experienced time. Time flew by for Einstein, and he was unable to do a proper accounting of it. He called this inability "the old song with the old motif and many variations" explaining to a friend: "The days and weeks slip by just so, without my noticing it except toward the end of the month . . . the old song with the old motif and many variations, a few of which you too probably know how to sing."[14] A few months later he again expressed how "my time slips by."[15] Comments to this end continued throughout his life.

Well into the first decades of the twentieth century, clocks competed against alternative time-measurement techniques. In the nineteenth century, one could use the time taken to recite an Ave Maria as a measure of time or, more profanely, simply refer to "a pissing while."[16] When Bergson wrote about our sense of time, he described it in terms reminiscent of older timekeeping methods—techniques that were only and with great difficulty and resistance replaced by clocks. His descriptions were varied, pointing to the multiple possibilities available for understanding time. Mathematical time elided the differences between instants (tied to experiences, feelings, memories, and meaning) by considering all moments on the same plane. He sought to recover significant ones. "If I want to mix a glass of sugar and water, I must, willy-nilly, wait until the sugar melts," he explained. "This little fact is big with meaning," he added. Why? "The time I have to wait," he explained, "is not that mathematical time." Rather it was a notion of time that "coincides with my impatience." That time "is no longer something thought, but lived." Depending on how impatient one was, one could chose to stir and thus see time "bite" into the future.[17] By focusing on apparently insignificant moments, Bergson showed how even those instants could have an actual impact in shaping time to come. Time would cease to be just another independent variable: "*modern science must be defined pre-eminently by its aspiration to take time as an independent variable.*" Yet that mathematical and scientific *t* variable was very different from actual time. "But with what time has it to do?" he asked.[18]

"THE FIRST AUTOMATIC DEVICE TO BE USED FOR PRACTICAL PURPOSES"

Humans and clocks had been compared against each other for centuries. Clock beats were measured against heartbeats, and vice versa. Galileo used the time of his pulse and heartbeat to determine a pendulum beat, not the other way around. In the work of René Descartes, mechanical bodily functions were also compared to clockwork technologies. In a letter to the Marquis of Newcastle, he equated the mechanical mechanisms of living bodies to those of clocks: "They act naturally and by springs, just like a clock." But for Descartes, it was clear that clocks were better timekeepers than human *judgment*. "A clock," he explained, "shows time in a better way than what our own judgment reveals."[19]

During the Industrial Revolution clocks became increasingly popular in industrial settings. According to Marx, they were essential for establishing new factory production methods. "The clock was the first automatic device to be used for practical purposes, and from it the whole theory of the production of regular motion evolved," he explained to his collaborator Friedrich Engels.[20] During these years, the use of clocks for regulating labor was violently contested by numerous craft workers and Luddites, who opposed the mechanical routine meted out by owners. A worker's sense of time and the time marked by a manager's clock competed forcefully.

Bergson refused to accept the tyranny of clocks. The philosopher argued against Einstein by considering clocks as servants in *Duration and Simultaneity*. They were meant to "serve us," he insisted. Forgetting their subordinate status could have dire consequences for philosophy and for science.

Under what circumstances could clocks be considered servants? Certain clocks started to be considered "servant" machines in the seventeenth century, when the German philosopher Gottfried Leibniz described new machines that he compared against older types in terms of the differences between slaves and servants. He contended that certain new instruments, which he called servant automatons, were unique: "There are [automaton] servants so well primed that they do not need signs. They get in ahead of them. Chiming watches, for example, and alarm clocks are servants of this kind. Far from waiting for signs from us, they give signs to us." Although they were meant to serve, very much like an engine would, a machine of this sort was not slavish. It "would not alter itself to fit with its master's thoughts."[21] The benefit of an alarm clock, for example, resided in that it would awaken its user even when its master would very much prefer to continue to sleep. But how could we evaluate the benefits of clocks when they were not used willingly as helpful servants but were instead forcefully imposed by masters regulating factory workers?

Bergson and Einstein were exposed to similar timekeeping technologies, yet they experienced and described them differently. As networks for determining and distributing time became ever more pervasive during the twentieth century, the two men (and their respective advocates and enemies) appeared even more divided in their understanding of time.

While Einstein, Bergson, and their interlocutors referred repeatedly to clocks and watches, the topic of *coordinating* clocks at a distance only created further divisions. Wristwatches were set on time by consumers who depended on an expanded network of coordinated clocks. Both men agreed on the pragmatic details of clock coordination. They concurred on how it should best be done and why it was needed. Yet Einstein and Bergson differed about the *significance* of the most common and best procedure. Until the last decades of the twentieth century, Einstein's special theory of relativity suffered from a persistent criticism that considered it as *procedural* and *technical* rather than universal. Clocks frequently ran late, and this did not mean that time was slowing down. "It is not because clocks go more slowly that time has lengthened; it is because time has lengthened that clocks, remaining as they are, are found to run more slowly," explained Bergson in *Duration and Simultaneity*.[22]

WIRELESS TIME

"Suppose that at a time T_0 a light ray is sent from the origin of the system," and that this light ray is used to synchronize a distant clock.[23] What time would the clock mark? Consider the velocity of this light signal as constant. Einstein was nearly done with his argument. With a few additional simple calculations, he arrived at one of the most astounding claims of his theory: that a clock in motion would mark time differently than a stationary one. In practice, distant clocks, connected via radio wireless signals, were coordinated in exactly the manner described by Einstein in his famous paper on relativity.

Bergson described the same clock coordination procedure as Einstein. Clocks are set "by an exchange of optical signals, typically electromagnetic," he explained during his meeting with the physicist.[24] Bergson, following Poincaré and others, argued that Einstein offered a particular *procedure* of time-coordination based on a number of *conventions*. Einstein, instead, argued that it was, "objectively" time.

Well into the twentieth century, Einstein's critics accused him of defining time by reference to a practical principle, one "based on the *practice* of physicists, that the physicist's judgment of simultaneity should be determined by the characteristics of light." Was this redefinition justified? This "point has been something of a 'cause célèbre' in recent years," explained a writer intent upon weighing in on the question.[25]

Paris was at the very center of a rapidly expanding time-coordination network in which astronomers sent time to the world via radio waves.[26] How did new radio-based signaling technologies affect the meaning of time? During the meeting on April 6, 1922, accurate time was determined via a complex timekeeping and distribution network that relied on many assumptions, including presuppositions about the behavior, constancy, and speed of light in relation to the Earth's rotation and orbital velocity.

In 1887, the German physicist Heinrich Hertz had seen a spark jump as if by magic in a coil of wire when he passed electricity through another, close-by, coil. Did the electricity traveling through one coil cause the spark to go off in the other one? What caused these strange at-a-distance effects? Einstein, who was a young boy of fifteen at the time of Hertz's discovery, had clear reasons for concern. Would his family's electric distribution business, which was based on transmission by wires, be imperiled or aided by these new technologies? When he was only sixteen, he wrote his *first* research paper after reading the "reports in the popular science magazines about Heinrich Hertz's recent discoveries" on wireless transmission.[27]

"The bewildering possibility of telegraphy without wires, posts, cables, or any of our present costly appliances" intrigued a growing number of scientists and engineers, and researchers worldwide raced to explore the new effects.[28] In 1894 Guglielmo Marconi transmitted dots and dashes across his garden. At first, the label "wireless telegraphy" seemed most appropriate, but soon researchers started to note important differences with traditional telegraphy. Like the semaphore, wireless was based on the transmission of light, but unlike the semaphore, the signal could not be seen by the naked eye. Like telegraphy, wireless involved electricity, but unlike telegraphy, transmission occurred without cables. Like X-rays, wireless was able to penetrate walls and travel through bones, barriers, and all sorts of obstructions. Édouard Branly, a Parisian researcher working at the Catholic Institute in Paris, coined the word "radio" from the Latin *radius*, meaning "ray of light." What would happen if traditional timekeeping methods were combined with these new technologies? How would our understanding of time change?

The potential of early wireless telegraphy was at first limited to sending and receiving simple signals across short distances. Its use was primarily military.[29] At the time Einstein published his famous paper, wireless transmission was still in its infancy, but it soon became a practical means for sending *time* signals.

In 1908 new antennas placed at the top of the Eiffel Tower increased the range of wireless transmission from 400 km to 6000 km. Experiments on the transfer of voice started soon thereafter. Civilian use of wireless time signals sent from the Eiffel Tower became regular in Paris after 1910. With the invention of the triode (three-electrode lamp), the transmission of radio waves improved so much that by 1922 it was possible to reach north and south to the poles and even get the waves to Australia. The increase use of telegraph and wireless prompted investigations into the time of transmission taken by light signals to reach a distant destination. Before the advent of those technologies, the speed of light concerned only astronomers, who, accustomed to dealing with large distances, were forced to take its transmission time into consideration. When in the 1850s clocks started to be coordinated by telegraph wires across longer distances, scientists wondered if they needed to take into consideration the time of signal transmission. Investigations into transmission delays became paramount, especially when it was time that was being sent to and fro.

One of the advantages of coordinating clocks via light signals was due to light's tremendous speed—so fast than most instruments could not even measure it. For this reason, scientists and engineers could assume that it was infinitely fast, and they often did. They had no evidence that faster means of signal transmission existed; they had no evidence of it being affected by the motion of the emitting body; they had no evidence of it changing if the waves were directed north or south instead of east or west. They could consider it as constant in all directions. Were these assumptions justified? The time that was sent could not possibly be exactly the same as the time that arrived. Or could it? An eastbound time signal was bound to arrive at its destination faster than a westbound one, as the Earth's rotation brought its eastern target closer while it dragged the western target farther away. Or not? How did time need to be adjusted?

Scientists expected that the speed of the Earth's velocity might be a distorting factor in the time distribution network. Max Planck, the scientist responsible for publishing Einstein's seminal 1905 article, explained how the speed of the Earth's rotation affected this time system. "A time signal is sent out from a central station such as the Eiffel Tower by means of wireless telegraphy, as proposed in the projected international time service," he began. If the Earth was considered stationary, then stations at the same distance from the central station should receive the signal at the same time as the others, with an adjustable delay due to travel time. But if scientists considered the Earth in motion, "it is clear that those stations which, seen from the central station, lay in the direction of the Earth's motion, will receive the signal later than those lying in the opposite direction, for the former move away from the oncoming light waves and must be overtaken by them, while the latter move to meet the waves."[30] Ideally, scientists sending time from astronomical observatories needed to calculate delays in time signals sent across the globe. For this they needed to know the exact speed of light, the velocity of the Earth, and the effects of any ether wind that could retard a wireless signal. Until these questions were solved, this system could hardly be considered perfect. Yet scientists continued sending and receiving time signals while the public continued to set their clocks on time.

In practice, possible delays due to the finite velocity of the speed of light and of any effects on it coming from the Earth's velocity were simply ignored. A time signal traveling at the speed of light would be delayed only by a minimal fifteenth of a second when traveling half way around the globe. Reichenbach, who had been drafted into the German radio corps at the Russian front during the Great War, explained how radio engineers considered this delay as negligible. Knowing fully well that "the time it takes the wave to travel from Nauen to New York is only about 1/50 second," he explained to his readers why this short delay was frequently ignored. It was such a small quantity that "for that reason, we do not need to take into consideration this time, and we can say with good sense that the waves arrive at the same time in New York that they are sent in Nauen."[31] His experience during the war had taught him that light was special: "Only the speed of light is so great."[32]

CHAPTER 22

Telegraph, Telephone, and Radio

How did advances in electromagnetic communication technologies inform Bergson's understanding of Einstein's work? Despite fully accepting "the invariance of the electromagnetic equations," which were frequently seen as the pillars of relativity theory, Bergson did not agree with Einstein's conclusions.[1] He simply did not believe that these equations, even though tied to recent discoveries and contemporary technologies, should lead scientists to adopt Einstein's interpretation.

The philosopher carefully considered the connection between a stationary observer and a traveling one in terms of electromagnetic communications. He imagined what a dialogue between "Peter" and "Paul" would be, as they sped away from each other. Bergson wrote it down, as if it were a script. Peter, in Bergson's account of the story, says to Paul: "The moment you separated from me . . . your time swelled, your clocks disagreed." According to Bergson, it was "obvious," what "Paul would reply": that everything was normal for him (Paul) and that it was Peter's system that had gone awry.[2]

The conversation between the two observers, as narrated by Bergson, led nowhere. It consisted in back-and-forth repetition between the travelers. It was characterized by misunderstanding and mistrust. For this reason, it was senseless to defend Einstein's theory by reference to it. Yes, he understood the connection between two clocks in terms of the exchange of electromagnetic signals. "How do we synchronize two clocks located at different places?" Two operators in charge of setting the clocks "communicate" with each other about the time by means of "optical signals, or more generally, electromagnetic ones," he explained. "A person in O sends a person in A a ray of light destined to be returned back," he continued. This procedure was equivalent to that of the Michelson-Morley experiment "with the difference, however, that mirrors have been replaced by people."[3] But neither of these two cases (one based on two twins communicating with each other at a distance and the other one on the Michelson-Morley's experiment), according to the philosopher, led to Einstein's conclusions.

To fully investigate the topics of his interest—of how science migrated from the concrete to the abstract, and of how "a mathematical representation was transformed into transcendental reality"—Bergson stressed aspects of the twin paradox that could *not* be explained simply by recourse to new electromagnetic technologies.[4] He did not believe that the common technical account of the twins' reciprocal time dilation effects solved the questions at stake.

Could the debate between Einstein and Bergson be solved by reference to new technologies for long-distance communication? When discussing the validity of Einstein's work and the reasons why time could be defined by the behavior of light waves, commentators referenced a bevy of new light-based technologies populating the world around them. They evaluated Bergson's arguments against Einstein in terms of telegraphy, telephone, and radio. Some argued that these technologies showed the need to accept Einstein's theory. They even imagined new improvements, such as television, to prove Bergson wrong. But others thought of these questions in terms of the possibility of a *meaningful* connection between two observers at a distance from each other. For the first group, electromagnetic communication technologies sufficed as examples showing the validity of Einstein's work; for the second, they were sorely deficient. New technologies and references to them only exacerbated the debate.

Why light? Light was not only used for illumination—it was also, and primarily, used for communication. Since ancient times, people at a distance had sent signals across space by using torches. Code systems were later developed for maritime and military communications. The semaphore, primarily a military technology, was used in the eighteenth century to send all the letters of the alphabet as well as numbers across long distances via visual signals. With the development of telegraphy, these complex sign systems were reduced to simple dots and dashes characteristic of the Morse telegraph system. By the time Einstein wrote his paper, light signals were optical (torches and semaphores), electrical (telegraphs), and electromagnetic (wireless). They were used not only to set clocks at a distance, but also and more important, to send voice, audio, images, and even money. The answer to the question of *what time is* changed as technologies for the transmission of light signals proliferated and were transformed from technologies for sending simple signals, including time, to those used for wider communication. Scientists and philosophers saw the meaning of time itself change along with these media transformations. The discussion between Einstein and Bergson became a disagreement about the nature of long-distance communication more generally. Physicists' and philosophers' understanding of communication at a distance only grew further and further apart.

TELEVISION AND WIRELESS: LANGEVIN AND BECQUEREL

In his colorful presentation of relativity in Bologna of the "voyager on a rocket ship" now known as the twin paradox, Paul Langevin thought of new ways in which the two clocks could be compared while remaining at a distance from each other: "It is fun to imagine how our explorer and the planet Earth would see each other mutually live, if they could stay in constant communication by light signals or by wireless telegraphy, during separation, and thus understand how the asymmetry between two measures of time is possible." These speculations made sense in light of the surge in the development of wireless technology from 1905 (the date of Einstein's paper) to 1911 (the date of Langevin's). Langevin "imagined" a scenario where a voyager sent on a rocket ship traveling at a speed close to that of light and a stationary observer would communicate wirelessly, using "hertzian signals," or wireless telegraphy. He also imagined a scene, anticipatory of *television*, where "our explorer and the Earth could see each other live," through the exchange of luminous signals.[5] For Langevin, the possibility of seeing or communicating the effects of time dilation was used to illustrate the reality of relativistic effects. Langevin imagined ways not only of sending, receiving, and comparing time signals but also of actually seeing any temporal processes dilate.

To explain *time dilation* in his 1905 publication, Einstein had to imagine what would happen if one of the two clocks in his theory would be "transported" back to meet the other one. Although he did not yet have the tools to account for the acceleration necessary for changing the direction of motion, he nonetheless ventured to claim that the clock would slow down during the voyage and that it would be behind the other. Half a decade later he no longer had to think about actually transporting the traveling clock back to Earth. Langevin's explanation, which relied on a technology similar to television decades before it was invented, showed him ways of thinking about time dilation without having to include the topic of acceleration into the discussion.

The physicist Jean Becquerel used the example of radio to prove Bergson wrong. Ultimate proof of Einstein's theory, and of the equal validity of different times, Becquerel argued, lay in the possibility of exchanging time signals via telegraph or wireless signals. He used the example of two fictional observers named Pierre and Paul, famous characters of the twin paradox. According to the theory of relativity, if one twin would travel outside of Earth close to the speed of light his clock would run slower compared to that of the twin who remains on Earth. But how could the comparison actually be made, asked a number of skeptics? Becquerel proposed a solution: "One can imagine, for example, that Pierre sends to Paul, minute by minute, wireless telegraphy signals that indicate to Paul the hour of the clock."[6] In his attack on Bergson and defense of Einstein, Becquerel described two cases that would prove the reality of time dilation, both of which led to the same conclusion. One of them described the whole trajectory traveled by the speedy twin strewn by clocks synchronized to Earth through electromagnetic waves that could then be used to compare at every moment his watch against that of the stationary clock. In this way, he could see his clock slowing down vis-à-vis that of the twin remaining on Earth. The second scenario involved the twins exchanging time signals by electromagnetic means (telegraph without wires): "T.S.F [*telegraph sans fils*] signals."

The possibility of actually comparing the differing times of a moving clock against a stationary one became an important proof of Einstein's relativity theory. For the most part, that proof eliminated preexisting doubts that the effects described by the theory would not affect time in general. References to wireless signals in discussions of the twin paradox proved to many, after Becquerel's formulation, that they were clear examples proving why the time of the traveling clock should not be considered as secondary. Becquerel disagreed with Bergson's interpretation, arguing that both clock times were equally real and therefore both twins should be considered equally live and conscious.

Many commentators followed Becquerel in understanding Einstein's theory by reference to contemporary telecommunications technologies. Some of them were directly involved with these technologies, such as Poincaré, Langevin, de Broglie, and Reichenbach. Poincaré was a professor at the École professionnelle supérieure des postes et télégraphes. In important articles published in the technical journal *Éclairage électrique*, he made "the telegraphers' equation" mathematically more "general." He developed techniques so that scientists and engineers could deal with "all the cases" involving "receiving apparatus" that reacted to changing "limit conditions." Research published in 1904 allowed him to work with telegraphy, telephony, and wireless with the same mathematical tools.[7]

A number of prominent thinkers understood Einstein's work by imagining a very distant telephone communication—so distant that the delay of the transmission would no longer be negligible. In a symposium where Whitehead was asked to be the final respondent, the philosopher Wildon Carr described the "time of transmission" in the theory of relativity by reference to a telephone conversation: "Suppose two persons A and B conversing with one another by telephone over a distance, say, 500 miles, for a duration of say, three minutes." Both people would have conversed for a total of three minutes, which could be expected. But another counterintuitive effect also followed: "But the three minutes lived by A is not simultaneous with the three minutes lived by B because the transmission of over five hundred miles of connecting wire time."[8] Even more astounding was the additional claim that there was no way of knowing what absolute simultaneity would mean. Absolute simultaneity could not be established either between the two people on the phone or in any other case.

Reichenbach used the example of the "telephone" and "radio telephone" to illustrate how we could grow "accustomed" to the reality described by Einstein. "If a telephone connection with the planet Mars were established," he explained, "we would have to wait a quarter of an hour for the answer to our questions." If our communication technologies functioned with a similar delay in that way, then "the relativity of simultaneity would become as trivial a matter as the time difference between the standard times of different time zones today."[9]

By the end of the First World War, commentators on Einstein's theory did not need to have been working directly with communication technologies in order to see it in those terms. Western Union radio clocks, according to one philosopher, had proven that one of Bergson's strongest defenders, Alfred N. Whitehead, was mistaken: "Any one who checks his clocks by radio is determining simultaneity at a distance in this [Einstein's] way." Did Einstein's critic have an excuse? "If it be objected that when this statement was made radio was not in very general use, the reply is" simply no. Why? Because "'Western Union clocks' have been in use in America for more than twenty-two years."[10]

BERGSON'S CONNECTED WORLD

In contrast to Einstein, Langevin, Becquerel, and many other scientists who stressed how the effects of relativity could be explained simply by reference to contemporary communication technologies, Bergson did not think that any of these examples led to those conclusions. Hidden within their technical explanations lay certain assumptions about what meaningful communication really was. Debates about Einstein's work, he argued, could not be solved by recourse to technical explanations involving procedures for sending and receiving "optical signals."[11]

In 1913, when the philosopher was invited to become president of the Society for Psychical Research in London, he wrote one of his strongest statements defending the possibility of some sort of telepathy or clairvoyance between observers at a distance. Bergson explained how even as simple a fact as their interest in having him speak there proved that the span "across the two hundred and fifty miles of space" between Paris and London was not a barrier preventing a different sort of nonphysical and indirect connection to occur between them.

> I suspect that there is in this a case of telepathy or clairvoyance, that you felt from afar the interest I was taking in your researches, and that you perceived me, across the two hundred and fifty miles of space, attentively reading your *Proceedings* and following with keen curiosity your work.[12]

Bergson was very clear that he did *not* understand by this reference to "telepathy or clairvoyance" a type of communication that was comparable to either face-to-face, correspondence, telegraph, telephone, or any other form of communication that involved only the simple causal transmission of a message. It was based on other types of relations, such as environmental similarities, fortuitous resonances, and circumstances that led to the possibility of having different people in different places have similar thoughts. Much of his work was concerned with describing the power of our minds to travel to realms beyond the reach of causal communication understood narrowly, to imaginary, symbolic or virtual worlds, and back. He then brought these insights to bear on a much narrower and more common understanding of communication. Did not all forms of meaningful communication, even face-to-face, one-on-one, and of the simplest possible kind, have to include imagination and interpretation? What made communication meaningful?

During these years, telepathy was often invoked as an example of forms of communication that could not be explained in terms of the simple causal transmission of signals. But its meaning changed rapidly. When compared directly against new communication technologies based on electromagnetic transmission, telepathy was for the most part discredited. Numerous experiments had clearly disproved it. Yet many scientifically literate researchers continued to believe that not *all* forms of communication should be considered in these limited terms. Bergson, who was well aware of the limitations of telepathy, at times invoked it in a broader sense: to elucidate aspects of communication that could not be understood by reference to a simple causal-transmission model. Scientists' accounts of time dilation often took for granted the transfer of meaning.

In *Duration and Simultaneity* Bergson considered the case where the twins' could be seen by "a supreme consciousness" who was capable of "communicating telepathically with both." That "consciousness" would indeed see the effects of time dilation. But "from the point of view of physics, that argument does not count," since "no message could be transmitted, no causality could be exercised at a speed faster than that of light."[13] Throughout the rest of his book, Bergson explained how neither electromagnetic communication nor telepathy in the usual sense of the word nor positing "a supreme consciousness" would lead to Einstein's conclusions. He discounted the first case because it assumed a concept of communication so narrow that it sidestepped questions of meaning and agreement. Paul and Peter could go on disagreeing forever, never settling on the "paradoxes" of relativity, namely the "multiple times that flow more or less rapidly, upon simultaneities that become successions, and successions simultaneities, whenever we change our point of view."[14] He discounted the other two because they clearly fell outside the realm of contemporary science.

After he discounted these three cases, he tried a completely different rhetorical strategy in order to explain the "oddities that have led so many minds astray" and that so jolted the public's imagination.[15] He asked what would happen if one imagined that the stationary observer and the traveler were *completely* separated from each other. He underlined the differences between them to the point that he only referred to one of them as "live and conscious" while the other was reduced to a "mirage," a "midget," a "phantom," or a simple "fiction." Why? This portrayal, in his view, would better reveal how the paradoxes of relativity arose. They showed how inadequate Einstein's theory was if understood solely in terms of communication technologies and also how philosophically insufficient it was if considered in their complete absence. As Einstein jumped back and forth from the abstract world of mathematics to the concrete world of science and technology to that of the imagination, Bergson urged caution. "The philosopher, who should distinguish the real from the symbolic, must speak in a different way," he concluded.[16]

The examples of the telegraph, telephone, and radio created further rifts between Einstein and Bergson and their respective supporters. These technologies, ostensibly created to reduce distances between individuals and foster agreement by permitting communication, produced a completely different effect. The writer Franz Kafka was one of many writers who eloquently noted how their drawbacks often outweighed their benefits. In the same year Einstein met Bergson, he compared light-based communications technologies to traditional ones. A new system based on "the telegraph, the telephone, [and] the radiograph" replaced old practices of "writing letters" that could only be sent by relying on transportation networks such as "the railway, the motor car, the airplane." Kafka, like nearly everyone else, had hoped that these new technological innovations would join with previous ones finally creating "a natural communication" in which "the peace of souls" would reign. But an opposite effect surfaced, nourishing more and more "ghosts" that disturbed communication and prevented "kisses" from reaching their destination. In a personal letter to his lover, he explained:

> Writing letters, however, means to denude oneself before the ghosts, something for which they greedily wait. Written kisses don't reach their destination, rather they are drunk on the way by the ghosts. It is on this ample nourishment that they multiply so enormously. Humanity senses this and fights against it and in order to eliminate as far as possible the ghostly element between people and to create a natural communication, the peace of souls, it has invented the railway, the motor car, the airplane. But it's no longer any good, these are evidently inventions being made at the moment of crashing. The opposing side is so much calmer and stronger; after the postal service it has invented the telegraph, the telephone, the radiograph. The ghosts won't starve, but we will perish.[17]

In Kafka's world, one that was populated by the same technologies surrounding Einstein and Bergson, new means of electromagnetic communications seemed to foster an ever-growing separation between individuals. In this very same context, physicists' and philosophers' understanding of simultaneity and of communication "at a distance" grew further and further apart.

CHAPTER 23

Atoms and Molecules

Timing is everything. Nineteen five was Einstein's annus mirabilis and the date when—alongside his revolutionary relativity paper—he published at least three other remarkable texts. The second one dealt with molecular motions. This paper was so important that the physicist Max Born later remembered how "at the time atoms and molecules were still far from being regarded as real." Einstein's work changed the landscape. "These investigations of Einstein," continued Born, "have done more than any other work to convince physicists of the reality of atoms and molecules."[1] How was the atomic view of nature defended by Einstein connected to a particular notion of time?

"We do not walk backward or digest before we eat," noted the philosopher Émile Meyerson on April 6, 1922.[2] With this apparently self-evident assertion, Meyerson proceeded to ask the physicist a more difficult question. How did the theory of relativity account for processes, such as eating and digesting, which always took place in a certain order? Meyerson's comment stemmed from a much more general impression that Einstein's theory did not adequately explain our sense of the irreversibility of time.

Although Einstein did not deny that "orientational sense of time" was important for us, he repeatedly wrote it off as a phenomena disconnected from the fundamental laws of nature: "It is entirely correct that this temporal basis of events finds no expression in the fundamental laws we use as a basis."[3] Could Bergson explain what these fundamental laws did not? Einstein's friend, the mathematician Hermann Weyl, who would eventually join him in Princeton, explained how Bergson's philosophy focused on the flow of time and our resistance to it:

> Order and organization are the hallmarks of life. This gives the impression that life as it develops on Earth resists falling into the pit of heat death imposed by the entropy of inorganic matter. Bergson coined the great word *élan vital* for this resistance.[4]

"The 'timelessness' of Einsteinian physics" and "the time-obsessed flux of Bergson" have "gained an undisputed ascendancy in the intellectual world," noted Wyndham Lewis.[5] How could such contradictory perspectives both be right?

Einstein's theory was perfectly consistent with physical theories that explained why certain events occurred only in a specific order. It was also perfectly consistent with psychological theories that explained our sense of the flow of time. But on its own, it described a universe where our sense of time passing by us was an illusion. According to Einstein and many of his supporters, the universe at its most basic molecular level was perfectly reversible. Sure, physicists admitted that at *our* scale—the macroscopic scale—events *were* irreversible for us, but this did not mean that its elementary processes could not be reversed. Irreversibility was only a statistical effect affecting aggregate particles. Under the statistical explanation defended by most physicists at the time, it was possible (although unlikely) to "walk backward or digest before we eat." Einstein and Bergson accepted the statistical interpretation of our sense of time passing by, but while for Einstein it entirely sufficed as an explanation of all phenomena—including our lives, for Bergson it was wholly insufficient and begged a complement.

THERMODYNAMICS AND REVERSIBILITY

We may think that atoms were an established concept in physics, but atomic theory was contested on many fronts well into the first decades of the twentieth century. Physicists of the caliber of Ernst Mach, Pierre Duhem, and Wilhelm Ostwald refused to accept it partly because atoms could not be seen. Why posit unobservable atomic entities while other approaches, such as Ostwald's "energeticism," which was based entirely on visible effects, could be used? One of the reasons why atoms became so controversial during this period had to do with how they were used to answer the broader question of why time seemed to flow in one direction, the "arrow of time."

From the middle of the nineteenth century to its last decade—for more than half a century—scientists had struggled to understand two related yet seemingly contradictory phenomena. The first one was related to the first law of thermodynamics, the principle that energy was neither created nor destroyed but simply transformed. The second one was related to the second law of thermodynamics, through which we know that work and heat processes flow in a certain direction.

The first law suited reversible phenomena; the second explained irreversible processes. For many years, they seemed to contradict each other. How could the first law claim that energy was neither created nor destroyed, while the second one described dissipation? Scientists noticed a paradox that they soon baptized the reversibility paradox. James Joule, a British gentleman who ran his family brewing business and whose first experiments on electricity involved giving electric shocks to his servants, developed the first law with precision. Joule's famous experiments consisted in submerging a paddled wheel into an insulated barrel of water and measuring the increase in temperature in correlation to the moving wheel: "I shall lose no time in repeating and extending these experiments, being satisfied that the grand agents of nature are, by the Creator's fiat, indestructible; and that wherever mechanical force is expended, an exact equivalent of heat is always obtained."[6]

The second law developed out of the work of the French scientist Sadi Carnot, who experimented with steam engines and looked for ways to improve their efficiency. Carnot related engine efficiency to the amount of heat that was lost in the process, regardless of the materials used to power the engine. Carnot's theory could be used to explain why perpetual motion machines could never be built. It contrasted starkly with Joule's, in that Carnot's work focused on lost heat, or heat dissipation.

In classical Newtonian physics, mechanics was perfectly reversible. The same mathematical techniques explained the pulling or pushing of a pulley, the lifting and lowering of a lever. If explained solely in terms derived from classical mechanics, the work performed by an engine would be perfectly reversible; but in practice, it was not. The proliferation of engines after the dawn of the Industrial Revolution made it increasingly necessary for scientists to focus on irreversible processes and to develop new theories to account for them.

Many scientists, most prominently among them Rudolf Clausius, William Thomson, and Hermann Helmholtz, tried to reconcile these two seemingly opposing laws, but the relation between the two laws remained unclear and frequently contested until the twentieth century. How could Joule's experiments on energy conservation be reconciled with Carnot's theory of dissipation? The reversibility paradox, as formulated by William Thomson and Josef Loschmidt, stumped scientists for nearly half a century.

James Clerk Maxwell, Josiah Willard Gibbs, and Ludwig Boltzmann developed a molecular theory of heat that explained the processes of heat diffusion using the reversible laws of classical mechanics. Did heat flow always only in one direction? Statistically speaking, yes. If heat was explained in terms of time-reversible mechanics, how could a physicist explain its directionality? Boltzmann clarified these paradoxes in a probabilistic fashion, giving to the second law a statistical interpretation. If successful, his work would solve not only one of the central riddles of physics, that of reconciling reversible mechanics and increasing entropy, but also provide the basis for a kinetic and molecular theory of heat, which would displace the old and problematic caloric theory by explaining it instead in terms of the energy of particles in motion.

If the second law was only statistical, where were the exceptions? Where were those cases when events seemed to be running in reverse? Boltzmann answered that highly improbable cases could simply be neglected: "One may recognize that this is practically equivalent to never," he concluded.[7] In the twentieth century, the physicist Richard Feynman explained the paradox in this way: "Things are irreversible only in a sense that going one way is likely, but going the other way, although it is possible and is according to the laws of physics, would not happen in a million years." In other words, according to modern physics it was perfectly possible for a soap bubble to form rather than burst; it was just statistically improbable.

CRITICS OF THE STATISTICAL INTERPRETATION

Critics of the statistical explanation argued that if heat was merely molecular motion, then there was no reason why it would not equally well flow from cold to hot as from hot to cold. This criticism threw into doubt the viability of the atomic view of nature on which it was based. Despite the best efforts of Maxwell, Gibbs, and Boltzmann, it was hard to reconcile clear evidence of irreversibility with a mechanical and molecular theory of heat. These criticisms were hurtful and important, launched most forcefully by Ernst Mach, who firmly opposed the atomic and molecular view of nature on which these theories were based, as well as by Ostwald, who sought to found thermodynamics on an entirely different basis. Ostwald rejected a molecular theory of heat because of it seemed incompatible with macroscopic irreversible processes. Under that theory "the tree could become a shoot and a seed again, the butterfly turn back to a caterpillar, and the old man into a child," he argued. For him "the actual irreversibility of natural phenomena thus proves the existence of processes that cannot be described by mechanical equations, and with this the verdict on scientific materialism is settled."[8]

Free will and life itself seemed in blatant contradiction to the second law of thermodynamics. Thomson excluded living beings from being subjected to the second axiom.[9] At first he even excluded vegetative action and chemical action, distinguishing living creatures by their possession of free will. He differed from diehard materialists (Helmholtz, du Bois Reymond, Huxley, and Tyndall), who wanted to subject everything, even free will and life, to thermodynamic laws. Certain aspects of life seemed to flourish in ways that could only be explained as statistical exceptions.

Effects that seemed to run counter to the decay associated with entropy and the arrow of time captivated scientists: dust floating in the air or pollen floating on puddle of water. Einstein investigated the movement of tiny particles in sugared water, while Bergson famously described our experience of waiting for sugar to dissolve in a glass of water. The philosopher also looked closely at the tiny droplets that seemed to hover in the air in the path of a teakettle's steam. Were these examples of those events that could, statistically speaking, happen but that rarely did? In their laboratories, scientists isolated representative examples of these effects. Many of them fell under the label of Brownian motion.

BROWNIAN MOTION

Brownian motion was strange. Small particles suspended in liquids or gases seemed to move and dance as if alive—as if they would never tire or slow down. They were named after the Scottish botanist Robert Brown, who in 1827 investigated the movement of pollen floating on water. When he used dust instead of pollen, he saw that their movement did not change, disproving the usual explanation that pollen moved because it was alive. Two investigators noticed that the movement persisted for a whole year, even when the floating particles were perfectly sealed in a container.[10] "Brownian motion never ceases. It is eternal and spontaneous," explained Jean Perrin, a friend of Einstein who would forcefully back his molecular explanation of their movement.[11]

The strange movement of these dustlike particles had captured the imagination of thinkers since ancient times. Lucretius famously described the strange "dancing" movement of dust floating in the air:

> Observe whenever the rays are let in and pour the sunlight through the dark chambers of houses: you will see many minute bodies in many ways through the apparent void mingle in the midst of the light of the rays, and as in never-ending conflict skirmish and give battle combating in troops and never halting . . .

What caused them to move?

> Such tumbling imply that motions also of matter latent and unseen are at the bottom. For you will observe many things there impelled by unseen blows to change their course and driven back to return the way they came now this way now that way in all directions round . . . and step by step issues forth to our senses, so that those bodies also move, which we can discern in the sunlight, though it is not clearly seen by what blows they so act.[12]

Like Lucretius, Einstein would famously link their visible movement to the subtending force of invisible atoms.

How did Brownian motion affect current theories of time and its flow? These particles did not move in the manner dictated by the second law of thermodynamics and entropy. Where did their inexhaustible energy come from? Experience showed that all other physical phenomena eventually slowed and cooled down, but not Brownian motion. Why not? The particles' motion did not settle into a dominant direction that would permit them to eventually end in a state of equilibrium. "If this motion with all its regularities is really to be observed, then classical thermodynamics can no longer be viewed as strictly valid at the microscopic level," explained Einstein.[13] Their movement went against the usual flow of things, implicating the direction of time itself.

Boltzmann's statistical interpretation was only partially successful, and Einstein was well aware of its weaknesses. The first problem was that it only applied to gases. Aware of these limitations, Einstein tried to improve on it. "The Boltzmann is absolutely magnificent," he wrote to his girlfriend. He was "firmly convinced of the correctness of the principles of his theory, i.e., I am convinced that in the case of gases we are really dealing with discrete particles of definite size which move according to certain conditions."[14] But the theory did not work in cases that did not involve gases, such as liquids or solids. Could it be expanded to those territories? Einstein saw a clear opportunity: "Great as the achievements of the kinetic theory of heat have been in the domain of gas theory," it failed everywhere else. Einstein set to work "to close this gap."[15]

In the year 1905, one of Einstein's revolutionary papers described an effect that could be related to Brownian motion. Einstein was initially cautious about linking his work directly to these strange motions: "It is possible that the motions to be discussed here are identical with the so-called 'Brownian molecular motion,'" he speculated. But in a letter written shortly after the paper was sent, he was bolder, explaining how "physiologists have observed <unexplained> motions of suspended small, inanimate, bodies, which motions they designate as 'Brownian molecular motion.'"[16] Einstein argued that the molecular theory of heat should cause small particles floating on liquids to move very quickly on its surface: "bodies of microscopically-visible size" or "small particles" should "perform movements" that "can be easily observed in a microscope."[17] Einstein drafted the mathematical formulas for their motion.

For many scientists, Einstein's particular contribution was decisive for the establishment of modern physics. It helped scientists confirm the molecular and atomic theories on which the statistical interpretation of thermodynamics was based and to solve the reversibility paradox of the last century. An increasing number of physicists inspired by Einstein's contributions to Boltzmann's theories considered the motion of molecules as perfectly reversible and foundational for our understanding of the concept of time in physics.

Einstein's solution nonetheless entailed significant problems. If Einstein's theory of relativity *and* Brownian motion were both right, there would be no essential distinction between the past and the future in the same way that none was then known to exist between left and right. Einstein's theory of relativity did not account for time flowing in one direction, while his theory of Brownian motion was eventually used to prove that reversibility was a fundamental property of atoms. Yes, Einstein admitted, there were many cases where the arrow of time dominated, even in the microscopic world. If, for example, one used a diluted solution with too few particles, the Brownian motion phenomenon largely disappeared, and the arrow of time phenomenon took over. But irreversibility at this molecular level was, according to Einstein, due to "highly improbable" cases. Reversibility and multidirectionality was the rule, whereas flow (brought about by irreversible and unidirectional tendencies) was a rarity.

IRREVERSIBILITY VIA REVERSIBILITY

After Einstein's contribution received additional experimental proof, and after it was furthered by his supporters such as Jean Perrin, most scientists espoused a statistical interpretation of entropy and an atomic and molecular theory of matter. Yet the paradox of explaining irreversibility via reversible laws of nature remained.

Was the quest to find what differentiated the past from the future as misguided as those attempts to find a difference between left and right, which mythical thought had labored over? "There can be no doubt that all natural laws are invariant with respect to an interchange of left and right," explained Weyl. Why should time be any different?[18]

By the time Boltzmann's statistical interpretation became widely accepted, Bergson felt that materialistic and deterministic theories were a tight noose constraining, as never before, current beliefs about the world. He reviewed "the two most general laws of our science." One was the law of the conservation of energy, by which the "total energy remained constant," and the other one was the "second principle of thermodynamics" of the "degradation of energy." Bergson considered the statistical interpretation given by Boltzmann, but since it was "equivalent, practically, to absolutely never," he felt compelled to look outside of physics for an explanation of the exceptions he saw were rampant in the living world.[19]

Bergson defined the élan vital as a retarding force that worked against degradation. It was "attached" and "riveted" to matter but not entirely part of it. According to him, living bodies were part matter—they were "riveted to an organism that subjects it to the general laws of inert matter." But they "tried to rid themselves from those laws." They "did not have the power to reverse the direction of physical changes, such as Carnot has determined them." But they were nonetheless "a force that, left on its own, works in the opposite direction." Incapable of "stopping the march of material changes, it nonetheless is successful at retarding them."[20]

Bergson illustrated this counterforce "riveted" to matter, using the example of the steam coming out of a pressure cooker. He explained, with poetic detail, the jet of vapor coming out from the pivot. All liquid drops eventually came down, but careful observation revealed that "a small part of the vapor remains . . . for a few instants, and it makes an effort to lift the drops that fall; it is able, what is more, to delay their fall."[21] He warned that his comparison was inadequate, since all vapor would condense and the drops would eventually descend, but it sufficed for the purposes of imagining those droplets caught in a perpetual state of motion. Bergson's descriptions of movement emerging from a force "riveted" to matter and "retarding" its inevitable fall and decay ran counter to Einstein's entropy-based molecular explanation.

Scientists' solutions to the reversibility paradox remained highly unsatisfactory during these years. The philosopher Martin Heidegger continued to note how the paradox, although solved by scientists, remained pertinent philosophically. "Though one allows 'non-reversibility' as a distinct predicate of time, one does so on the understanding that one would much rather reverse time's direction, that is, that one would very much like to repeat and retrieve time and have it completely available in the present moment as something present-at-hand."[22] Other researchers similarly protested against scientists' efforts to base knowledge on reversible processes. The differences between past, present, and future, which Einstein considered mere psychological illusions, were central for others.

CHAPTER 24

Einstein's Films: Reversible

Could films show us how time advanced? At the close of the nineteenth century, the Lumière brothers had presented a new camera. It recorded and displayed moving scenes on a screen. Had scientists finally discovered a means for capturing time—as it unfolded? Could it help settle the disagreement between Einstein and Bergson?

Clocks were not the only timekeeping instrument against which Einstein and Bergson compared their own experience of time. Increasingly they would understand time by reference to cinematographic cameras. When Einstein, Bergson, and their numerous interlocutors debated about the nature of time, they frequently mentioned film. What did they learn from it?

Bergson had started writing about the cinematographic method around 1900 and continued to refer to the camera throughout his life. He would mention cinematography again in his polemic against Einstein. Why did Bergson think it was relevant to talk about film when discussing physics? The philosopher protested that if one flattened Einstein's universe and arranged one instant after another, the result would end up looking "like a screen upon which the cinematography of the universe would be run off." The only difference between a film of the universe and Einstein's model, alleged Bergson, was one: "with the difference however that here there is no cinematography external to the screen, no photography projected from without; the image takes form on the screen spontaneously."[1]

Bergson criticized "cinematographic" conceptions of the physical universe as well as "cinematographic" conceptions of time. By assuming that temporal events in physics succeeded each other in the same way as successive frames in filmstrips, science and film shared an underlying, deceptive notion of time. Although they both *seemed* like they conveyed events unfolding in time, this effect was a mere illusion, he explained repeatedly. Cinematography (as a technique of representation *and* as a model for how sequential events succeeded each other in the universe) shared the same faulty notion of time: one that could be divided into its constitutive parts, that could be represented spatially, that was homogenous, and that in principle, could be grasped all at once, from beginning to end.

BERGSON'S CRITIQUE OF THE CINEMATOGRAPHIC METHOD

Bergson's critique of the cinematographic method was based on the conviction that something essential escaped from the small gaps, or frame lines, bordering successive film stills.[2] Nobody was looking at frame lines: "As to what happens in the interval between the moments, science is no more concerned with that than are our common intelligence, our senses and our language: it does not bear on the interval, but only on the extremities."[3] The philosopher explained how cinematography provided viewers with an illusory kind of movement that differed from real movement: "Suppose we wish to portray on a screen a living picture. . . . How could it, at its best, reproduce the suppleness and variety of life?" By taking "a series of snapshots" and projecting "these instantaneous views on the screen, so that they replace each other very rapidly," movement could be reproduced. A note-taker in one of Bergson's courses explained the philosopher's point:

> The photographs that we take of a galloping horse are not, in reality, the elements of the gallop from which they were taken. And the cinematographic machine, that with these series of views, recomposes their trajectory, does not give us the illusion of movement other than by adding to these views, in the form of a certain mode of succession, the movement that in them they cannot contain.[4]

An *illusory movement* resulted from shifting *real movement* elsewhere. In the case of cinema, it resulted from shifting it inside the apparatus: "In order that the pictures may be animated, there must be movement somewhere. The movement does indeed exist here; it is in the apparatus."[5] For the illusion to work on screen, real movement had to be moved somewhere else.[6] Bergson exhorted his followers to peer into the apparatus to find real movement hidden there; to find, inside, a moving universe that could never be divided into separate, constitutive elements.

Bergson continued to assail the cinematographic method relentlessly. It was a pervasive, constraining, and infirm approach used to pass off illusory movement *as if* it was real. Referring not merely to the modern cinematographic camera, but also to the proclivity of the human mind for arranging temporal images spatially, he criticized its restrictiveness and urged scientists to "set the cinematographic method aside" and search instead for a "second kind of knowledge."[7] The passing of time, he insisted, involved the creation of the new and the unforeseeable. Time was uncontainable. Every instant bit into the future.

Bergson's criticisms of cinematography appeared in print in *Creative Evolution* (1907). The explicit intention of the book was to combat a mechanistic view of evolutionary theory. Herbert Spencer, known for coining the term "survival of the fittest" and chief popularizer of Darwin's theory, conceived of evolution as a step-by-step mechanical process that could explain the emergence of the human mind and even our ethical choices. Bergson disagreed wholeheartedly with this aspect of Spencer's work and this kind of Darwinism. To convince his readers, he questioned Spencer's mechanistic evolution by *first* questioning the basic techniques of representation used in the sciences by everyone ranging from evolutionary scientists to astronomers. According to him, essential aspects of the world (especially those connected to life and its unfolding through time) could never be grasped or captured entirely by either the intellect or a machine.

References to photographic and cinematographic cameras formed part of Bergson's fight against materialistic theories of life and mind. He analyzed time's relation to optical (photographic and cinematographic) and auditory (phonograph) storage technologies to point out how much they missed. Memory, he argued, was much deeper and multilayered than any record of it (such as in a photograph, film, or phonograph record) could ever be. In his 1911 lectures at Oxford University, he explained current theories of memory by reference to photography. It was wrong, he claimed, to perceive memories in the brain as a storage mechanism of visual memories as if they were "photographic clichés" and of auditory ones as if they were simply "phonograms" allowing playback. After "a material object made an impression on the eye and left in the mind [*l'esprit*] a visual memory," this memory was much richer than a single photograph could ever be, much deeper than even a thousand photographs, and much more complex than even a film. Even the shortest impression would leave on the mind "as many and more [images] than those of a cinematographic 'film.'"[8] Bergson never accepted common attempts to describe memory in terms of these technological analogues.

FROM A TO Z, OR FROM Z TO A

How was film actually used by scientists at the time of Bergson's critique? Cinematographic studies of microparticles had led scientists to firm conclusions about the nature of time in the universe. Immediately following the publication of his theory of Brownian motion, Einstein's theory was tested with the new cinematographic technologies.[9]

According to statistical and molecular theories of thermodynamics, our sense of the "arrow of time," and of temporal irreversibility, was actually based on reversible effects at the microscopic, molecular level. Brownian motion showed reversibility at this molecular level. Physicists used Brownian motion films to study this basic reversible form of movement, one not marred by the "arrow of time."

Brownian motion films were much more than representational techniques for a certain branch of physics. They were tied to how physicists understood the concept of time much more generally. Late in life, in his attempts to convince his friend Michele Besso of his point of view against Bergson's, Einstein used the example of a Brownian motion film. Although in our macroscopic dimension we sensed time flowing in a certain direction, scientists, including Einstein, found no evidence of such directionality in the microscopic realm. Einstein explained to his friend why the subjective feeling of time flowing was just an illusion by asking him to consider playing a movie of Brownian particles in reverse:

> Imagine that one has filmed the Brownian movement of a particle and kept the images in the correct chronological order with respect to the neighboring images; only they forgot to note if the correct temporal order went from A to Z, or, well, from Z to A. The shrewdest man in the world would not be able to find the arrow of time from that material.

A Brownian motion film in reverse was indistinguishable from an unreversed one. Einstein continued his letter by extending the logic of this cinematographic Brownian motion film to the rest of the universe: "In the elementary realm, all processes have an inverse. Too bad, then, if the theory of relativity sinned against the theory of the arrow of time."[10] Einstein's reference to a Brownian motion film was neither fortuitous nor merely illustrative—cinematography was an essential technique for understanding the passage of time across a wide spectrum of fields, from the microphysical to the macrocosmological.

When Einstein saw a Brownian motion film, he saw in it a model for how the laws of physics worked, one that was *not* characterized by irreversible unidirectional temporal flow. Evidence for the arrow of time was present in nearly all regular films, but not in this one. Yet Einstein prioritized it as a model for understanding the universe.

BROWNIAN MOTION FILMS

Brownian motion was easily seen with a microscope. It could also be displayed on a screen for public viewing. In his famous paper on the subject, Einstein agreed that these particles could be "easily seen," but seeing their motion and reaching agreement about their movement and nature turned out to be quite different propositions. As with many of his other publications, this one took years to garner the full support of the community of physicists. Film played an essential role in securing it.

Why film? Part of the challenge of studying Brownian motion was due to the limited information about it available to researchers. "The data available to me . . . are so imprecise that I could not form a definite opinion on this matter," Einstein complained.[11] The physicist understood that one difficulty in answering the riddles of Brownian motion had to do with problems in observing very fast, irregular motions of very small particles. An "observer operating with definite means of observation in a definite manner" would have great difficulty determining how a certain particle moved from one place to another in a very, very short period of time.[12] It was important for Einstein, as for other Brownian motion researchers, to be able to track their movement at specific and fixed time intervals.

Soon after Einstein's paper, the physicist Max Seddig was able to photograph Brownian motion particles at intervals of one-tenth of a second. A few years later, the French scientist Victor Henri used a different cinematographic setup that was capable of halving that time, to one-twentieth of a second. Henri's studies emerged from a laboratory tradition at the Collège de France led by the physiologist François-Franck, who helped him and lent him his laboratory. During those years, the Gaumont Company, along with their competitors Pathé, worked closely with scientists.

Henri remarked on the need for studying Brownian movement quantitatively, noting how this was difficult due to "the rapidity and the weak trajectory of movement [of the very small particles]." After preliminary successes, Henri concluded that the value predicted by Einstein (and corroborated by the physicist Marian Smoluchowski and Paul Langevin) was "four times weaker than the value found experimentally." He concluded that Einstein's theory was wrong: "It thus results from our experiments that Einstein's formula does not give the exact displacement of the Brownian movement of the grains studied by us." But it was not all hopeless. Henri noticed that if he measured the movement of the particles every four images (equal to a separation time of one-fifth), his results matched perfectly with Einstein's formula. Still, even with his cinematographic equipment, it was not easy to come to any definite conclusions about molecular motion since "the trajectory varies from one grain to the next, and is absolutely independent for each grain."[13] Another problem had to do with coagulating effects that slowed down the particles.[14] Henri measured the trajectory of the particles by comparing one cinematographic frame against the other and then represented their movement in graphic form. He focused on a few particles, rarely more than ten. He ended his paper claiming that the questions pertaining to Einstein's work were "a point that can be resolved by a cinematographic study."[15]

In a key text, Jean Perrin—who did more than most scientists to verify and promote Einstein's statistical work on Brownian motion—claimed that Henri's cinematography methods were trustworthy. He also found a way to reconcile them completely with Einstein's results, by explaining that the initial discrepancy found by Henri between their movement and Einstein's predicted value was actually due to a minor error in the measurement of the particles' diameters. These successes led Perrin to publish a book about molecular motion, titled *Les Atomes* (1909), in which he finally dispelled many scientists' doubts about the benefits of microphysical studies and statistical approaches to thermodynamics.

Perrin studied and published images of particles in Brownian motion mostly in graphic form. His investigations were based on careful and laborious observation through a microscope, with which he tracked the changing positions of the particles at fixed intervals of time and drew them on square-grid paper. Even when the technologies employed by Perrin were not actually cinematographic, he considered his observations as fitting with a more general concept of cinematography—the practice of tracking movement across fixed intervals of time. If studied "cinematographically," he argued, Einstein's victory—and that of modern microphysics as the dominant approach in physics—was clear.[16]

Beginning in 1911, Perrin used Brownian motion films to reach a broader public and popularize these theories, commissioning them for almost a decade.[17] Scientific films, many of them enlarging microscopic phenomena or using time-lapse photography, became widely known to the public starting in the 1910s. During this period, they started to emerge as a distinct genre that was increasingly separate from commercial movies.

FILMS IN REVERSE AND THE ARROW OF TIME

While Brownian motion films led scientists to certain conclusions, movies that showed how different the world looked in reverse led various other thinkers to side with Bergson. Soon after shooting their first films, the Lumière brothers filmed the slow demolition of a wall and then showed it in reverse. Fallen boulders, scattered rocks, and clouds of dust magically gathered back together to form a wall. They elicited a lot of laughter with another film in reverse: *Charcuterie mécanique*. After recording the successive steps of a pig being butchered and made into sausage, they played it backward: from the sausage came out a pig. What did these films reveal about the nature of time? What role did they play in cementing further divisions between Einstein and Bergson?

When scientists discussed time in terms of reversibility, they frequently considered it in the context of these new technologies. Einstein and Bergson held different views about irreversibility, just as they held different views about film. Films in reverse, particularly sound films, showed to many viewers that the feeling of time flowing in one direction could not be brushed aside as a mere illusion. Satosi Watanabe, a quantum physicist who sided with Bergson, was one of many authors who explained the laws of entropy by reference to playing films in reverse. "Imagine that a natural phenomena is filmed and that the film is projected in the inverse sense of the course of time." By then, he could speculate that "you may have seen" these effects: "For example, you may have seen on the screen a diver resurface from the water, her feet before her head, and spring in the air to arrive at the diving board." The definition of reversibility used by most physicists was based on the possibility that these reversals might, statistically speaking, occur: "A phenomenon is said to be reversible if its movement in reverse is possible according to the laws of nature."[18] Watanabe proceeded to explain that reversible phenomena and irreversible ones should not be distinguished by the mere *possibility* of reversibility but rather by the frequency with which reversals actually occurred. Scientists should not focus only on reversibility as a statistical *possibility*, they should focus on it in terms of its actual realization—and this almost never happened. Because of these limitations in current theories of thermodynamics, Watanabe took Bergson's philosophy as more useful than Einstein's for studying the passage of time.

In the 1950s Hans Reichenbach used the same films-in-reverse example to illustrate thermodynamic processes. "The relation between irreversibility and time order is well illustrated by the series of pictures which we see when a motion picture is run in reverse," he explained. These movies showed "the strange aspect of cigarettes becoming longer and longer while they burn, or of pieces of broken pottery that rise from the floor to the table and assemble into neat dishes and cups."[19] Why were these special effects so bizarre? Reichenbach did not have a complete answer. Recalcitrant to turn to Bergson for inspiration, he died before being able to complete the last chapter of his last book, *The Direction of Time*.

Descriptions of thermodynamics by reference to movies in reverse became classic. In the 1960s, the famous physicist Richard Feynman explained how thermodynamics was frequently demonstrated in the classroom by playing a movie backward: "The demonstration of this in lectures is usually made by having a section of moving picture in which you take a number of phenomena, and run the film backwards." Feynman continued: "The moving picture should work the same going both ways." The paradox remained at the cornerstone of modern physics: irreversible processes seemed everywhere, yet the laws of physics were unable to explain them: "That is, in all the laws of physics that we have found so far there does not seem to be any distinction between the past and the future."[20]

Experiments with gramophone records played in reverse brought additional lessons. "This unidirectional character of time," explained a writer in 1926, was "exhibited" by reversing the motion of the turntable: "In plain words, playing a phonograph record backwards takes the meaning out of it."[21] Increasingly, scientists felt the need to consider entropy in terms of the transfer of information in addition to energy. New sound films, in contrast to earlier silent films, intensified the need to think about the arrow of time, and thermodynamics, in this new way.

How could the very same medium initially criticized by Bergson and employed by defenders of Einstein be used to justify the philosophy of Bergson and criticize the science of Einstein? Part of the answer has to do with the different ways in which time itself was conveyed in different kinds of films. Throughout the twentieth century, films of various kinds (ranging from laboratory Brownian motion films to blockbuster feature films) and used in various ways (played forward or in reverse) remained at the center of scientific and philosophical discussion about the nature of time. The development of film threw into sharper relief the differences between Einstein and Bergson. From Brownian motion to a movie of diver jumping into the water to the same image in reverse, cinematography was much more than an applied technique of representation. It played a central role in a world divided into two irreconcilable theories of time.

CHAPTER 25

Bergson's Movies: Out of Control

At the time of his first critique of Bergson, Bertrand Russell had never seen a film. It soon became clear that he had to go to a theater to actually see what everyone—particularly the French philosopher—was talking about: "When I first read Bergson's statement that the mathematician conceives the world after the analogy of a cinematograph, I had never seen a cinematograph, and my first visit to one was determined by the desire to verify Bergson's statement."[1] Russell understood fully well that Bergson's critique of cinematography was not only directed at the popular form of entertainment, but it also targeted some of the most valued principles of mathematical logic, touching on the very foundation of calculus.

Since ancient times, philosophers had debated about the relation of the *discrete* in the form of still instants, to the *continuous*, in the form of movement. How can a flying arrow both move and occupy a fixed length in space? Zeno's paradox showed just how hard it was to answer that question. In modern times, these ancient questions were debated not by reference to flying arrows but to cinematographic machines.

Logicians such as Russell tried to prove that discrete entities could indeed make up a continuum (if they were infinitely small, the continuum would be perfect) and that a series of (equally short) instants could make up movement. Bergson believed that they did not: a real continuum could never be made up of discrete entities, and real movement could never be made up of separate instants. Russell's seminal contributions to the foundations of mathematics, done in collaboration with Alfred

North Whitehead, depended centrally on the idea of an absolute continuum that was infinitely divisible and at the same time composed of discrete yet infinitesimal entities. Bergson's philosophy, in contrast, allowed for the division of a continuum into separate pieces, but it did not allow for its reconstitution through these discrete elements.

Unconvinced by Bergson's critique of mathematical logic, Russell proceeded to refine his theory by interpreting cinematography in a different way from the philosopher. In one of his first attacks on Bergson, he claimed that the philosopher's arguments did not hold because one could imagine a film camera operating with an infinite number of frames: "A cinematograph in which there are an infinite number of films [frames], and in which there is never a *next* film [frame] because an infinite number come between any two, will perfectly represent continuous motion."[2] Why did Russell turn to film? Could he settle longstanding debates about the nature of logic by imagining a perfectly continuous film machine?

During the Great War, Russell joined forces with Einstein in his denunciation of the war against Germany and in his defense of relativity theory. He attacked Bergson once again, who had patriotically sided with his country against Germany. In Manchester (1915), he delivered a talk that outlined some of the beliefs he dearly held for the rest of his life—including his hatred of Bergson. That day, Russell "suggested" that "the cinema is a better metaphysician than common sense, physics, or philosophy."[3] What did he mean by that? How could "the cinema" have powers greater than those of laymen, physicists, and philosophers?

"CINEMA IS A BETTER METAPHYSICIAN"

After seeing his first film, Russell became convinced of how appropriate Bergson's "favourite illustration of the cinematograph," really was. Yet he drew a completely different conclusion from the French philosopher. Russell explained that he "found [it] to be completely true," but for exactly the opposite reasons.[4] For Russell "the cinematographic illustration" was appropriate for solving the problems of mathematical logic and describing the world in materialistic terms, including life and time; for Bergson it stopped short of revealing the most important aspects of the biological and physical universe, those having to do with aging and duration. Russell realized that in a film the illusion of continuity was made up of "momentary" sections:

> When, in a picture palace, we see a man rolling down hill, or running away from the police, or falling into a river, or doing any of those other things to which men in such places are addicted, we know that there is not really only one man moving, but a succession of films, each with a different momentary man.[5]

That film comprised discrete frames was not a drawback for Russell, as it had been for Bergson. It was its strength.

> The illusion of persistence arises only through the approach to continuity in the series of momentary men. . . . The real man too, I believe, however the police may swear to his identity, is really a series of momentary men, each different one from the other, and bound together, not by a numerical identity, but by continuity and certain intrinsic causal laws.[6]

For Russell there were no "real" entities to contrast against those "on screen"—both were constituted in similar ways. Film showed the strength of "intrinsic causal laws" that linked one state to the next on the screen as much as outside of it. Russell saw no need for Bergson's cautions against the machine's illusions, not even when thinking about the world at large, since "what applies to men [on the screen] applies equally to tables and chairs, the sun, moon and stars."[7] Delighted, he had found in cinema an ally. During these years, cinematography fit uneasily with Bergson's philosophy. Its effects were much smoother than they had been when he first mentioned the technology. It could be used, as Russell did, to disprove the philosopher's conclusions.

Cinematography changed radically in the years following the Great War. Early techniques were soon abandoned in favor of new instruments and filming methods. Scientific films would soon do much more than prove the essential reversibility of the world at the molecular level. They would be used to study the essence of life as much as that of the physical universe. In these new contexts, cinematographers fought against the camera's timekeeping regularity, speeding up the recording and projecting speeds and using cuts and edits to skip over or zoom into long time spans.

Einstein and Bergson were situated in radically different places with respect to the visual culture of their time. Their use and experience with images, both moving and static, drew them farther and farther apart in a century that saw film take off from its early origins in scientific laboratories to become much more. For those who followed their work, certain kinds of films furnished clear evidence in favor of one man, but other kinds of films proved exactly the contrary.

Early cameras and projectors were driven by hand. Later, automatic clockwork mechanisms were installed in commercial cinematographic cameras to record and display images at fixed intervals. These changes affected how scientists thought about film. Étienne-Jules Marey was one of the first photographers to stress the need to record and display images with clocklike precision. He protested that in the famous horse-in-gallop photographs of his contemporary Eadward Muybridge, the intervals between the frames were neither fixed nor properly determined. Muybridge's cameras went off every time the horse broke with his stride a series of strings strewn across the running path. The resulting photographs were therefore taken according to the horse's speed, not according to clock time. In contrast to Muybridge, Marey wanted cinematography to become a time-recording technology as much as an imaging one.

Should scientists always opt to keep the intervals fixed in order to better display movement, as Marey advocated? Certain topics, like the horse in gallop, seemed to call for expanding or contracting the intervals between frames or regulating their speed according to the speed of the filmed object. Otherwise, interesting visual effects were lost. Which technique was better? What did these competing methods reveal about the nature of time? As the film industry grew, an increasing number of filmmakers abandoned the strict timekeeping standards advocated by Marey and employed by astronomers and physicists. Where did Einstein's and Bergson's conception of film fit within these debates? While Bergson had been extremely critical of the medium, his philosophy would be revived in the interwar years for the inverse purpose: to explain the wonders of cinema.

UNREGULATED FILMS AND EARLY BIOLOGICAL FILMS

Did films capture and reveal movement? According to Bergson they did not. They only showed an illusory kind of movement because of an essential trick: they hid the real movement "within the apparatus."[8] The on-screen spectacle worked only because one could find real movement elsewhere. To understand the temporality that concerned Bergson, viewers should think about how technologies displaced real movement by hiding it. They should think twice before considering the movement on screen as real.

Film techniques and technologies changed radically after Bergson's first negative comments on cinematography. Bergson would eventually refer to certain aspects of film that *did* resonate with his philosophy of time. These were not regulated films. In fact, they were precisely unregulated ones.

In a lecture on dreams delivered in 1901, Bergson explained that our lived sense of time echoed how we experienced it in our dreams.[9] It was a sense of time comparable to that of a broken clock. Clocks, he explained then, were kept on time by a "balance wheel that slows down and breaks into regular sections the indivisible and almost instantaneous tension of the spring."[10] The balance wheel of a clock, he had already explained, permitted the division of time into intervals, making it possible to treat it as something homogenous and analogous to space. In connection to that technology arose "the mixed idea of a measurable time, that is, of a time which is space through its homogeneity and duration because of its succession"[11] Dreams did not have such a mechanism: "It is the balance wheel that is missing in dreams."[12]

When Bergson reedited this article for publication in 1919, he had a more precise and different description. The time of dreams was like that of "a cinematographic film if one did not regulate the unraveling." Ordinary well-regulated cameras did not have the ability to capture this sense of our temporal experience. But out-of-control ones did. "In a few seconds, the dream can present to us a series of events that would last whole days during our waking lives," he explained.[13]

Two senses of time, their difference and opposition, emerged alongside transformations in film technologies and techniques. A subjective sense of time, often associated with Bergson's notion, was frequently associated with cuts, edits, and changes in frame speeds. Einstein, during one of his busiest years, described his life in these terms: as a cinematographic camera running fast. During a stressful trip to Zurich, it appeared to him like a sped-up film. "The final days of my Zurich stay resembled a runaway motion-picture projector," he wrote to a friend.[14]

His descriptions of the universe, in contrast, were widely considered to be cinematographic in a different way: perfectly cadenced, regulated, and without cuts or montage.

Scientific and epistemological debates about how to portray movement and about the nature of time and causality involved direct references to filming techniques. How should cinematographers portray simultaneous events? How should they portray them when they were so distant from each other that they could not be filmed by the same camera? Could filmmakers use two cameras and then reassemble the scene as in a montage? Or should they opt for fast cuts moving from one locale to the other? The film critic Bela Balázs argued against split-screen "simultaneism" or the showing of "a number of simultaneous events," especially in cases when "there is no causal relationship between these events themselves." Showing simultaneous events with no causal relationship to each other was simply wrong, in his view: "They give a film a *false dimension*." Balázs protested against a technique used by the filmmaker Abel Gance, where at the same time that the viewer would "follow the fate of his hero in Paris," he included shots of things that were "not relevant to the plot" but that "represent a simultaneous reality," such as "villages, people working in the fields, or a girl at a window."[15] Balázs's "rule" echoed Einstein's call for considering as invalid the simultaneity of events that were not actually connected to each other. For the most part, the scientific documentary and pure cinema, based on long takes, shunned the montage and editing techniques of fictional film and the psychological film genre, as the two dominant techniques for portraying time and simultaneity became increasingly differentiated.

BERGSON VINDICATED

Innovations in the temporal, narrative structures of films (often associated with the "psychological " genre) showed the pertinence of Bergson's philosophy for film theory and practice. "Film will be Bergsonian or it will not be at all!" exclaimed the writer Émile Vuillermoz, adding that "everything that Bergson applied to sculpture and painting is realized in the cinema that will become, in the hands of artists, the most powerful of the plastic arts."[16] A 1914 interview of Bergson by

Georges-Michel Michel, a painter and journalist, revealed the philosopher's positive assessment of cinema—if it was understood in a certain way.[17] Later discussions between Vuillermoz and the writer Paul Souday showed how Bergson's philosophy was considered in terms of movies, as an increasing number of viewers saw films reveal a Bergsonian notion of time. How could a philosopher who built his career denouncing the cinematographic method end up as one of the most important thinkers of the Seventh Art?

Unedited, mechanical film techniques seemed to capture the passing of time so imperfectly that Arthur Eddington, the man who had made Einstein famous, referred to cinematography in lectures delivered in Edinburgh in 1927. In addition to referring to Bergson, he explained to his audience that he no longer believed with Einstein that the illusion of time passing by was simply our mental construct. Our proclivity to conceive of temporality as becoming and evolving rather than given all at once had to be taken seriously. "Unless we have been altogether misreading the significance of the world outside us—by interpreting it in terms of evolution and progress, instead of a static extension—we must regard the feeling of 'becoming' as (in some respects at least) a true mental insight into the physical condition which determines it," he concluded. That sense of time—which he associated with Bergson's notion—had to be considered by scientists. It could not be simply disregarded. "It is clearly not sufficient" to argue "that the change in the random element of the world should deliver an impulse at the end of a nerve, leaving the mind to create in response to this stimulus the fancy that it is turning the reel of a cinematograph," he explained. Our minds did much more than simply register stimuli in a mechanical and automatic way; they faithfully registered an aspect of time connected to a legitimate and irregular sense of "becoming."[18]

BIOLOGICAL TIME AND LIVED TIME

Film critics were not the only ones who found Bergson's philosophy pertinent for understanding the powers of cinema. Biologists who used film in order to understand cell cultures did so as well. While physicists used precisely timed and cadenced exposure times and intervals between frames, biologists favored different techniques. Biological films were "projected at the ordinary rate of 16 [frames] per second or less" but they were all filmed at variable speeds varying "according to the activity of the culture." Some at "every 10, 15, 20 or 30 seconds, during periods varying from 24 to 72 hours."[19] The need to introduce variations in recording speed, in its evident contrast with other regularly cadenced films, threw into relief the difference between "biological" and "physical" time and between living and dead matter. The notion of lived time, as one in opposition to physical time, obtained stronger legitimacy in a number of scientific disciplines during these years.

Variable-interval films became essential for cell biology. Key exponents of the "new cytology" (cell-based biological research) noted how these films revealed a different aspect of time, biological rather than physical. When a book with the title *Biological Time* appeared in 1936 citing Bergson positively, the philosopher was thrilled. He discussed it with a friend, saying it was "serious, since it deals with interesting facts" that "are not in any way incompatible with my representation of the concept of duration."[20] The critic André George, who had previously found fault with Bergson, now contended that Bergson's philosophy was corroborated by this new research into physiological time.[21]

The controversial surgeon, microbiologist, and Nobel laureate Alexis Carrel, along with his student the biophysicist and philosopher Pierre Lecomte du Noüy, developed the concept of biological time in the interwar years and considered it a vindication of Bergson's work.[22] Carrel and du Noüy were, coincidentally, using the very instrument that Bergson had so potently criticized. In 1931 Carrel described the cells he filmed in terms of Bergson's philosophy of life: "The present of a living organism does not pass into nothingness. Bergson has clearly shown how the past persists in the present."[23]

Carrel paired with du Noüy after World War I broke out. Du Noüy was a lieutenant of the 61st Reserve Division at the Hôtel du Rond Royal, which was in the process of becoming Front Hospital No. 21. Under Carrel's leadership the hospital became a research center where part of the research pertained to a study of infected wounds funded by the Rockefeller Institute of New York. Du Noüy started working on a project that traced the healing of wounds across time and tested the effects on cicatrization of different antiseptics, the initial size of the wound, and the age and state of health of different subjects (ranging from soldiers to dogs). Some of their work was only cinematographic in the sense that it was based on graphs tracking changes across specific time periods (as in Perrin's graphs of Brownian motion), but others were actually recorded and projected as films. Their biological films had a drama of their own, permitting "an indiscreet onlooker" to witness the "birth, nourishment, battles, and death" of cells. The cells acted like "children let loose in a school-yard" with "the meetings, the bumps, the flights, the struggles."[24] "Cinematographic films of cell-cultures have revealed totally unknown and unforeseen facts," he explained. The cell spectacles were identical to the spectacle of "everyday" life: "We observe, in short, at the scale of a thousandth of a millimeter, all that we are accustomed to see everyday around us."[25] In cicatrization, as in the reproduction of cells in vitro and in vivo, du Noüy found that the time of organisms was not cadenced like that of the physicists. Clearly it was *this* time that was important for us, du Noüy argued, both poetically and practically: the time for the healing of wounds of war. "Everything," du Noüy argued, "occurs as if sidereal time flowed four times faster for a man of fifty than for a child of ten."[26]

Du Noüy was inspired by the concepts Bergson had introduced in *Creative Evolution* (1907), but he was particularly admiring of the text that confronted Einstein: "In pages 58 and 61 of *Duration and Simultaneity*, he arrives, by pure reasoning at conclusions very similar to ours."[27] Elsewhere in his book, du Noüy cited whole passages of Bergson's text as evidence of how pertinent it was to focus on aspects of time that did not flow like the physicists'. "Bergson," he argued, "disengaged very clearly the necessary but vague notion of physical time from the more precise notion of duration."[28] Biological time, according to him, was particularly useful since it bridged the impasse between time as described by physicists and as it was intimately felt: "All that we can say at present is that our crude language, lacking appropriate words, translates this knowledge into improper, inadequate expressions such as: 'There are two species of time,' or 'Physiological time does not flow uniformly like physical time.'"[29]

In his later work, du Noüy continued to cite Bergson. He approvingly cited how he was read by the quantum physicist Satosi Watanabe. *L'Homme devant la science* (1935) cited a text in which Watanabe mentioned Bergson: "Our psychological life is a continuous duration: it is, as Bergson says, 'the continuous progress of the past, which gnaws at the future and swells up as it advances.'"[30] Bergson served as a thread tying together microcinematography, cell biology, and quantum physics.

FILMS, PHYSICS, AND MORALITY

Bergson did not change his mind about the limitations of cinematography he had painstakingly described, but in later years it became clear that he despised only specific aspects of the medium that he associated with a "static" conception of the universe and of life within it. In the 1930s, he expanded his cinematographic critique to morality. In *The Two Sources of Morality and Religion* (1932), he explained the difference between the "dynamic morality" that he advocated and the "static morality" that he deplored by reference to the cinematographic camera. Static morality was one that "has become ingrained in customs, ideas, and institutions; its obligatory character is to be traced to nature's demand for a life in common." Dynamic morality, on the contrary, was "impetus, and . . . related to life in general, creative of nature, which created the social demand."[31] Just as real movement could not be obtained through static images that succeeded each other rapidly, so dynamic morality could not be reached through the fulfillment of static social obligations. In discussions about morality, Bergson returned to the "illusions we have time and again denounced." The most important of them was related to his critique of cinematography.

> [It] consists in the conception of movement as a gradual diminution of the space between the position of the moving object, which is immobility, and its terminal point considered as reached, which is immobility also, whereas positions are but mental snapshots of the indivisible movement: whence the impossibility of reestablishing the true mobility, that is to say, in this case, the aspirations and pressures directly or indirectly constituting obligation.[32]

Static morality—a mark of "closed societies"—was as deficient as the illusory portrayal of movement by cinematographic means. Dynamic morality, in turn, characterized "open societies."

The philosopher urged readers to think twice about their "social" and "moral" obligations, including obligations to their families and nations. These moral obligations often amounted to more harm than good. They might help some people, but they hurt the greater majority: "Who fails to see that social cohesion is due, in large part, to the need of a society to defend itself against others, and that it is first against all other men that we love the men among whom we live?"[33] Bergson urged his readers instead to open their souls even more, beyond the social to the human, and beyond humans to nature: "What does it allow in? If one says that it embraces all of humanity, we would not be going too far, but we would also not be going far enough, because its love extends to animals, plants, and all of nature."[34]

"You must because you must."[35] Bergson cautioned readers to be particularly suspicious about such a command. Obligations were meted out to us. Which ones should we follow? Bergson rebelled against having a blind allegiance to them, and not only for moral reasons.

Following moral orders blindly was as dangerous as following physical laws uncritically—and both were connected: "But if physical laws seem to us as a form of commandment when they reach a certain generality, so a [moral] imperative directed to everyone is presented to us as a law of nature." Scientists failed to note that every law, "be it physical, social or moral," was a command that could—and should—be thoroughly examined: "The scientist himself can hardly prevent himself from believing that the law 'governs' facts and consequently is prior to them."[36] He urged scientists to consider the reverse: not to follow laws that governed facts, but to focus on the facts that governed laws. *The Two Sources of Morality and Religion*, Bergson's last monograph, finally tied together his three previous critiques: cinematography, morality, and physics.

CHAPTER 26

Microbes and Ghosts

When Charles Nordmann, a Parisian astronomer greeted Einstein at the border station in Jeumont, he was immediately struck by the physicist's "captivating physiognomy." "Einstein was tall (approximately 1 m 76)" and had surprisingly "wide shoulders." "The head, that head from which the world of science emerged anew, attracted and drew attention." The astronomer identified Einstein's skull as "extraordinarily brachycephalic," with a "wide" forehead of "exceptional length." Wide, flat heads were hardly a characteristic of geniuses, who according to "phrenologists and certain biologists," should be "dolichocephalic" with deep craniums and narrow foreheads. Einstein was a clear exception. His traveling companion was immediately intrigued, analyzing even more carefully the physicist's appearance as he accompanied Einstein on the four-hour-long train ride to Paris. "A small very short black moustache" sat on top of a "sensual mouth." "The nose is sharply delineated and slightly aquiline," he remarked.[1] The physicist Louis Dunoyer found Nordmann's physical descriptions unnecessary, even propagandistic. Who would "judge the excelence of a theory from the nose of its author"? he asked. Not "a single French scientist," he answered.[2]

Other observers had been similarly attentive to Bergson's appearance. The philosopher had "a broad and powerful forehead with eyes that looked a bit like those of nocturnal birds accustomed to seeing in the dark, bright eyes, curious about everything, hiding under the shade of eyebrows."[3] His "powerful brain" appeared disproportionately large on top of a "slender body." His "wide forehead" housed "sunken blue eyes under thick eyebrows." His nose was "like the beak of an eagle," and a blond mustache was "cut flush" above "thin lips."[4] Whereas an admirer marveled at Einstein's "very black hair, mixed with silver, undisciplined, curls falling towards the neck and ears, before standing very upright, like an immobile flame of that large forehead," one of Bergson's enemies described the philosopher as looking like a "bald fetus."[5] Appearances mattered on April 6, 1922. The status, hierarchy, and worth of different kinds of beings was discussed even when debating about the nature of mathematics and multidimensional geometries.

Did time look different from the ground level of the lowliest of mice as from the all-encompassing point of view of a superman? Did it look different through the eyes of the tiniest microbe than from those of a giant, spread apart in an immense forehead? These examples—mice, flat fish (lampreys), supermen, microbes, and giants—appeared over and over again in seemingly abstract debates about the nature of space and time.

Did the radically different perspectives of these beings shed information about the nature of time? And if so, in what measure? Did the perspective of scientists suffice, or should animals be included? If so, which ones? Why should physicists decide about these questions, instead of philosophers? Was our knowledge of the universe inescapably anthropomorphic, or could the limitations of our all-too human perspective be overcome? These questions became even more relevant as racist theories against Jews and other minorities gained popularity during those years. "I begot children with a physically and morally inferior person and cannot complain if they turn out accordingly," Einstein explained to a friend during a particularly stressful period of his life.[6] He rashly thought it was "urgently necessary that physicians conducted a kind of inquisition for us with the right and duty to castrate without leniency in order to sanitize the future."[7] He did not spare his Serbian wife and children from these reflections.

Einstein disregarded the perspective of microbes; Bergson sought to include them. Discussions about sense organs should have absolutely *no* bearing on his conclusions, stressed the physicist. Microbiology had nothing to do with microphysics. His theory was based only on the laws of physics, and these laws were not apt to change according to how they were viewed. Bergson disagreed.

Time for whom? asked Bergson. "Scientific microbes," explained the philosopher that April evening at the Société française de philosophie, would perceive an enormous distance between the two locations that Einstein considered as being roughly at the same place.[8] "A thinking microbe," he argued, "would find an enormous interval between two 'neighboring' clocks."[9] If two microbes were positioned slightly to the left or to the right of each other, each microorganism would get slightly different readings of time (since the time taken by a light signal to reach either of them would be shorter or longer depending on exactly where they were positioned between two clocks). This tiny being, argued Bergson, would know that its ascertainment of simultaneity would not coincide with that of his microbial friend beside him, and certainly not with a human's. It would reach an entirely different conclusion about the nature of time than Einstein. This microbe would be "more Einsteinian than Einstein."[10]

Bergson elaborated these thoughts further in *Duration and Simultaneity*. The example of the microbe, according to him, proved that the concept of local simultaneity described by Einstein was not apt to stand for a universal definition of time, since it was simply an extension of one particular—and very partial way—of understanding time. To do justice to the concept of time (including how it was perceived by different beings) required a lot more work than Einstein had done.

MICROBIOLOGY AND MICROPHYSICS

Bergson speculated about beings who were simply sized differently, not only small, but also large. He asked Einstein directly what relativity would be like for "a superman with a giant's vision."[11] Would Einstein's theory of relativity still hold for a very large observer? How could we prevent it from perceiving absolute simultaneity? Why would we be justified in setting a limit to this potentially infinite chain of observers ranging from the infinitesimal to the infinite? Bergson reminded his readers that a different-sized observer would not draw the line between the distant and the local in the same way that Einstein did. From this, he drew the conclusion that we were simply *not* justified in drawing such a sharp distinction (between the local and the distant) as Einstein had done in his famous paper.

Bergson was not the only philosopher to ask these questions. The objection that Einstein's definition of simultaneity assumed a human-sized observer was frequently brought up. Was Einstein describing events at a uniquely human scale and selling them as universal? No. "Physical propagation velocities have nothing to do with the character of our sensory organs," he retorted as he sought to sideline an inquisitive attendee at one of his lectures. The participant agreed that "a velocity greater than the velocity of light is out of the question for the organs available to humans."[12] But he objected that it was perhaps possible for other beings with different sense organs to access the world at speeds faster than light. An infinitely large eye would *meet the phenomena at their location*, no longer requiring them to propagate in space at a finite speed. The speed of light would be irrelevant for this observer, who would effectively perceive the event *before* a distant smaller being, who would have to wait for light to arrive at him. If infinitely large, the effect would be instantaneous.

FLATLANDERS

While discussing gigantic and miniscule beings, scientists also discussed abnormally thick and thin ones. Paul Pierre Lévy, a mathematician and professor at the École polytechnique, raised a question during the discussion. He asked the physicist what he thought of "infinitely flat beings on the surface of a sphere."[13] Einstein, in his popular account of his theory, had referred to "flatland" beings confined to live in two-dimensional space. The example of flatlanders showed how our sense of three-dimensionality was dependent on our limited point view.

The example of flatlanders was widely popular at the time. A novella written decades earlier, titled *Flatland: A Romance of Many Dimensions* (1884), had described in convincing detail the drama of living in a world of two and more dimensions.[14] Bergson used the example of flatlanders as well. In *Duration and Simultaneity*, he contrasted imaginary "flatland" beings confined to live in a two-dimensional space with a "superman" who was not confined to our three-dimensional world. He reached the expected conclusion that the reality of a superman, as that of a flatlander, was radically different from that of humans living in three-dimensional space. But he drew different conclusions from this fact than did Einstein. His argument echoed Poincaré's. The mathematician had insisted that there was something special in our three-dimensional understanding of the world because of how it related to our particular, human, constitution. In *The Value of Science* (1905), Poincaré confronted authors who claimed that our understanding of the world as three-dimensional was a mistaken inference stemming from our particular physiological constitution.

Scientists evaluated the value of non-Euclidian mathematics in terms of recent research in animal physiology. Poincaré studied the most up-to-date information about how different species experienced the space around them. Japanese mice and flat bottom-feeding lamprey fish were thought to experience the world in two dimensions or less: "Japanese mice have only two pairs of canals; they believe, it would seem, that space has only two dimensions."[15] Physiologists knew that semicircular canals in the ear functioned as motion sensors that oriented us in space. The number of dimensions experienced by different species was largely considered to depend in some way on these organs. Humans had three canals, Japanese mice had only two, and flat fish had only one.

Japanese mice and lamprey fish were actual examples of the flatlanders frequently discussed by mathematicians and physicists. While Poincaré pored over the latest research on mice that could not lift up their bodies from the ground and bottom-feeding fish scurrying along the floor of our oceans and lakes, he continued to refer to "the fiction . . . of beings who, having been educated in a world different from ours, would have been led to create a non-Euclidean geometry," adding that "beings still more strange may be imagined." How did the perception of space and time by these beings relate to the actual nature of space and time?

Einstein and Bergson drew different conclusions from the examples of different beings and flatlanders. For the physicist, they showed why the common conception of space as three-dimensional arose from a limitation due to our particular bodily constitution. For him, our ability to think led us to discard this limitation and reach broader conclusions. This ability was a mark of our superiority vis-à-vis inferior beings. It was even a mark of Einstein's own superiority. When his younger son, Eduard, once asked him why he was so famous, Einstein replied: "When a blind beetle crawls over the surface of a curved branch, it doesn't notice that the track it has covered is indeed curved," he said. "I was lucky enough to notice what the beetle didn't notice."[16] The reason why "the universe of these beings is finite and yet has no limits," is because they—blind beetles—simply did not know any better.[17] When during his trip to Chicago he explained to a journalist why he had noticed aspects of the universe that other beings had been unable to see, the reporter felt utterly insulted:

> It began to trickle into his brain that the two-dimensional organism referred to was himself, and far from being the 13th Great Mind to comprehend the theory he was condemned henceforth to be one of the Vast Majority who live on Main Street and ride in Fords.

The journalist bitterly noted the growing gap between "Main Street" and science. He felt that the inferior organism mentioned by the physicist was not unlike the "Vast Majority" of people who "ride in Fords."[18]

Einstein used the example of flatlanders to show why our three-dimensional perspective should be abandoned and why the universe should be described in multidimensional terms. For Bergson, the same example (of flatlanders) proved why no scientific conception—and certainly not Einstein's—was free from certain assumptions about how different living consciousnesses related to the world. Einstein, Bergson, and their followers and detractors held different views about the relation of humans to other sentient beings—even ghostly ones.

GHOSTS, PHANTOMS, AND FICTIONAL BEINGS

"Only ghosts,' Einstein argued, can hear the sounds of "an eternally uniformly occurring tick-tock." Who believed in these ghosts? Almost everyone, argued Einstein. Why? Because almost everyone believed in the notion of time associated with a universal clock. "Ask an intelligent man who is not a scholar" what time is, Einstein continued, and you will see that he takes time to be this ghostly "tick-tock."[19] The physicist quickly chased away these ghosts. "There is no audible tick-tock everywhere in the world that could be considered as time," he concluded.

While Einstein wrote about ghosts, Bergson turned to phantoms. He accused Einstein and his followers of introducing into science fictional, fantastical scenarios and portraying them, uncritically, as real. *Dilated* time, the philosopher would not tire to insist, was the time of *phantoms*.

It would be tempting to think that scientists could tell what was fictional and what was real on April 6, 1922; what applied to flesh-and-blood and what was ghostly; what was of this world and what was otherworldly. It would be tempting to think that the boundaries between fiction and science were clear and distinct at that time and had been so at least since the time of the Scientific Revolution. Yet scientists and philosophers that evening debated about *who* should draw the distinction, and how. Throughout *Duration and Simultaneity*, Bergson disagreed with how Einstein drew the boundary between fiction and reality in his work. More important, the two men disagreed about how the boundary between fiction and reality was drawn more generally.

Would physicists be the ultimate arbiters about what was realistic and what was fantastical in our world? In 1923 Einstein published a short introduction to a small book containing wondrous stories.[20] He had encountered these stories as a child, remembering them for the rest of his life.[21] "The theory of relativity," wrote Einstein "saves us" from some of the "bizarre" conclusions of book. What conclusions? Why was Einstein so concerned with setting the record straight and pointing out its mistakes? Physics, in Einstein's short introduction, occupied the position of a judge vis-à-vis other forms of knowledge.

The book, *Die Gestirne und die Weltgeschichte* (The Stars and the History of the World), contained stories of beings that could travel faster than light. From their perspective, the history of the world looked completely different. Einstein had good reasons for concern: faster-than-light speeds invalidated the most important conclusions of relativity theory. What is more, these beings were gifted historians. By positioning themselves far away from Earth in strategic places where light waves of past events where just passing by, they could see key episodes of history. Einstein first read these stories as recounted by Aaron Bernstein—a favorite author of his during his youth.[22] Bernstein recounted how

> in one point in space, the light of the scenes of the French Revolution is just coming into view. And even further away, the invasion of the

> barbarians has just become the order of the day, Alexander the Great is still conquering the World. . . . And even further away in space, the representation of Earth's past by way of light will just be advancing into the future, historical events that have long been dead for us will just be coming to life.[23]

For Einstein, science could check these fictional scenarios and the extravagances of overactive imaginations; for Bergson, the relation of science to fiction and the role of the imagination in both needed to be understood philosophically. Why ignore the fact that in the famous voyager of relativity one could count on the law of gravity as surely breaking the spaceship in half because of the strain of the speed? Why take other effects, such as time dilation, as scientific facts? These questions were raised after some of Einstein's most prominent defenders filled their popular accounts of relativity with examples that even the most sober of readers would probably consider fantastical.

When Paul Langevin described the "voyage à boulet," he explicitly described the ship as "the projectile of Jules Verne."[24] Arthur Eddington described smoking cigars that could last twice as long and asked his readers to visualize the indicator dials on aviator's watches going at half speed.[25] Bertrand Russell explained the theory of relativity by narrating it from the perspective of a drugged balloonist seeing the fireworks display during the Fourth of July. He wrote of flies landing on stagnant pools of water and bending the previously flat surface toward their bodies. He vividly described passengers eating at railway carts where "dinner plates which you see as ordinary circular plates, will look to the outsider as if they were oval."[26] What role did these narratives play in science? Were they merely creative techniques for explaining complex science to nonspecialists, or was there more to them?

The topic of disagreement between Einstein, Bergson, and their respective defenders had repercussions not only for science but also for the arts and for art criticism. When first reading about relativity, Bergson was reminded of the work of H. G. Wells, author of *The Time Machine*:

> We read in one of the first works on the theory of relativity, by Silberstein, that Wells had wondrously anticipated this theory when he had his "time traveler" say that "there is no difference between time and space except that our consciousness moves along time."[27]

He was not the first or the last to draw that connection. Many others who read Einstein and his popularizers were "reminded of . . . the remarkable novel *The Time Machine* by the Englishman Wells, who a dozen years ago," in a "remarkable anticipation of future research" spoke "of time as of a fourth dimension which is of equal validity with our ordinary three dimensions of space, indeed is even interchangeable with them."[28]

Why was Bergson intent on comparing Einstein with H. G. Wells? Clearly, the physicist's contributions went far beyond those of the English writer. It was evident that each man was working in completely different areas for different purposes and different audiences. Why draw a connection between the two?

Scientists, including Einstein, frequently used thought experiments, known as Gedankenexperiment, in their work.[29] These thought experiments were different from other products of the imagination in that they, ostensibly, *could* be reproduced under certain circumstances. It did not matter that some aspects of the thought experiment may not be practically realizable at that moment. They were legitimate as long as they could one day be realized; as long as their potential realization did not contradict any known laws. When engaging in thought experiments, scientists and philosophers frequently distinguished between imagined scenarios that were potentially realizable and imagined scenarios that were considered to be ultimately unrealizable, and they understood that a certain territory of the not-even-imaginable also existed. The differences between these three cases were essential for scientists, who used the first method frequently to decide about the validity and scope of their theories and who shunned the other two.

In the eighteenth century, the Enlightenment thinker Denis Diderot studied the particular qualities required to transform an imaginary vision into a "realistic" scene. Similar questions concerned scientists and philosophers, who debated about what should be considered "realistic" in science. At the time of Bergson's and Einstein's debate, the line between realistic thought experiments and improper ones was as unclear as beliefs about what could be realistically possible and what would remain completely fantastical. For this reason, their initial confrontation was soon transformed into a much larger disagreement about which imaginary scenarios should, would, and could fit (or not) within the realm of scientific possibility.

IMAGINATION

Bergson insisted that scientists should be allowed to "imagine" that two events taking place at a distance from an observer were simultaneous even though he acknowledged that the observers described by Einstein could not actually determine their simultaneity. Granted that scientists did not currently know how to determine it, and even the most talented engineers could not implement such a comparison, should the possible existence of such simultaneity be denied? Bergson insisted that scientists should be allowed to think of simultaneity at a distance in terms of how they understood local simultaneity, and they should be justified to treat the first in the same way they conceived the latter. We should not, he argued, prevent our minds from thinking in this way: "But why prevent our imagination, and even our understanding, from treating the simultaneity of the readings of two very widely separated clocks like the simultaneity of two clocks slightly separated, that is, situated 'at the same place'?"[30]

What was at stake? Disagreements between Bergson and Einstein about what scientists could legitimately imagine as possible had implications that veered from the metaphysical to the theological. Bergson wanted to defend the power of imagination and understanding to think in infinite terms: "How can we prevent our imagination, and even our understanding" from thinking in this way? Why prevent our imagination from conceiving the possible existence of a being large enough to grasp events all at once? asked Bergson. "A superman with a giant's vision," he wrote, "would perceive the simultaneity of two 'very distant' instantaneous events as we would perceive two 'close by' events."[31] Bergson refused to exclude intuitions of simultaneity-at-a-distance from the toolbox of scientists.

A number of Bergson's defenders agreed that they should be able to imagine the possible existence of such powerful (infinitely fast or infinitely large) beings as part of their scientific investigations. Maritain argued that one could logically think that "at the instant that a clock shows this time in New York such event occurred in Paris" even though "the same observer could not be at the same time in New York and in Paris to see the clock here and the event there." This natural, anti-Einsteinian conception of simultaneity, Maritain argued, approached that of angels and God. While humans could only obtain it through a process of "abstraction" connected to an "ideal existence" using "pure notions," angelic intelligences could reach it "intuitively" through their "actual existence."[32] We should strive to reach these perspectives, he argued, and not write them off from science. Whitehead, similarly, refused the injunction against thinking of simultaneity at a distance in an intuitive sense.[33] These imaginary alternatives were considered in geopolitical and ethical terms. Eddington used the examples of Peru and China to explore these questions.[34] Would we deny the simultaneous existence of a faraway being in a space just because it was physically unreachable?

Einstein's injunction against considering such possibilities as scientifically realistic was particularly radical because of how it contrasted with Poincaré's classic descriptions of the role of imagination in science. For Poincaré, the power of science lay in its capacity to imagine radical changes in the small and the large, the outside and the inside, and—even more pertinent to the case against Einstein—the slow and the fast. These extrapolations of the imagination were natural for humans, he argued. We could, for example, easily imagine giants: "I shall imagine what a giant would experience who could reach the planets in a few steps." We could easily imagine our selves as gigantic "or, if we prefer, what I should feel myself in the presence of a world in miniature, in which these planets would be replaced by little balls, while on one of these little balls there would move a Lilliputian that I should call myself."[35]

Discussions about giants and Lilliputians in physics were amusing, but they had serious repercussions. In his discussions on geometry that Einstein later contested, Poincaré discussed the example of a man trapped in a prison. "Suppose I am enclosed in a chamber between the six impassable boundaries formed by the four walls, the floor, and the ceiling," he wrote. He could always imagine a world outside, as well as the possibility of escaping, by naturally extrapolating from the possibilities of movement inside the prison to its outside. Poincaré stressed how "this act of imagination would be impossible for me if I had not previously constructed my restricted space and my extended space for ordinary use."[36] Because of how we ordinarily engaged with our spatial surroundings, *this* act of the imagination (one connected to the idea of liberty) was possible, he insisted. Why would considering scientifically the real possibility of one day obtaining ever-faster transmission speeds be any different?

If fiction could be realistic and reality fictional, where exactly did the boundary between them lay? Bergson forcefully insisted that the beings in Einstein's cars, ships, boxes, and cabinets were fictional and phantasmagoric. As Einstein's science moved from Earth to space, from clocks to time, and from dead matter to living beings, Bergson and his followers claimed that when the physicist substituted one for the other, he ended up with neither—only with a ghost of the two. In the end, Einstein and Bergson, their followers and critics, not only debated about what was already possible according to the most up-to-date knowledge of physics, they disagreed about what *could* be possible, now and forever.

CHAPTER 27

One New Point: Recording Devices

Two years after the debate, Bergson reconsidered all the arguments his polemic had set off. After passionate and lengthy discussions, the philosopher found that "only" one "new point" had emerged. This new point centered on the example of automatic inscription devices used for timing, recording, and transmitting events in time.

Bergson's critics and Einstein's defenders claimed that to understand time "automatic registering devices suffice, without the need of an observer to see what they mark."[1] Einstein used the word "observer" merely for convenience, but these "observers" could simply be recording instruments:

> It is true that for the sake of convenience authors of works on this issue have often introduced "observers" who are assumed to perceive the phenomena. But these observers can just as easily be replaced by recording instruments, and whether or not they are "living and conscious" is irrelevant.[2]

Bertrand Russell, in his explanation of relativity, insisted on this same point: "It is natural to suppose that the observer is a human being, or at least a mind; but it is just as likely to be a photographic plate or a clock."[3]

After considering this "new point" Bergson was *almost* ready to concede to Einstein.[4] He had already admitted that—in those circumstances—Einstein was entirely right: "One could naturally say . . . that [clocks traveling at different speeds] cannot run in synchronicity. . . . In effect, time slows down when speed increases."[5]

Bergson considered the possibility of completely eliminating the human observer from Einstein's theory. *If* clocks could replace humans, then Einstein would be entirely correct. But, for Bergson, the complete elimination of the human observer from the world was itself a philosophical riddle. The universe, as we know it, was one that housed consciousness *and* was conceived by conscious beings. For this reason, Bergson continued to claim that the time marked by the clocks described in relativity theory was not Time. To consider these slowed-down clocks as Time itself was tantamount to adopting a metaphysical stance in which humans and automatic registering devices were interchangeable.[6] Where did this metaphysical stance come from? Why was it so seductive?

On repeated occasions Einstein claimed that the laws of nature he described remained valid even if no one perceived them. They described the universe as if no one had made them. But scientists and philosophers were divided about these propositions, just as they were divided about the merits of Bergson's philosophy versus Einstein's science.

BETWEEN CLOCKS AND HUMANS

Einstein was at first criticized for moving from examples involving clocks to those involving people without acknowledging any difference between the two cases. In a lecture given in January 1911, the physicist claimed that time dilation would affect "a living organism in a box."[7] After learning about Langevin's presentation in Bologna—with its references to Jules Verne and space travel—Einstein increasingly thought about how his theory affected not only clocks but also humans. What happened to an atom or a person—or anything else for that matter—inside moving vehicles, cabinets, or boxes that sometimes had windows and sometimes did not; that sometimes moved up and at other times down, or forward or backward; or that accelerated, or remained at constant speeds, on the surface of the Earth or in a remote corner of the universe? The movement of these vehicles and boxes, argued Einstein, affected time in general just as much as the time of anything caught within. "What kind of being," we place in these contraptions, he explained, is "immaterial to us."[8] Whatever they be, their time would dilate if moving at speeds close to that of light.

Would the effects of time dilation affect biological beings? Yes, answered Einstein. Not so quick, replied Bergson. Throughout *Duration and Simultaneity*, Bergson expressed the need to think differently about how relativity affected flesh-and-blood conscious beings than it did nonliving entities. He chastised Einstein for moving so quickly from descriptions involving the first to those involving the second. Why such differences in interpretation?

Would changes in astronomical time directly transfer to changes in human time? The relation between astronomical time and human time had become an increasingly relevant question for various thinkers since the middle of the nineteenth century. After realizing that the clockwork universe seemed to be winding down, thinkers across fields started to inquire into the consequences of the Earth's changing velocity on humans. Would the speed at which living beings functioned change accordingly? The jurist and writer Felix Eberty, who had conjured the stories about infinitely fast beings that so concerned Einstein, explained how life processes would change in direct proportion to changes in the acceleration of the Earth's rotation. If its velocity was doubled, "the drawing of our breath, and the stroke of the pulse would proceed with double their usual rapidity, and our new period of life would appear to us as the normal length." Because of intimate connections between biological life and astronomical processes, these changes would likely occur "without our being able to perceive the change" since our perceptual capacities would change at exactly the same ratio.[9] Unless a person knew enough about astronomy and thermodynamics, a common observer simply had no other way of knowing that these changes were occurring. This effect was similar to that which affected Gulliver's Lilliputians who considered themselves to be "perfectly grown men" because they had had no one against which they could have compared their miniscule statures.[10] Speculations about how the temporality of biological systems was connected to astronomical ones were continued by Ernst von Baer, Carl du Prel, Bernstein, Poincaré, and others. Einstein and Bergson followed the work of some of these authors closely.

What would happen to our sense of time during moments of rapid changes in the time of the universe? For Bergson, they would likely produce a proportional change in our "intra-cerebral molecular movements." If so, we would remain oblivious to these changes. Scientists who talked carelessly about the reality of dilated times could only do so by assuming that a "watchful consciousness" that was itself *not* subjected to these changes existed independently of them and served as a point of comparison.[11] If they did not want to posit this hypothetical consciousness, then scientists would have to think more carefully about how the time of the universe was connected to the time of conscious beings, the time of science to the time of experience, and changes in acceleration to differences in time.

"FROM OUR HUMAN POINT OF VIEW"

Accounts that threw into relief the potential tension between clock time and lived time led scientists to rethink the status they should attribute to each. They threatened to upset current hierarchies between the biological and physical sciences. Perhaps the notion of time connected to living consciousness should be prioritized? Charles Richet, the scientist most closely associated with the concept of physiological time and "a close friend of Bergson," was one of the most prominent scientists to insist that Einstein's conclusions about time in the universe could not be so uncritically extended for humans.[12] Richet revisited his lifelong work on "the unit of psychological time" in 1921. In a volume of the prestigious *Comptes rendus de l'Académie des sciences* that was filled with new essays about the meaning of time in Einstein's theory of relativity, Richet summarized work that had commenced as early as 1870s. Richet concluded that the smallest time-unit of consciousness was 0.08 seconds. This number could be associated with many others, such as vibrations of light, sound, or an electric spark. It was "indissolubly tied to the function of our organs and the workings of our conscience. It is an integral part of our humanity."[13]

Richet argued that "psychological" time was better than other measures, including astronomical time. To prove his point Richet turned to an example reminiscent of the "voyage à boulet," in which a man was sent to space for ten years to travel around the Earth in twenty-four hours going from east to west. This man "astronomically remained in the same day, but for that same reason he did not age any less that 10 years." Richet concluded that "time passed in our conscience and organs, independently of everything that is not in our conscience and organs, regardless of the measure adopted."[14] Just because astronomical time was halted in this way, that did not imply that time stopped.

Richet brought his example even closer to relativistic concerns. "If we were impelled through space by an even more prodigious translation speed," the time of the traveling observer and the stationary one could appear "much longer or much shorter." This however, "matters little" since "from our human point of view, the succession of physiological phenomena would remain the same."[15] After citing Richet's work, a writer to *L'Écho de Paris* turned to "that which Einstein wanted to say." He sought to understand Einstein's contributions in light of current research on physiological time—a topic to which the "press referred frequently."[16] "Following Charles Richet," explained the journalist, the "physiological unit of human time" is "1/12 of a second." This unit, according to the writer, arose from "a long adaptation to our milieu" and thus "could not vary brusquely in astronomical milieu." Even if clock time or astronomical time dilated, time would not instantly change for an observer traveling at great speeds. A change could only occur if the observer had "followed a long evolution in a milieu whose speed was more rapid than ours."[17] In later works, Richet would reach even more radical conclusions that went against many aspects of the science of his time. He dedicated *The Sixth Sense* to Bergson, calling him "the most profound thinker of modern times."[18]

DIRECT-READING INSTRUMENTS

Was it necessary to talk about actual humans when discussing physics? What if instruments replaced humans? Bergson's obsession to refer back to actual observers could be bypassed by replacing them with clocks.

The use of automatic inscription devices in science increased as the nineteenth century progressed. Yet in Bergson's view, they were unable to replace human observers in their entirety. Just as scientific notions of time could not be freed from certain assumptions about how these related to human consciousness, recording devices by themselves would never solve these questions. Why? For the simple reason, answered Bergson, that an observer would always be needed to see and read the record.

At the time of the debate, reading most scientific instruments was usually not more complicated than reading a clock and noting a number indicated on a dial. By then, observation practices in science had been simplified by direct-reading scales and automatic inscription devices.[19] For the most part, taking a measurement was as simple as comparing one signal (such as a clock hand) against another one (such as a number on a dial) at a single glance. In contrast to other forms of observation or reading practices, the use of these instruments required almost no interpretation from the part of the observer.[20] Scientists' "reading" activities had been largely reduced to noting numbers.

Since the fifteenth century, numbers started their long exodus from the alphabet of letters.[21] Printed scales on bars and balances, grated leveling instruments, thermometers, barometers, chronometers, galvanometers, and photometers—all permitted scientists to use direct-reading methods to take measurements in ways that could be easily recorded and transmitted. Some of these instruments left a trace on a scale that could then be consulted at leisure by the experimenters; others used needles to indicate numbers on a dial. With these devices, scientists could simply measure the phenomena under investigation with an instrument that permitted the reading off of a certain quantity, usually as a number. Scientists rarely had to smell and feel substances or describe them in verbose poetic detail.[22]

Numerical measurements were hardly limited to length and time. Temperature, electricity, pressure, light, color, and even facial traits and expressions could be assigned a clear quantitative value. These techniques were used not only by physicists but also by teachers, doctors, anthropologists, and even policemen. Standardized charts and simple numerical codes were used to send numerical data telegraphically, even permitting the transmission of something as complex as a description of a criminal's appearance. The astronomer Arthur Eddington noted these dramatic changes: "The whole subject-matter of exact science consists of pointer readings and similar indications," titling an entire chapter of his book "Pointer Readings."[23] Edmund Husserl, in his influential book *The Crisis of the European Sciences*, complained how "the visible scales, scale-markings, etc. are used as actually existing things, not as illusions."[24] A few years later, Susanne K. Langer, a philosopher and a student of Whitehead, described the pervasiveness of inscription devices in science:

> The men in the laboratory have departed so far from the old forms of experimentation—typified by Galileo's weights and Franklin's kite—that they cannot be said to observe the actual objects of their curiosity at all; instead, they are watching index needles, revolving drums, and sensitive plates. . . . Observation has become almost entirely indirect; and readings take the place of genuine witness. The sense-data on which the propositions of modern science rest are, for the most part, little photographic spots and blurs, or inky curved lines on paper. . . . Instead of watching the process that interests us, that is to be verified—say, a course of celestial events, or the behavior of such objects as molecules or ether-waves—we really see only the fluctuations of a tiny arrow, the trailing path of a stylus, or the appearance of a speck of light.[25]

When Einstein and Bergson met, the act of taking a measurement was not much different from that of reading a clock.

READING SIGNALS: TWO SIGNALS ARRIVING AT THE SAME PLACE

The concept of measurement in science changed in step with these media transformations. For most of the nineteenth century, measurement required scientists to compare two objects directly, or one of them against a scale, by aiming or leveling an instrument against an object, as with a theodolite, or when using rulers, by aligning two different objects against each other. Precision measurements were frequently done with a micrometer, requiring users to carefully bring two marks in line with each other or to find the exact center of circles and dots.[26] In many cases all an observer needed to do was compare the arrival time of the two signals.[27]

Einstein's work emerged in the context of profound changes in the meaning and practice of observation and measurement more generally.[28] His theory of relativity was based on a particular notion of scientific observation, one that considered it in terms of light signals sent and arriving at a source and that contrasted starkly with how it was defined at other historical periods.[29] He described "simultaneity" in terms of the behavior of light rays meeting at one source. One of his interlocutors described this conception as "a coincidence at once spatial and temporal on the retina of the observer."[30] The experimenters described by Einstein did not go to the objects themselves to measure them and they did not draw, feel, or smell them. Einstein's descriptions of simultaneity flourished in a culture increasingly characterized by simple responses to successive stimuli—responses to the arrival of light signals at a single place.

Did scientists read the book of nature? Did they read it *as a* book? How literally did they understand their work as reading? In the sixteenth century, investigators understood both in quite interchangeable ways: reading nature was an activity similar to that of reading the Bible. But by the time that Bergson and Einstein debated, scientific practices had changed so significantly that scientific work no longer seemed comparable to interpretative reading. Galileo was an important figure marking this change. Because "the book of nature" was written in the language of mathematics, he argued, scientists did not need to interpret it in the same manner as they had to interpret other texts.[31] Science, in his view, was not an interpretative (that is, hermeneutical) activity. For this very reason, scientific work should be considered to be different from that of the humanists.

The debate between Bergson and Einstein on time involved an argument about direct-reading and automatic inscription devices, about instruments where a needle pointed directly to a number or that left a legible mark on a roll or a piece of ruled paper. When he debated with Einstein in person, Bergson insisted that automatic inscription devices alone could never measure time by reminding readers that these instruments—in order to serve us—had to be read. He argued for the need to think more carefully about these instruments. "It is to them that we must turn to criticize" Einstein's conception of time. The philosopher closed the meeting by reminding listeners that every time humans "read an instrument" to find out what time it was, human concerns affected this "reading." Philosophers, he believed, were free—even obliged—to explore this area of contact.

To the very end, Bergson differed with Einstein by focusing on what "reading" meant—either "reading" a clock, "reading" a scientific instrument, or "reading" the book of nature. At the time of the debate, concepts of "reading"—both inside and outside laboratories—were rapidly changing as new recording technologies—graphic, photographic, and phonographic—started to intersect in new ways with traditional print culture. Both Bergson and Einstein lived in the midst of these profound media transformations.

Bergson considered instrument-reading practices—even ones as simple as reading a clock—as *philosophically* essential. They needed to be considered if we wanted to understand how scientists gained knowledge of the natural world. They showed how this knowledge depended on the active intervention of a human observer. Einstein disagreed. For him, the universe—and our knowledge of it—did not depend on any observer, human or otherwise. It would go on ticking just as efficiently, and exactly as his theory described it, without us and as if we never existed. Clocks, by themselves, showed how time advanced.

Why did a debate that started about time end up as a debate about reading? And why were those involved discussing such a peculiar form of "reading"—the reading of a clock or a scientific instrument? The debate between Einstein and Bergson exacerbated just as instrument "reading" practices differed ever more radically from other types of reading. New laboratory instruments obviated the need for any complex analysis of sensations or signs, symbols, letters, words, sentences, paragraphs, or books. If reading an instrument was not like reading a book, was it still necessary to refer to it? Yes, answered Bergson.

An inescapable human component always appeared, Bergson reminded listeners. Without this element, "it would not be possible to read an instrument." Scholars needed to think about what enabled its reading to take place and how it involved an "intuitive" sense of simultaneity and succession. "Without it we could not read an instrument," he repeated again in his book.[32] Bergson knew fully well that these practices, when understood philosophically, were hardly straightforward. He concluded his comments that evening by stressing that "we need to return to psychological assessments" in order to understand how we could "read an instrument"—even one as simple as a clock. With this last sentence, the discussion came to a close.

PART 4

THE WORDS

CHAPTER 28

Bergson's Last Comments

DESCARTES CONGRESS, PARIS

Bergson mentioned Einstein one last time, in writing, in 1937. He was seventy-eight years old. In a long note he sent to the Descartes Congress, he first apologized: he was too sick to attend. He then explained how, when thinking about Descartes, he was immediately reminded of Einstein. He proudly wrote that he was old enough to remember the first congress. But almost four decades later, his health was so fragile that he could barely write. "Rheumatism has made writing become for me a veritable suffering," he explained to a friend during those years. "And besides, I have never been able to dictate."[1] Making a valiant effort to overcome his debilitating arthritis for this particular occasion, Bergson wrote about Einstein one last time.

The note described Einstein as brilliant, savvy, and ambitious. But it provided an image of Einstein that differed markedly from the one the physicist promoted of himself. According to Bergson, Einstein was driven as much by discipline as by pleasure. There was no denying that in his early years he had been a soldier with a mission, but things changed in his later ones. Einstein was a man who had "practiced *grand tourism*, covering, first as a soldier [for science] and then for his own pleasure, Germany, Hungary, Switzerland, Holland, and even more countries."[2] But Bergson then accused Einstein of having used the League of Nations *not* for its intended purpose of promoting relations among scientists and intellectuals, but primarily as a networking forum used for his own advantage—to "get in contact with scientists all over the world, corresponding with a princess, lecturing to a queen." Yes, sometimes, Bergson pictured Einstein deep in thought. But mostly he pictured him as an action hero: "I also see him on a ship where the crew conspire to steal and to throw overboard, anticipating them, and drawing his sword to hold back the bandits."[3] The scene described by Bergson was like those that could be seen in the new blockbuster movies and propaganda films that were gaining more and more audiences during those years. "Einstein," explained Bergson, always tried to produce a "maximum effect" from his efforts. His whole life was organized for this purpose, argued the philosopher. The physicist had positioned himself in America in order to "organize his life to draw maximum effect from it."[4]

Bergson did not commend a life as active as Einstein's, but he did not preach passivity either. Bergson urged his reader to strive to connect thought with action more tightly. Delivering one of his most celebrated and oft-quoted phrases, he concluded: "One should act like a man of thought, and think as a man of action."[5]

FAME AND LIGHTHOUSES

The image that Einstein promoted of himself was radically different from Bergson's descriptions. In a famous 1933 speech, now remembered as the "lighthouse" lecture, Einstein described scientists as pure, isolated thinkers. He compared their lonely work to those who worked "in the service of lighthouses and lightships."[6] Which image of the scientist's work, and of Einstein's own life, was most accurate? Einstein was hardly isolated. The selfsame address vaunting the virtues of a lonely scientific existence was delivered at the Royal Albert Hall in front of some 10,000 people and recorded so that it could be used as a soundtrack for a newsreel. Einstein's public portrayal of the work of scientists hardly fit with his own work.

Einstein was not the first scientist to carefully promote himself to the public. Darwin followed his own image obsessively, collecting thousands of news clippings of himself. But Einstein was the first scientist to obtain a worldwide reputation through mass media. He was, bar none, the main representative of the modern scientific genius—a figure that emerged alongside that of the Hollywood star. In 1917 the *Motion Picture Magazine* included the question "Who's Your Favorite Film Star?" yielding more than one hundred candidates. The American beauty Mary Pickford emerged as number one.[7] But Einstein rivaled her in terms of face and name recognition. During an embarrassing moment while traveling in California in 1931, the famous movie star introduced herself to Einstein, who was caught off guard, having no idea who she was.[8]

Shortly after the widespread news coverage of the eclipse expeditions that confirmed his theory, Einstein noticed how in Berlin "every child knows me from photographs."[9] "I must serve as a famed bigwig and decoy-bird," he wrote to a friend, and "let myself be shown around like a prize ox," he explained to another one.[10] His wife Elsa charged a modest fee for every photograph taken of him and every autograph he gave. Critics were quick to accuse: "Everything is publicity for Einstein."[11] Einstein consciously went on an antipublicity campaign, concerned that critics accusing him of being a propagandist were being taken too seriously. His friend Max Born urged him not to "authorize publication of *Conversations*," a laudatory book about Einstein, "after having been accused of seeking 'publicity.'" Additional publicity would be "offering your opponents more ammunition."[12] In 1922 the *Einstein Film* premiered to the delight of a growing public.[13] Complete with "the illusions of trick shots," it was one of the most watched educational films of the time.[14] After work on the film started, Einstein was careful to ensure that his personal involvement with the film would not be made public.[15]

The 1930s saw the development of a full-blown film industry increasingly based on full-length blockbuster feature films. Einstein loved the movies. One of the first things he did when he returned to the United States was to go to the movies. In California, he saw Sergei Eisenstein's *Que Viva Mexico* and Charlie Chaplin's *City Lights*, prominently appearing next to the director, who was also one of the world's most famous celebrities, at the premiere. He also went to the Warner Studios, meeting Jack Warner, and visited Carl Laemmle's Universal Studios (who filmed the scientist with a secret camera). He became close to the socialist writer Upton Sinclair, who promoted the Soviet film director Sergei Eisenstein in Hollywood, and wrote a preface to one of his books. After he was invited to see the film adaptation of Erich Maria Remarque's *All Quiet on the Western Front*, he became an important advocate of this film, which had been censored in Germany because of its brutal depiction of the war.[16] Einstein had a strong interest in propaganda films. In 1938 he went to the preview of *The Fight for Peace* in New York City and later watched *Where Will You Hide?* an antiwar animation film.[17] After the war, Einstein continued to go to the movies.

Einstein was as much a subject of mass media as he was an avid spectator of it. Even his sons were buying and reading biographies of their own father to try to figure out who he was. Einstein cautioned them that they might not be getting the full picture.[18] A full-length biography tellingly titled *The Drama of Albert Einstein* (1945) described the man in terms of a tension between his private and media persona.[19] The aim of this biography, like so many written since then, sought to disclose the "real" man underneath the public face. In 1950 he was filmed at his Princeton home for the television program *Today with Mrs. Roosevelt.* His seventieth birthday was considered a "radio occasion."

Because some of the historical events in which Einstein had participated occurred before the widespread use of recording technologies, Einstein embarked on various reenactments of them. In 1932 he agreed to record a version of his earlier "What I Believe" (1930) speech so that it could be played back with a phonograph. He was also filmed reenacting an event that had taken place six years earlier: the signing of the famous August 1939 letter to President Roosevelt urging him to support nuclear weapons research. The reenacted film was produced for the weekly newsreel series tellingly titled *The March of Time*.

EINSTEIN ON IDENTITY AND SELFHOOD

Einstein was keenly aware of the different roles he had to play in life and how "playacting" was an essential part of it. Being a scientist was one of many other roles. When complaining about how "all but one" of his colleagues were trying hard to "poke holes" in his work to "refute" his theory, he considered their competitiveness as a comedic farce: "If one is pressed into playing one's role as an actor in this farce, one is richly compensated for the pain and effort by being able to watch as a spectator the others' playacting."[20] The physicist was aware of the power of being a media personality years before his theory became a topic of world-wide interest. Only days before the Great War exploded, he already referred to himself as "the great animal in the illustrated newspapers."[21] Afterward, when his fame increased even more, this feeling only became more acute.

New innovations in printing technologies permitted the press to print news as fast as the telegraph and telephone delivered it, greatly outpacing communication by correspondence. By the 1920s, largely due to the development of the printing telegraph and typewriter keyboard, every individual could become a miniature printing press. Bergson and Einstein took advantage of these new technologies in different ways. Einstein shunned these mechanical technologies for his own writing but welcomed those who used them to write about him.[22]

Einstein's wife complained about an imbalance between the writings *from* Einstein and those *about* Einstein that she received. She obtained less news from him than about him: "Your agenda is revealed to me always first by newspapers, before your messages arrive."[23] So did his most intimate friends, who first read in the newspapers Einstein's scandalous intentions to leave Germany after the anti-Semitic events of the Philharmonic Hall anti-relativity lectures and to quit the League of Nations. "The *Neue Freie Presse* reports," started a letter from his friend Fritz Haber.[24] "In the newspapers I read," started another letter by Lorentz.[25] Many other letters, all starting with similar phrases, followed. "Various papers published the announcement that you were giving up your Berlin job. Is that true?"[26] His friend Paul Ehrenfest did not recognize the man whom he read about in the newspapers, finding an essay by the physicist "so chock-full of un-Einsteinisms that I could not believe you had written it yourself." The print media, explained Ehrenfest to Einstein, "really did succeed in making you, for a short time, 'beside yourself.'"[27] During that time Einstein described himself as a *media* King Midas, who turned everything he touched into news: "Like the man in the fairy tale, whose touch turned everything into gold, thus it is with me, with everything turning into banner line news."[28]

During the war, Einstein had imagined a person's inner mind like a "house" full of "furniture": "When I peer into the mind of a decent average citizen, I see a dimly lit comfortable space." That house contained a "shrine" in which there was "written in huge letters the word 'patriotism'" and that "contains the moral requisites of bestial hatred and mass murder." Einstein asked "the man of the house" to find "a more fitting piece of furniture" to substitute for this shrine. He should "consider placing a piano or a bookshelf in that same corner."[29] At the time that Einstein wrote these lines, it was natural to consider a bourgeois room furnished by pianos and books, which were only starting to compete with the headlines "written in huge letters" of the house-delivered newspaper. Einstein associated the rise of print media with the rising tide of nationalism that he tried to combat.

From the very moment that relativity theory became headline news, Einstein became acutely conscious of having at least two roles in life: one connected to the media press, which he referred to as "my public life," and a second private side, always in contradistinction to this public one.[30] His public life was mythical and otherworldly: "I have become an idol due to the clamor of the press. The role I play is similar to that of a saint's relics that a cathedral absolutely has to have."[31] His private life was the opposite—so bare that he simply described it as "life in the raw."[32] "Behind the conventional mask of behavior and speech," he argued, there lay a "real person."[33] In public forums, Einstein frequently referred to a private side; in private media, he described a public role. In both he referred to a core self that (he believed) was not shaped by any form of media. These different roles, including his understanding of "life in the raw," were connected to different media technologies and cultural techniques, including writing, speaking, listening, and even performing for his colleagues, for the public, and for the camera.

In 1921 an embarrassing incident revealed tensions in Einstein's public and private persona. On returning from a visit to the United States, Einstein described an incident when "he struck up a casual conversation" with a "young Dutch woman" who turned out to be a journalist. She then wrote an "extemporaneous account . . . prepared from memory" that did not "correspond exactly to what I actually said."[34] The account claimed that Einstein had described Americans as "uncommonly bored" and that men were "toy dogs for their wives." Americans, Einstein had allegedly told the journalist, could be easily duped into believing anything if "one tells them about something tremendous that will influence all future life, and of a theory that is within the realm of comprehension of only a select group of the very learned, and famous names are mentioned of predecessors who also made discoveries."[35] As damage control for this media fiasco, he proceeded to explain why this printed account was "extemporaneous" and hence unable to capture the events as they had really transpired. Einstein attempted to correct this distortion by writing a personal follow-up response titled "What He Really Saw."[36] In his autobiography, Einstein described his "essential being" as defined principally by his thinking: "The essential in the being of a man of my type lies precisely in *what* he thinks and *how* he thinks, not in what he does or suffers."[37] He believed that these three elements could be separated.

Bergson resisted these divisions. More important, he condemned modern societies for augmenting them and technologies for fostering them. His call to defend a philosophical conception of time was part of a much larger effort aimed at mending the growing rift between public and private roles in contemporary society.

BERGSON ON IDENTITY AND SELFHOOD

Bergson and Einstein thought carefully about how media transformations affected them. They were hardly alone in considering the effects of these vast changes on their own lives. Both were affected by them as public figures, in terms of their reputation and influence. What they said was often recorded. Almost anyone who ever met them or talked to them wrote down and later published their memories and interactions. Lectures were frequently stenographed and almost immediately reproduced in print.[38] But Einstein and Bergson experienced these media transformations in different ways. Einstein's life was subjected to the new mass-media culture even more starkly than Bergson's. Outliving the older philosopher by fourteen years, the physicist lived to see himself broadcasted on television.

In 1916 Bergson extended his thoughts about the nature of time while writing about common divisions of the self. He argued that philosophers since antiquity had divided the self into two parts, which they described in different ways. He noticed how certain commonalities remained across all their descriptions and how hierarchies were quickly assigned. One element was always described as closer to the "eternal," while the other was caught in actual time. One part of the personality was always more real and more essential than the other, which was often described as a "shadow" or "projection" of the first. One was like "gold," while the other was like "money." Eventually, Bergson claimed, the labels of subjectivity and objectivity became the most common way to describe these divisions.

Bergson argued that the self did not need to be divided in this way. He insisted that those who believed in this division suffered from the same misconceptions of those who believed in the illusions of the cinematographic camera, which portrayed a form of moving reality through static images. But just as real movement could never arise from discontinuous elements, a person was not made up of these two separate parts. This way of conceiving ourselves was "an illusion analogous to that which a cinematographic machine conscious of itself would feed on."[39] The philosopher resisted dividing the self in two parts, especially into subjective and objective parts. These divisions—which originated in antiquity and obtained full force in the world of mass media—had to be resisted. Bergson wrote these lines at a time when the figure of a film star was just gaining importance.

Bergson, noted orator and man of letters, noticed how new technologies were affecting the role and standing of intellectuals in the wider world. Even the cinematographic camera, which he had so potently criticized, turned on him. The philosopher was captured in film as he joined the Académie française.[40] Bergson was the first speaker to "a series of philosophical and social interviews" from Radio Paris. His inclusion was part of a broad effort to use radio in order to educate "an immense mass of cultivated people, who, until now consider it an undignified amusement."[41] But Bergson's involvement with mass media paled in comparison to Einstein's. As media technologies based largely on electromagnetic telecommunications came to dominate the airwaves, both men increasingly adopted distinct roles as public intellectuals.

Should intellectuals embrace mass media? Should scientists take advantage of its persuasive powers for reaching consensus and assent? Science in the age of mass media had to take advantage of these new forms of communication. At the same time that Einstein and Bergson debated about the significance of the exchange of light signals for physics and philosophy, they reconsidered the role of the scientists and intellectuals in the age of long-distance electromagnetic communications.[42]

THE LAST MENTION OF EINSTEIN

The last time Bergson mentioned his book on Einstein was in a long footnote in his final book *La Pensée et le mouvant* (1934). The note stuck out like a sore thumb, since footnotes "of such length are rare, if not exceptional, in Bergson."[43] For nearly a decade, the revised second edition of his *Duration and Simultaneity* had been republished numerous times without any changes. But when Bergson referred to his work on relativity one last time, it was clear that certain aspects of his old argument against the physicist were not worth repeating. Yet some points were worth clarifying. His critique of relativity at this later point was even more succinct, less technical, and more integrated into his philosophy than ever before. He restated points that he had developed years earlier about the relation of science to philosophy and of both to time, for which he would remain famous.

Bergson reminded his readers that Einstein gave us a notion of time that was essentially incomplete. "With regard to Time attached to Space, to a fourth dimension of Space-Time, it has no existence . . . other than *on paper*."[44] For the very last time, he argued that time could never be measured completely and that reality should not be confused with measurement. He concluded: "The reality of [Einstein's] Space-Time is purely mathematical, and one cannot transform it into a metaphysical reality, or into 'reality' tout court, without giving to this word a new meaning."[45] The mathematical models that scientists impressed on paper could never capture the vibrancy and diversity of the concrete world.

Bergson explained how his book had been "frequently misunderstood," especially by "those who, in transporting themselves from physics to metaphysics," went back and forth from the world of "calculations" to the world of "perceptions" without mentioning the large gap that lay between the two. For this reason, they lost the very "essence" of time. Bergson granted to Einstein the honor of having found a "mathematical expression of the world," which was a feat of "capital importance," but as soon as the physicist went back from mathematics to the world of concrete things, something was lost.

This new note complemented the arguments advanced in *Duration and Simultaneity*. Science not only divided events cinematographically, and in the process lost time, but it was inevitably tied to certain media (from film to paper): "There is no physics, no astronomy, no science possible if we refuse the scientist the right of laying down schematically the whole of the universe on a piece of paper."[46] Time on paper, one should not forget, was not time itself.

Soon after publishing his last book, Bergson's health declined sharply. Living conditions in Paris worsened rapidly. During the economic crisis of the 1930s and after the fall of France to Nazi Germany, the philosopher did not use his fame or reputation to obtain special privileges from the government, refusing to ask Vichy officials for special treatment. Renouncing all privileges, he decided to wait his turn in line in the street in the inclement December weather and register with other French Jews. Newspapers recounted how he wore a simple robe over his pajamas. His feet were covered only with slippers.[47] He died in early January 1941, a few days after the New Year. He was eighty-one years old. "When they came to get the coffin," recounted the poet Paul Valéry, "we said our last goodbye to the greatest philosopher of our time."[48]

CHAPTER 29

Einstein's Last Thoughts

INSTITUTE FOR ADVANCED STUDY, PRINCETON, NJ

Has time ever advanced clockwise? Einstein outlived Bergson by nearly a decade and a half, but he did not stop thinking about the philosopher. He was prompted to renew conversations about Bergson by his friend Besso. "I would like to formulate Bergson's desire as follows," his friend explained to Einstein on Christmas Eve 1951: "to turn subjective time into something objective."[1] In intimate personal discussions among lifelong friends that lasted until they died, they discussed the arrow of time, its passing, and Bergson.

"Here I sit in order to write, at the age of 67, something like my own obituary."[2] With this first sentence, Einstein started an autobiographical text about his life and work. Besso read Einstein's "autonecrologie," back in Europe and decided to write a letter to Einstein asking him some difficult questions. The "forced passage of time," Einstein would respond, "presents itself to us in an imperative fashion." But this sense of time passing by was due to a subjective "baggage of consciousness," which scientists could correct for. No longer young, Einstein remembered well just how disputed his view of time had been—especially, he wrote to his friend, by Bergson.[3]

Far from the old continent and living in a completely different setting at 112 Mercer Street in Princeton, New Jersey, Einstein still held his ground against the long-dead French philosopher. How could Einstein, during his advanced years, still maintain that the passage of time was merely subjective? How did he explain his own wrinkles, his own worsening health? Had he legated to physicists a "world without time," as his friend the mathematician Kurt Gödel accused him of doing?[4]

After having brought up the name of Bergson that Christmas Eve, Besso wrote him another letter the following summer asking him "whether the question of what is so strangely compelling about the passing of time is ever on your mind?" Would the question upset Einstein? Besso finished his letter asking for forgiveness: "Forgive your dear Michele, who is now truly old" for bringing up the topic again.[5] Perhaps Einstein would now answer this question differently? These musings seemed fit for old men forced to take stock and look back instead of ahead. Einstein replied, acknowledging that he "was also already quite musty." He understood what motivated Besso's "allusion to subjective experience" in his later years. He proceeded to rephrase Besso's question: "You say that this passage [of time] is accompanied by suffering, which—if one interprets it as a physicist—is tied to irreversible processes." "I do not know how to help you," he first answered modestly, noting that the difference between how we experienced the past, present, and future was due to our limited, subjective perspective. "This is was what bothered Bergson the most," he replied.[6]

The rest of the physicist's response in that letter was as strong then as it had been in his youthful years. Einstein was proud to have eliminated these subjective elements from the "conceptual construction of the objective world." To the very end of his life, Einstein disagreed with Bergson—yet Bergson, dead or alive, emerged over and over again as a potent critic. Was the passing of time—as these two men experienced it during those fateful years—a mere illusion? In subsequent letters, it became clear that Einstein could not convince his friend. Einstein cited Bergson once again in their correspondence, drawing comparisons between Besso's and Bergson's views: "You cannot get used to the idea that subjective time with its own 'now' should not have any objective meaning. See Bergson!"[7]

Two years after explaining to his friend the illusory nature of the passage of time, Besso died. Einstein would die less than a month after that. Grieving, he wrote to his friend's son and sister offering his deepest condolences, underlining once again his unfaltering belief in the illusory nature of our sense of past, present, and future: "Here Besso has once more preceded me a bit in leaving this strange world," he explained to them, knowing fully well that his own time would soon be up as well. But "that does not mean anything. For us, physicists of faith, the separation between the past, present and future, holds nothing more than the value of an illusion, however strong it may be."[8] For Einstein, their youth (just like their old age), their births (just like their deaths) were simply moments whatsoever, slices of time that they had not been able to see or foresee but that had always been there and that had, in the physical universe, no special importance. The sequence of events which they had experience was an effect arising from their own limitations in conceiving them otherwise—limitations that could be overcome when they thought and acted as physicists "of faith."

EINSTEIN RECONSIDERS BERGSON'S POINT: "THAT WOULD BE WONDERFUL!"

Had Einstein learned anything from Bergson? Had he gleaned any insights from any of the philosopher's numerous followers? The philosopher Filmer S. C. Northrop tried to find out. He traveled from Yale to Princeton, knocked on Einstein's door, and asked him this very question. Northrop had been a student of Whitehead's at Harvard University. He wanted to know what Einstein thought of his mentor's objections. "A few weeks ago," sometime in 1941, recounted Northrop in an article that appeared alongside the physicist's autobiography, he had "a lengthy discussion" with Einstein about these old questions.[9]

Northrop was aware of the strong "Bergsonian influence" on his teacher Whitehead. He traced Whitehead's infatuation with the Frenchman's work to the "impressionable war years" of 1914–1918. His teacher, he argued, had been "continuously conversing" with his colleagues in Britain about "the French philosopher." Northrop was now ready to relay the results of what Whitehead had learned from Bergson back to Einstein. He personally explained Whitehead's objections to the brilliant physicist who had failed to grasp them for most of his adult life. All the paradoxes of relativity, he explained, arose from an epistemological weakness. It consisted in maintaining a strict difference between the local and the distant. Since we could not, he argued, firmly determine the difference between what was local and what was distant, we could also never determine the difference between what was actually perceived (locally) and what was determined rationally. "Whitehead," Northrop patiently explained to the physicist, denied an essential difference between local "phenomenal events" and distant "postulated physically defined public events." For this reason, he fought against Einstein's redefinition of time in terms of the simple timing of two events at a single place.[10]

"That would be wonderful! So many problems would be solved were it true!" exclaimed Einstein, after having for the first time understood this particular criticism against his theory.[11] To Einstein's delight, it seemed as if nearly two decades of conflict could have a fairytale ending.

The reason why Whitehead opposed Einstein's conclusions, Northrop explained, was because he was unwilling to differentiate that which was sensed from that which was deduced mentally. Whitehead refused "to distinguish very sharply between the very limited part or aspect of simultaneous nature which is disclosed in direct sense awareness and the concept of simultaneous nature at it is prescribed in one's deductively formulated scientific theory."[12] The philosopher criticized Einstein's successes as residing in an essential "fallacy" that he named "the fallacy of misplaced concreteness." It referred to the tendency of establishing a strict distinction between "sense impressions" and "ideas." Whitehead, like Bergson before him, believed this distinction could never be absolute; that we could never establish a fixed boundary between matter and mind. This fallacy, continued Whitehead, led to the "bifurcation of nature" that divided the universe into two main categories, one physical and material and the other psychological and mental, forgetting what lay in between them, their connection, and their constant interplay.

Einstein first replied with surprise after understanding the criticisms launched against him, "Oh! Is that what he means?" But on second thought "after a moment's silent reflection," he remained unconvinced: "Unfortunately, it is a fairy tale. Our world is not as simple as that," he concluded.[13]

THE OFFICIAL HERE AND NOW

Einstein and Bergson differed about what they considered to be "local" compared to what was "distant." But they also differed about the importance of this difference. Einstein prioritized a "local" definition of time, arguing that it was the only "objective" measure of time. Because it worked well "for the place at which the clock is located," it could serve as a firm base for evaluating the time of "places remote from the clock."[14] For setting faraway clocks, scientists could simply take into consideration the time of transmission of light signals back to this base. The problem appeared to have been solved. Or not?

Bergson quickly noticed that Einstein's definition depended on making a sharp distinction between the local and the distant. Bergson was not convinced: "Where does proximity commence and where does the far end?" Telescopes and microscopes, telegraphs and telephones, and even simple reading glasses would all affect our perception of something as being far or near, distant or local. What is more, a slight change in our assessment of the far and the near would lead to a change in our perception of something occurring now or later. Distance affected temporal perceptions, as every thunderstorm easily confirmed. An observer standing midway between two distant flashes would see them as simultaneous, while one standing closer to one than to the other would see the closest one go off before the other one. Einstein knew this lesson well, but Bergson drew additional conclusions from it. Our sense of the far and the near, and therefore of now and later, *always* varied "according to the point of view, the terms of comparison, the instrument or perceptual organ."[15] Einstein's apparently solid definition of "local" in distinction to the "distant" depended on these factors as well.

Why did Einstein and Bergson hold such different views about something as apparently simple as what was "here" and what took place "now"? A host of new technological developments affected the experience of the here and now. More important, they created profound rifts in how different people perceived and thought of these assessments—repercussions that ranged from the metaphysical to the theological. Photography, phonography, telegraphy, cinematography, and radio changed the visual and auditory reach of observers as words, images, and sounds were transported in new ways. During the years of rapid media transformations, disagreement about something as apparently simple as the here and now was rampant. The "objectivity" of Einstein's definition arose in a specific historical and material culture where certain ways of noting the here and now gained currency, connected to new ways of communicating and experiencing presence.

Einstein redefined time in general by referring it back to a simple "local" procedure. He defined simultaneity by sharply dividing it into two components, local simultaneity and simultaneity at a distance. One definition applied for the local case, and another one (which had to include the speed of light transmission) worked for the distant case. The first definition served "for the place where the clock is located; but it is no longer satisfactory when we have to . . . evaluate the times of events occurring at places remote from the clock." This first definition was colloquial, explained by the commonsense assumption that "that train arrives here at 7 o'clock" meant that "the pointing of the small hand of my clock to 7 and the arrival of the train are simultaneous events."[16] But for determining if distant events were simultaneous, Einstein argued that observers needed a different understanding. They needed to take into consideration the extra time of propagation of a signal from the event to the place of the watch, due to the extra travel distance. Einstein prioritized a certain sense of the "local" as a privileged procedure for determining time. His understanding of what was near and what was far depended on what was immediately *ready at hand* for him or completely *out of reach*.

Why did Einstein consider the comparison of an event against a watch as perfectly sufficient for defining simultaneity while the other comparisons at a distance were deficient? "The simultaneity of two events taking place (at approximately) the same location" was, according to Einstein, a sufficiently clear process. Any ambiguities in this determination were simply issues that Einstein decided to "not discuss here."[17] That specific procedure, he argued, was free from the many other complications haunting the determination of simultaneity of faraway events. Bergson was not satisfied by Einstein's call to "not discuss" those problems.

Back in 1905, Einstein knew he would have difficulties convincing others of the benefits of his "local" procedure. A footnote to his revolutionary paper referred to "the inexactitude that lurks in the concept of simultaneity of two events at approximately the same place." This inexactitude, Einstein argued, "can only be removed by an abstraction."[18] He knew fully well that the "imprecision that is inherent" in local determinations of simultaneity could be brushed aside by considering it abstractly instead of concretely. He was thus able to base distant determinations on local ones.

Bergson disagreed. Einstein, he argued, was not justified in considering "the concept of simultaneity of two events at approximately the same place" as an unproblematic case. He was not justified in considering simultaneity as the comparison of one event against a watch. He was not justified in removing these difficulties by abstraction. Bergson objected, arguing that Einstein introduced an artificial break between "local simultaneity" and "simultaneity at a distance." The philosopher insisted that the physicist's differentiation was spurious: "The distinction between 'small' and 'large,' 'not far apart' and 'very far apart,' has no scientific validity," he concluded.[19]

By the 1940s some commentators on Einstein thought that the physicist had shifted the problem of determining local simultaneity further into the brain, where neuroscientists now had to face it. They had to deal with precisely the question that Einstein bypassed. How did our mind match two events to determine that they occurred "at the same time"? Neurophysiologists faced the challenge of relating a conscious assessment (of two events taking place at the same time) to processes taking place in brain cells and carrying the information about these events.[20] There they once again faced riddles about the relation of human consciousness to the material world.

A particular insight that Bergson wanted to defend contra Einstein was that there could be no fixed boundary and therefore no essential difference marking when local events ended and distant ones arose, or between physical ones and mental ones. How powerful was his critique? In the eyes of some, it was so powerful that the entire basis of "empirical" science was called into question. Initial criticisms launched against Einstein for drawing an unnecessarily sharp distinction between the local and the distant grew into a much more damaging argument. They were soon connected to a broad critique of the claim that knowledge was built up from a firm basis of sense-data and extended mathematically, logically, and rationally.

Neither Bergson, Whitehead, Martin Heidegger, or Walter Benjamin thought that a sharp distinction could be drawn between the assessment of an event taking place "here" rather than "there," or "now" rather than "later." Heidegger argued that scientists obtained accurate measurements only by first assuming that a certain moment (the moment of measurement) could be separated from the rest of time.[21]

Benjamin, in his *The Work of Art in the Age of Mechanical Reproduction*, continued to think about how media affected these assessments, focusing mainly on the repercussions for art and history. Other thinkers did not accept Einstein's "local" procedure for determining time as the only or the most objective one because they did not accept his definition of locality. In 1935 the physicist and philosopher Moritz Schlick asked why Einstein's specific procedure for time should be used instead of others. He had a hunch. His best guess was that it worked better than others because a large number of people could agree on its results, and it could be corroborated with the largest number of senses—visual, auditory, and tactile: "The only correct answer is, because of its objectivity, that is, because of its inter-sensual and inter-subjective validity."[22] Although clearly the here and now was not experienced in the same way by everyone, everywhere, and forever, certain assessments were more valid than others. Were these conditions a sufficiently strong basis for science? In the nineteenth century, Marx had argued "that the forming of the five senses is a labor of the entire history of the world down to the present."[23] If conditions sustaining the validity of these assessments were historically determined, how could they be universal?

ORIGINAL SIN

The text that prompted Besso to write to Einstein and ask him about Bergson referred to an original, metaphysical "sin."[24] The reference appeared in Einstein's autobiography. It did not mention Bergson. What "sin" did he refer to in those pages? Einstein's sin, as he described it, had been to define an "ideal clock" as both a physical thing (an actual clock) and an ideal entity (based on the speed of light *defined* as constant). Einstein would never be able to free himself entirely from critics who accused him of being unable to justify his reasons for considering a "light clock" as an "ideal clock." For this very reason, "an ideal clock" appeared to be different from "all other things." Einstein could not forget that this led to an "inconsistency." But if he had not defined a "clock" in this dual way, his work would have either had to "forego a physical interpretation," with the disastrous result of "reducing physics to geometry," or perhaps worse, it would have remained a mere technical treatise about coordinated clocks. To avoid this danger, Einstein chose to embark on this inconsistency and treated a "clock" both ideally and concretely. His "sin" was "justified," he continued, because he had accepted the "obligation . . . of eliminating it at a later stage."[25] But even by the end of his life, he was still unable to find a solution.

The physicist also confessed that he remained uncomfortable with how his theory depended so centrally on a particular number, the value of the speed of light. This fundamental constant seemed arbitrary and too directly tied to contemporary telecommunications. He stressed how the seeming arbitrariness of the number *c* for the speed of light could be eliminated. If the unit of seconds from the equations were replaced by "the time in which light travels 1 cm" it could be made to equal one.[26] Its disconcerting "numerical value" could be embellished by metrological institutes, which were less visible to the public and could be asked to shift units in order to end up with the nice number one.[27] This, he believed, could be a step in the right direction, but it was hardly sufficient. Would it solve debates about the relation between Einstein's universal conception, contemporary technologies, and practical procedures? Would it solve even broader debates about the relation of science to technology?

By then, his theory was wondrously successful. Many new phenomena and effects were described by reference to it—but this initial contradiction remained. Einstein admitted that he could still not describe an "ideal clock" as emerging from a foundational theory of "moving atomic configurations." The "postulates of the theory are not strong enough to deduce from them . . . a theory of measuring rods and clocks." His "faith in the simplicity, i.e., intelligibility, of nature" led him to hope to find sometime in the future "such strongly determined laws that within these laws only rationally completely determined constants occur."[28] But to date, the sin persisted. Did this confession matter? His theory and his reputation were secure.

Einstein described his contribution to physics as one that had ostensibly taken "sense impressions" as a point of departure to understand more complicated "ideas," revolutionizing our theories of time and space. Yet toward the end his life he no longer believed that these basic components—sense impressions—could be clearly distinguished from the rest of the theoretical structure he had painstakingly built. Einstein confessed that there was "no such thing" as a clear distinction between physical "sense impressions" and mental "ideas." But believing in their distinction was a necessary "metaphysical 'original sin'" without which we could not advance knowledge.[29] If science did not stand firmly on a solid empirical ground, from where did it draw its power? Mathematics and logic, unless firmly connected to empirical observations, would remain empty abstractions.

With four groundbreaking papers that appeared in 1905, the physicist had revolutionized physics to a degree unmatched since the time of Newton, creating one of the most successful and comprehensive theories ever. Yet in 1949 he stood "guilty" as charged, so he said, because deep down nobody could draw a firm distinction between the world of sensations and the world of the mind. Could Einstein have done things differently? By the end of his life, he was much more circumspect than he had been as a younger man.

Contrast this confession with Einstein's words the evening he confronted Bergson, when he firmly insisted on a clear separation between subjective and objective factors, ascribing to psychology the study of the subjective realm, to physics the study of objective events, and to philosophy simply nothing—at least when it came to the study of time. Yet later in life he admitted that he did not think that the division between the subjective and objective could be established once and for all, or even that between physics and metaphysics. He no longer denied the role played by metaphysics in science, famously complaining that "a fateful 'fear of metaphysics'" had become "a malady of contemporary empiricistic philosophizing."[30] By then, he was similarly skeptical about the difference between objectivity and subjectivity. "For this conceptual distinction there is no logical-philosophical justification." It was merely a "presupposition" that permitted "physical thinking." Its "only justification lies in its usefulness," he wrote.[31]

What prompted Einstein's late-in-life confession? From the moment he debated Bergson in 1922 to when he confessed his "original sin" in 1949, Einstein successfully fought against philosophers who denied to science its grounding in objectively pure sensations freed from the distorting influences of the mind. But by the end of his life, Einstein offered a mea culpa. Bergson, much earlier, had insisted that this was a problem in Einstein's work.[32]

Bergson had been dead for almost half a decade when Einstein "confessed," but many of his insights were kept alive and passed down by colleagues and students who had worked with him or with his allies. A new generation of thinkers accepted the numerous difficulties in founding science on sense impressions and logic alone. They started to reevaluate the role of philosophy in science more generally and reconsidered some of the objections they had earlier brushed aside: "The metaphysician is treated no longer as a criminal but as a patient: there may be good reasons why he says the strange things that he does."[33]

MALE AND FEMALE

The twentieth century closed with the differences between Einstein and Bergson as vast as those between male and female.[34] It also ended with a physical understanding of the nature of time firmly ensconced. This scientific achievement dragged with it significant consequences in its wake. Critics decried that certain aspects of time, those connected to life, had been forever lost. "The more perfect the instrument as a measurer of time, the more completely does it conceal time's arrow," lamented Eddington.[35] Others denounced scientists for taking conventional standards too evidently tied to contemporary technologies as universal ones. Scientists, critics noted, focused too narrowly on measurement and instrumental results, forgetting the very conditions that permitted their measurability in the first place. These debates resonated beyond laboratories and universities, as Einstein and Bergson actively pushed contradictory political and cultural projects in a century of escalating violence.

In his personal letters, the physicist worried about the paradoxes of time in ways that were absent in his scientific work. In an intimate and sentimental letter to Besso from Princeton mourning the recent death of his wife Elsa, he confessed that he could not wrap his head round certain aspects of time. Reminiscing about the "thirty years" that had passed since he started his friendship with Besso at the patent office in Bern, Einstein calculated that they amounted to "almost 10^9 seconds":

> I now frequently and with pleasure think of our old time at the patent office and I cannot put it in my head that I left it almost thirty years ago. That makes up almost 10^9 seconds and one is surprised not to have been able to do more good during that time.[36]

Where had all those seconds gone? Accounting for them seemed impossible.

"Intellect" fought against "instinct," explained Bertrand Russell, as he dedicated himself to promoting Einstein over Bergson.[37] "Spontaneity" confronted "receptivity," stated Ernst Cassirer referring directly to each man as he demarcated competing philosophical projects.[38] "Science," explained Martin Heidegger, confronted "lived experience" so starkly that it led to a spike in the "irrational" as a necessary consequence.[39] When the physicist Percy Bridgman forcefully lamented the divide between the "universality of science" and the everyday "obvious structure of experience," readers could not help but accuse him of tacitly referring to both men.[40] Bergson was at first explicitly compared to Heraclitus of Ephesus; Einstein to Parmenides of Elea.[41] Bergson's philosophy was seen as similar to St. Augustine's; Einstein was repeatedly accused of being Aristotelian.[42] These ancient and modern thinkers were all associated with opposing ways of understanding of time—reinforcing dichotomies far vaster than those directly tied to the Einstein-Bergson confrontation. As the century drew to a close, the "time of the universe" and "lived time" appeared as irreconcilable as science and philosophy in ways that exceeded the discussion that took place that day.[43]

What had led Einstein and Bergson to hold such different positions in the first place? What caused the twentieth century to end up as divided as it did? These questions were neither asked nor answered; the two men and most of their numerous interlocutors simply took sides. Authority for speaking about time shifted considerably during the century, in step with major transformations in the hierarchical structure of society, which affected the relative standing of scientists, intellectuals, and the general public in a mass-media culture divided into experts and laymen. A particular understanding of time associated with scientific rationality and expertise gained prominence. Science emerged triumphant, lording it over the critical humanities and pushing artistic experimentation farther and farther away.

Time kept also slipped by. It cut through individuals who struggled to survive in a world divided into science and art, the public and personal, the objective and the subjective, the abstract and the concrete. The contradictions once represented by the two men gained independent lives of their own, appearing as timeless as time itself.

POSTFACE

Peter and Paul, the paradoxical twins of relativity theory, have produced fine young offspring. Bob, Alice, and Ted live, according to a recent account, in New Jersey and Manhattan. "In the language of quantum information, Alice can marry either Bob or Ted, but not both," explains the science writer Dennis Overbye. A new "high-octane debate" has engulfed Einstein's theory. The theories of Stephen Hawking, Juan M. Maldacena, and others vie to explain yet another paradox.[1] Shape dynamics, a branch of theoretical physics, has a global, locally undetectable, preferred frame. Recent experiments in quantum gravity have asked if the Lorentz symmetry can break at the Planck scale. My contention is that a preferred frame of reference has been hiding in plain sight: in our all-too-human world, in our everyday experiences, in art, and in philosophy.

The issues at stake today, and their social, institutional, and material context, are very different from the ones I have described—yet certain similarities persist. Questions pertaining to the relation of individual versus common experience and to information-transfer versus communication appear once again, albeit in a new guise. One difference is stark: scientists tackle these questions mostly on their own. Humanists are nowhere to be found in these conversations.

Bergson has not yet recovered from one of the most hurtful attacks launched at him at the end of the twentieth century. The physicists Alan Sokal and Jean Bricmont considered him as the main representative of a general malaise affecting radical philosophers and academics: postmodernism. Postmoderns, they argued, had a "historical connection with a philosophical tradition that emphasizes intuition, or subjective experience, over reason."[2] Bergson was their antihero. "One of the most brilliant representatives of this way of thinking is—no contest—Bergson, who pursued this mission until his debate with Einstein about the theory of relativity," they concluded.[3]

By the end of the century, various areas of science directly associated with Bergson frequently fell under the category of "postmodern science"—a label referring to unconventional research, generally considered legitimate but often remaining controversial, that stressed the indeterministic nature of the universe and the central role of philosophy within science itself.[4] In the 1970s the chemist Ilya Prigogine, who would go on to win the Nobel Prize a few years later, reviewed for the journal *Nature* a collection of essays on Bergson that included translated portions of the transcript of the April 6, 1922, meeting. Prigogine was extremely critical of Bergson's work on relativity: "Bergson's struggle with the Lorentz transformation in *Duration and Simultaneity* is as pathetic as it completely misses the point."[5] Although "Bergson was certainly 'wrong' on some technical points," Prigogine felt motivated to study those aspects of temporal development that had been left unexplained by Einstein.[6] As he studied the arrow of time in thermodynamics and thought about questions of indeterminism in the physical sciences, he became increasingly vocal in his defense of certain aspects of Bergson's work. Referring to "the famous discussion between Bergson and Einstein, which took place in Paris in 1922," Prigogine considered his own scientific work in terms of the historical legacy of the debate:

> Einstein gave a presentation of his theory of special relativity, and Bergson expressed some doubts about it. It is true that Bergson had not understood Einstein. But it is also true that Einstein had not understood Bergson. Bergson was fascinated by the role of creativity, of novelty in the history of the universe. But Einstein did not want any directed time. He repeated often that time, more precisely the arrow of time, is an 'illusion.' So, these ideologies seem to be irreconcilable.[7]

His scientific contributions attempted to reconcile Bergson's insights with the laws of physics.

The philosopher Isabelle Stengers, coauthor with Prigogine of *A New Alliance,* a book describing how science and philosophy could enter into a new partnership, returned to the Einstein-Bergson debate in her later publications: "Our scene is well known; it took place at the Société de

Philosophie de Paris, on April 6, 1922."[8] For her, the debate had marked "with the greatest force" the culmination of a project that had started during the time of the Scientific Revolution and lasted "for the last three centuries"—the project of eliminating change and diversity (and therefore real time) from science by reducing it to the "identical and the permanent." According to Stengers, modern physics had recently "rediscovered time," showing how "it will never again be able to be reduced to the monotonous simplicity" that Einstein gave it or to a simple "geometrical parameter that allows calculation."[9] Chaos theory and quantum mechanics proved that time was much more than Einstein had wagered.

> Physics, today, no longer denies time. It recognizes the irreversible time of evolutions toward equilibrium, the rhythmic time of structures whose pulse is nourished by the world they are part of, the bifurcating time of evolutions generated by instability and amplification of fluctuations, and even microscopic time, which manifests the indetermination of microscopic physical evolutions.[10]

Had science vindicated Bergson? Although a few thinkers were enthusiastic about the "new alliance" between science and Bergson, many others lamented that science, and the world, had become simply too postmodern.

Sokal and Bricmont were militant actors in the "science wars" of the 1990s, a confrontation that pitted physicists against humanists and culminated in mutual accusations of libel, incompetence, and even greed that reached the front pages of international newspapers. The issues at stake during the science wars were very different from those of the Einstein-Bergson debate, yet authors frequently drew connections between these two conflicts. Bergsonian philosophy and the large swaths of Continental philosophy connected to it were seen as the direct predecessors of a new enemy.

Sokal and Bricmont's *Impostures intellectuelles* (1997) set the "historical origins" of the growing rift between scientists and humanists in the Einstein-Bergson confrontation.[11] The closing chapter of the book, which was excluded from the English translation, was dedicated to exposing Bergson's error.[12] Bergson's "fashionable nonsense" had spread to "Deleuze, after passing through Jankélévitch and Merleau-Ponty," they argued.[13]

As professors of physics, Sokal and Bricmont framed Bergson's critique as mistaken with regard to *physics*. They continued to ignore Bergson's own claims that his book did not contest the results of relativity physics: "Bergson is mistaken," they insisted, adding that his "error is not a question of philosophy or interpretation, as is frequently thought; it bears on understanding the physical theory, and it enters, in the last analysis, into conflict with experience."[14] The authors repeated what had been said many times before, first by Einstein. They ignored Bergson's response. None of Bergson's claims were meant to bear on debates involving only physics: "The theory was studied with the aim of responding to a question posed by a philosopher, and no longer by a physicist." "Physics," he added, "was not responsible for answering that question."[15]

The science wars exacerbated the conflict between scientists and humanists by perpetuating the view that Bergsonian, Continental, and postmodern philosophy were antiscience. One of the main charges against all of them was that they were disconnected from empirical reality and that they fostered a dangerous form of relativism, promoting a view of truth as subject to endless debate and revision, with disturbing ethical consequences. But accounts that placed science, empiricism, and rationality on one side and, on the other side, Bergson, a disdain for empirical facts, and irrationality does not hold. Many scientists—including Poincaré, Lorentz, and Michelson—were either very close to Bergson or held views consonant with his work, refusing to accept Einstein's interpretation while wholly accepting all experimental results.

TIME IN THE AGE OF COMPUTERS AND PHOTODETECTORS

The controversy between Bergson and Einstein did not end as the twentieth century came to a close. Detractors and advocates continued to debate many of the same issues as before. In the 1950s the philosopher Adolf Grünbaum, still by reference to Bergson, once again asked if time could be understood "without including the human observer's retina or body in the analysis, let alone his stream of consciousness."[16] Yet important differences marked these later discussions. By the second half of the twentieth century, old arguments were relaunched by reference to *new kinds of instruments.* Early examples of telecommunication media, automatic inscription devices, and film cameras soon gave way to computers and photodetectors. To prove the philosopher wrong against the physicist, Sokal and Bricmont mentioned computers, reminding us that the actual observers described by Einstein (later commonly referred as Peter and Paul) could be replaced by machines: "Paul could be, for example, a photodetector coupled to a computer, and after the experiment everybody could consult the computer's memory and see which beam of light came in first."[17] Because proof of relativity could be obtained with automated computers, they argued, Bergson was wrong.

Did these new technological advances settle the debate? Defenders of Bergson also started referencing these new instruments to prove their (opposing) case. They continued to take a contrarian stance—arguing that scientists could not talk about new automatic machines as if no one had built them and as if they had not been designed for a specific purpose; nor could they ignore the fact that for the results to be meaningful they had to assume that someone, in the end, would have to witness them. The two groups continued to talk against each other, once again, by reference to the salient technologies of their respective eras.

A few writers, as diverse as Gilles Deleuze and Bruno Latour, joined a previous generation of thinkers (Whitehead, Heidegger, Benjamin, Merleau-Ponty) to offer new models for rethinking the relation of science, technology, and philosophy. They explicitly referred to Einstein or Bergson or to the categories of time frequently associated with them. Let me elaborate.

Merleau-Ponty asked Deleuze to write the entry on Bergson in a volume he was directing titled *Les Philosophes célèbres*, which appeared in 1956.[18] Deleuze was an ideal candidate for the task. A few years earlier, he had written his first essay on Bergson.[19] By 1966 he started a full-fledged revival of the philosopher by publishing *Le Bergsonisme*. Bergson, according to Alain Badiou, was Deleuze's most important influence. "The Bergson-Einstein debate opens up a veritable can of worms" for Deleuze, explained a recent philosopher.[20]

Deleuze saw in Bergson's philosophy an answer to the all-pervading scientism of his era by offering to philosophy a role that went simply beyond that of a critique and examination of science's foundations. He learned from Bergson why philosophy could remain as relevant as science itself. Deleuze included a key footnote to his book *Le Bergsonisme* about the relation of Bergson's philosophy and relativity theory. It indicated what he thought of Bergson's debate with Einstein. "It is frequently claimed that Bergson's reasoning implied an error with regard to Einstein," he admitted.[21] "But very often too," he added, "one has made an error with regard to Bergson's reasoning itself."[22] The error, according to Deleuze, resulted in taking Bergson's comments about time dilation too literally and forgetting that his main point was that the comparison of a stationary clock against a traveling one needed to be thought of as a comparison of a "quantity" against a "symbol." With this interpretation, Deleuze used Bergson's confrontation against Einstein as an example of why numbers and symbols, which were sometimes read uncritically by scientists as merely denoting quantities, should be understood as affecting our knowledge of the world more generally. But Bergson's attempt "to assimilate the scientific observer (for example, the cannonball traveler or relativity) to a simple *symbol*," was also not enough. Deleuze was not content to do "as Bergson does."[23]

Together with his collaborator Felix Guattari, Deleuze explained why these concerns stood at the center of their answer to *What Is Philosophy?* (1991). In their book—about the role of philosophy in the age of science—they took issue against Einstein's ally, Bertrand Russell, for believing in "sense-data without sensation" and for having "assimilated them to apparatuses and instruments like Michelson's interferometer or, more simply the photographic plate, camera or mirror that captures what no one is there to see."[24] In contrast to Russell, the two philosophers argued against the uncritical conflation of "data" gained with instruments as without them, stressing the differences between the two. One particular irony of "sense-data without sensation" they argued, was that it was always "waiting for a real observer to come and see." How was one to deal with the paradoxical relationship of real observers to recording devices? Recording instruments, they argued, only functioned because they "presupposed" an "ideal partial observer."[25]

In his earlier work, Deleuze considered that due to the high status of science in contemporary society, "we must ask why there is still philosophy, in what respect science is not sufficient." "It is not enough to say that philosophy is at the origin of the sciences and that it was their mother," he wrote. Deleuze considered two alternatives: "Philosophy has only ever responded to such a question in two ways, doubtless because there are only two possible responses." In the first one, philosophy completely gave up competing against science: "One says that science gives us a knowledge of things, that it is therefore in a certain relation with them, and philosophy can renounce its rivalry with science, can leave things to science and present itself solely in a critical manner, as a reflection on this knowledge of things." Its role, in this view, was essentially a reflective one. In the second case, philosophy continued to fight science as an old foe: "On the contrary view, philosophy seeks to establish, or rather restore, another relationship to things, and therefore another knowledge, a knowledge and a relationship that precisely science hides from us, of which it deprives us, because it allows us only to conclude and to infer without ever presenting, giving to us the thing in itself." Bergson, according to Deleuze, chose the second option. "It is this second path that Bergson takes by repudiating critical philosophies when he shows us in science, in technical activity, intelligence, everyday language, social life, practical need and, most importantly, in space—the many forms and relations that separate us from things and from their interiority."[26]

This second option required thinking about what made something "this rather than that, this rather than something else." It required focusing on singular, local, and concrete aspects of knowledge: "What science risks losing, unless it is infiltrated by philosophy, is less the thing itself than the difference of the thing, that which makes its being, that which makes it this rather than that, this rather than something else." Already in his early work Deleuze recognized how Bergson's work was marked by a number of dualisms: "Hence we see the meaning of the dualisms dear to Bergson: not only the titles of many of his works, but each of the chapters, and the heading that precedes each page, exhibit such a dualism. Quantity and quality, intelligence and instinct, geometric order and vital order, science and metaphysics, the closed and the open are its most known figures."[27] Deleuze's philosophy, and in particular his revival of Bergson, tried to combat these dualisms, forcefully criticizing the common interpretation of Bergsonian duration as merely subjective, interior, or psychological.[28] He called for the need to investigate the interplay between "two readings of time, each of which is complete and excludes the other."[29]

In *Logique du sens* (1969), Deleuze identified two dominant ways of understanding time, which he named Aion and Chronos. Naming neither Bergson nor Einstein directly in the text, the two senses of time nonetheless fit into the positions broadly associated—imperfectly—with their philosophies. Aion was a time made up of empty instants, whereas Chronos was "vast and deep." For Aion, "only the past and the future inhere in time and divide each present infinitely." For Chronos, in contrast, "only the present exists in time and gathers together or absorbs the past and future."

Deleuze's later work, in collaboration with Felix Guattari, explicitly returned to the figures of Bergson and Einstein. *A Thousand Plateaus* (1980), a book referred to as one of "the most important philosophical texts of the twentieth century," mentioned "the confrontation between Bergson and Einstein on the topic of relativity" as essential for understanding major divisions.[30] "Bergson thus brought to light 'two very different kinds of multiplicity,' one qualitative and fusional, continuous, the other numerical and homogenous, discrete," they explained. These divisions echoed many others: "We have on numerous occasions encountered all kinds of differences between two types of multiplicities: metric and nonmetric; extensive and qualitative; centered and acentered; arborescent and rhizomatic; numerical and flat; dimensional and directional; of masses and of packs; of magnitude and of distance; of breaks and of frequency; *striated and smooth*."[31] While at first Deleuze and Guattari explained how the "striated" and the "smooth" were two opposing ways of understanding the texture of space and time by direct reference to Einstein and Bergson, the philosophers soon used these terms as explanatory tools for understanding much broader divisions.

As an ambitious young philosopher, Deleuze recognized early on the need to go further, and that this meant going beyond those dualisms: Neither science nor philosophy should be limited to the exploration of only one of the two elements of these dualities. "But we must go further," he wrote.[32] How?

"THE MOST UNFAIR ACCOUNT OF SCIENCE"

The philosopher Bruno Latour referred to the debate as a locus classicus for thinking about the relation between science and other areas of culture:

> There is no better way to frame this question than the bungled dialog (well, not really dialogue, but that's the point) between Henri Bergson and Albert Einstein in Paris in 1922. Bergson had carefully studied Einstein's theory of relativity and wrote a thick book about it, but Einstein had only a few dismissive comments about Bergson's argument. After Bergson spoke for thirty minutes, Einstein made a terse two-minute remark, ending with this damning sentence: "Hence there is no philosopher's time; there is only a psychological time different from the time of the physicist."[33]

"Matters of concern" faced off against "matters of fact" by direct reference to Bergson and Einstein. These, like many other binary categories, have become ingrained examples of our pertinent cultural divides.[34] "Can we do better at the beginning of the twenty-first century?" he asked.[35]

A few years before the twentieth century came to a close, Latour studied Bergson's concept of time, calling Bergson's arguments against Einstein "the most unfair account of science."[36] Although he distanced himself from Bergson's critique, similarities between Bergson and Latour were readily apparent. "The crux of Bergson's argument is not really different from that of Latour," explained the sociologist Michel Callon.[37] Latour explained how Bergson's position was often dismissed because it was framed as one concerned exclusively with subjectivity: "Einstein argued that there was only one time and space—that of physics—and that what Bergson was after was nothing more than subjective time—that of psychology." According to Latour, Einstein's manner of dealing with Bergson became a typical way for scientists to deal with nonscience, including philosophy, politics, and art. While Bergson's account of Einstein's science had been "unfair," Einstein's account of philosophy was also tendentious: "We recognize here the classical way for scientists to deal with philosophy, politics, and art: 'What you say might be nice and interesting but it has no cosmological relevance because it only deals with the subjective elements, the lived world, not the real world.'"[38]

Latour's project consisted in asking: "is it possible to give Bergson another chance to make his case that, no, he is not talking about subjective time and space, but is rather proposing an alternative to Einstein's cosmology?"[39] But how? One could start by adopting a different metaphysical conception of time, concluding, with Latour, that it is not "coherent and homogenous."[40] One of the reasons why "we have never been modern," he argued, was because time—including historical time—was not something that could be lined up or summed up like simple integers on a ruler. By asking readers to "reject the idea of a coherent and homogenous time that would advance by goosesteps" he called on them to question the very possibility of modernity as we have known it.[41]

BEYOND EINSTEIN AND BERGSON

Divisions between Einstein and Bergson, between science and philosophy, and between opposing notions of time remain much vaster than the actual conflict between the two men. They precede it and surpass it. Calls for abandoning and moving beyond these dualisms have spread far and wide.[42] "Things that talk," "intangible things," "material semiotics," and "epistemic objects" are all labels used by contemporary scholars to capture the in-between territory of dualistic dichotomies.

What happens if we *get on with the job of doing the thing* and reread the debate in ways that no longer accept the binary terms associated with Einstein and Bergson as self-evident and inevitable? What happens to our understanding of science and of history if we shelve these binary categories—such as objectivity-subjectivity and nature-politics—and study instead how these categories strengthened at certain moments? For one, the outcome of the Bergson and Einstein confrontation no longer appears as clear-cut as before.

Our reasons for continuing to fight vanish. Instead of simply siding with one over the other, we can consider our universe filled with clocks, equations, and science as much as with dreams, memories, and laughter.

ACKNOWLEDGMENTS

"The entire Debate (and its over-arching significance) came as news to me," wrote one of the first reviewers of my manuscript, who then dutifully "checked the usual biographies of Einstein to see what had been reported there." He found almost nothing. Did the scarcity of references in the Einstein scholarship mean that this episode should continue to be "deservedly forgotten"? It has taken many conversations and exchanges with colleagues—and a yearlong delay—to explain why I believe it would be beneficial to take another look at these controversial events and write about them. I cannot thank these interlocutors enough, known or anonymous.

This book has given me more than I have given it. All historical accounts—explicitly or implicitly—make assumptions about what time is and how it progresses. Writing a historical account *where it is precisely the nature of time* that is debated presents additional complications and opportunities. I faced this challenge in my previous book. My hope is that, together, both tomes may help us recover aspects of the history of science that frequently get lost—its ability to show what ties scientific, technological, philosophical, historical, and everyday knowledge together. Throughout these pages, my focus has been on the connecting points between all of these categories—connections that help us move beyond the binary terms associated with Einstein or Bergson. I have followed previous thinkers who, like myself, had already worked hard to understand knowledge in this way: Bachelard's "abstract-concrete," Heidegger's "Zwischendinge," Whitehead's "un-bifurcated" philosophy, Benjamin's ever-changing "here and now," and Deleuze's understanding of "this rather than that." These approaches have helped me compose a historical text that questions common assumptions about science and technology in relation to philosophy and history.

Historians of science will also recognize that this book draws from contributions to the field appearing after 1986 that no longer privilege "social" explanations. For this reason, it differs from most social and cultural histories that accept the tripartite divisions of science, technology, and society. My aim has been to investigate the production of knowledge more broadly (epistemologically *and* ontologically, historical *and* scientific) and to write it a manner where its implicit use of chronology, causality, and agency throughout the text reflected my theoretical concerns.

I learned about the debate between Einstein and Bergson as I was finishing *A Tenth of a Second*, a book about time and modernity, centering on the period of the 1850s to the 1920s. Immediately after completing that book, I became interested in continuing to track some of its themes further into the century. It was a different world. My first book dealt with the rise of paradigmatic instruments of modernity (mainly telegraphy, photography, and early cinematography), focusing on the proliferation of keys and screens in the modern world and investigating the changing place of fingers and eyes. My second book had to deal with the telegraph and telephone as much as with radio, with early film as much as with the movies, and with clocks as much as with wristwatches. I started to think of this new project as a second volume extending the earlier one. Both aim to disclose the rise of authoritative knowledge in the modern world. While I had mostly written about lesser-known scientists, now I had to contend with one of the most well-known figures of our time. The more I immersed myself in the original sources the more evident it became to me just how starkly the events I read about differed from how they were generally portrayed. When I was told to ignore these vast differences, it was already too late.

For now, it is a distinct pleasure to thank all of you who have generously helped me along, from one moment to the next: my mother, Rocío, and my sisters, Lorea and Fernanda. Margaret Murray, Charles Barsam, and their sons William and James have let us share the joys of family and friendship. Sharon Harper, Dan de Gooyer and Adrianna, Tamara Griggs, Philippe Cluzel and Lou, and the third bird deserve special thanks; my friends and colleagues: Stefan Andriopoulos, Peder Anker, Ladina Bezzola, Hasok Chang, Graham Burnett, Yuval Dolev, Mahalia Gayle, Lisa Gitelman, Jeremy Greene, Jeanne Haffner, Orit

Halpern, Caroline Jones, David Kaiser, Alexei Kojevnikov, Christopher Krebs, Natasha Lee, Patrick McCray, Karen, Brian and Richard Mariscal, Massimo Mazzotti, Christopher Phillips, Sherri Roush, Henning Schmidgen, Hanna Shell, Oliver Simons, Matt Stanley, Mauricio Suárez, Fernando Vidal, and Heidi Voskuhl; students who patiently let me multitask as teacher and research-scholar, primarily Noam Andrews, Jérôme Baudry, Lisa Crystal, Stephanie Dick, Gregor Jotzu, Florin Morar, Kyle Parry, and Joseph Pomianowsky; it has been a clear privilege to share so much with Vincent Lépinay; in Spain, Manuel Lucena Giraldo and Javier Moscoso; in México, Carlos López Beltrán; and in France, Bernadette Bensaude-Vincent, Ioulia Podoroga, and Frédéric Worms. Élie During generously guided me through his favorite labyrinthic sci-fi novel. This book relies heavily on his scholarship.

I am deeply appreciative of the numerous invitations I have received that led me to rethink and revise: Northwestern University, Columbia University, Radcliffe, the University at Buffalo, the Bard Graduate Center, Yeshiva University, the University of British Columbia, the University of Rochester, the University of California, Berkeley, Georgetown University, the University of Pennsylvania, Boston University, New York University, Princeton University, UNAM México, the Van Leer Jerusalem Institute, ZfL and ICI in Berlin, Centro Cultural Matucana 100 Chile, the Fundación Juan March, CSIC and Medialab-Prado Madrid, and the École Normale Supérieure and the Centre Pompidou in Paris, among others. I would like to thank the warm hospitality of the history department at the University of Illinois at Urbana-Champaign and celebrate their defense of free speech.

Most of this book was written in Weimar thanks to a generous fellowship from the Internationale Kolleg für Kulturtechnikforschung und Medienphilosophie (IKKM). Conversations with Francesco Casetti, Lorenz Engell, Ben Kafka, Markus Krajewski, Bernhard Siegert, and Sigrid Wiegel impacted these pages thoroughly. A Charles A. Ryskamp fellowship from the American Council of Learned Societies (ACLS) permitted me to dedicate a full year to writing. I wish to thank everyone at The Park School for building a wonderful environment for my family and for being supportive of me as a writer and scholar. I thank my editor Ingrid Gnerlich for her timeliness and trust and Gail Schmitt for her careful attention to this project. Palle Yourgrau at Brandeis University encouraged me to continue with it, just as his biography on Simone Weil reminded me of the value of work even if left unfinished. At Harvard University, I would like to thank my colleagues and friends: David Armitage, Allan Brandt, Janet Browne (and her generous support of our corridor of excellence), Giuliana Bruno, Tom Conley, Ellen Guarente, Sheila Jasanoff, Evelynn Hammonds, Patrice Higonnet, Michael Kelley, Everett Mendelsohn, Charles Rosenberg, Linda Schneider, Diana Sorensen, and Deborah Valdovinos. Mario Biagioli has been a generous friend and mentor both at Harvard and beyond. With David Rodowick I had the pleasure of discussing how the histories of physics, philosophy, film, and media intersected, at the same time our lives did as well. I would have never been able to complete this project without the example and encouragement of Lorraine Daston, who read various versions of the manuscript and was there for me during the most difficult moments leading up to its publication. It has been a privilege to work with Peter Galison, who taught me how the history of physics was my history too. This book is dedicated to Billy, who is the reason why I have become ever more aware of time passing by and of my desire to stop it.

NOTES

For readability, some of the quotations used in this book have been abbreviated without the use of ellipses. Unless otherwise noted, most citations of Einstein's correspondence are from *The Collected Papers of Albert Einstein*, ed. Diana Kormos Buchwald, 14 vols. (Princeton, N.J.: Princeton University Press, 1987–). Most citations of Bergson's works refer to the first critical edition of his works, *Le Choc Bergson*, under the direction of Frédéric Worms. Unless otherwise specified, Bergson's letters are cited from *Correspondances*, ed. André Robinet (Paris: Presses Universitaires de France, 2002). Additional sources of correspondence included in the notes come from Albert Einstein, *Correspondances françaises*, ed. Michel Biezunski, vol. 4, Oeuvres Choisies (Paris: Éditions du Seuil, 1989); Albert Einstein and Michele Besso, *Correspondance 1903–1955*, trans. Pierre Speziali (Paris: Hermann, 1979); *The Born-Einstein Letters: Friendship, Politics and Physics in Uncertain Times* (New York: Macmillan, 2005); Angelo Genovesi, *Il Carteggio tra Albert Einstein ed Édouard Guillaume* (Milano: Franco Angeli, 2000); Angelo Genovesi, "Henri Bergson: Lettere a Einstein." *Filosofia* 49, no. 1 (1998): 3–39; Albert Einstein, *Letters to Solovine* (New York: Philosophical Library, 1987). When available, I have consulted previous translations. All translations of previously untranslated material are mine.

CHAPTER 1. UNTIMELY

1. "Dialogue": Marie Cariou, *Bergson et Bachelard*, Questions (Paris: Presses Universitaires de France, 1995), 90; "theater": Jean Langevin and Michel Paty, "Le Séjour d'Einstein en France en 1922," *Fundamenta Scientiae*, no. 33 (1979): 27.

2. Svante Arrhenius, "Presentation Speech," 10 December 1922 in *Nobel Lectures in Physics (1901–1921)* (Singapore: World Scientific, 1998), 479; also quoted in Abraham Pais, *"Subtle is the Lord . . .": The Science and the Life of Albert Einstein* (Oxford: Clarendon

Press, 1982), 510; Abraham Pais, *Einstein Lived Here* (Oxford: Clarendon Press, 1994), 75. Although Pais mentioned the reference to Bergson during the presentation, he did not atttribute any significance to it. "Why did Einstein not get the Nobel prize for relativity? Largely, I believe, because the Academy was under so much pressure to award him." "How Einstein Got the Nobel Prize," in Pais, "*Subtle is the Lord* . . . ," 502–512, on p. 511.

3. Isaac Benrubi, *Souvenirs sur Henri Bergson* (Neuchâtel: Delachaux & Niestlé, 1942), 104.

4. "La Théorie de la relativité: séance du 6 avril 1922," *Bulletin de la Société française de philosophie* 22, no. 3 (1922). Rephrased as "More Einsteinian than Einstein" in Henri Bergson, *Durée et simultanéité: á propos de la théorie d'Einstein*, ed. Élie During, 4th ed. (Paris: Presses Universitaires de France, 2009), 55.

5. Wyndham Lewis, *Time and Western Man* (Santa Rosa: Black Sparrow Press, 1993), 48.

6. "La Théorie de la relativité," 107.

7. Charles Nordmann, "Einstein à Paris," *Revue des deux mondes* 8 (1922): 26.

8. German Ambassador in Paris, Report to the Foreign Ministry, quoted in Siegfried Grundmann, *Einsteins Akte* (Berlin: Springer-Verlag, 1998), 212.

9. "Crowds in a craze from a theory that they manifestly do not comprehend," in Louis Dunoyer, "Einstein et la relativité (I)," *La Revue universelle* 9, no. 2 (15 April 1922): 180.

10. Ibid.

11. Ibid., 179.

12. Bergson, *Durée et simultanéité*.

13. For the theme of "intuition" against "reason" in Bergson's work see Bertrand Russell, *Mysticism and Logic* (Garden City, N.Y.: Doubleday and Co., 1957), 16; Arthur O. Lovejoy, *The Reason, the Understanding, and Time* (Baltimore: Johns Hopkins Press, 1961), 68; George Santayana, *Winds of Doctrine: Studies in Contemporary Opinion* (New York: C. Scribner's Sons, 1913), 63.

14. Alan D. Sokal and Jean Bricmont, *Impostures intellectuelles* (Paris: O. Jacob, 1997).

15. Paul Andrew Ushenko, "Einstein's Influence on Contemporary Philosophy," in *Albert Einstein: Philosopher-Scientist*, ed. Paul Arthur Schilpp (La Salle, Ill.: Open Court, 1949), 609.

16. Pais, "*Subtle is the Lord* . . . ," 510.

17. *The Collected Papers of Albert Einstein* (English translation supplement) does not include Bergson's commentary. See document 131, vol. 13, pp. 129–131.

18. Hermann Weyl, *Space-Time-Matter* (London: Methuen and Co., 1922), 174.

19. Hermann Minkowski, "Raum und Zeit," *Jahresbericht der Deutschen Mathematiker-Vereinigung* 18 (1909): 75. Lecture given at the 80th Naturforscher Versammlung in Köln on 21 September 1908.

20. For Bergson's influence on Virginia Woolf, see Birgit Van Puymbroeck "'Time in the Mind' vs. 'Time on the Clock': Modernism, Bergson and Woolf's to the Lighthouse," submitted in partial fulfillment for the degree of Master in de Taal en Letterkunde: Frans–Engels, Ghent University, Faculty of Arts and Philosophy. For the similarities between Bergson and Faulkner see Walter J. Slatoff, *Quest for Failure: A Study of William Faulkner* (Ithaca, N.Y.: Cornell University Press, 1960).

21. To some artists and writers, particularly later in the United States, Einstein would represent creativity more than mechanization.

22. Walter Arnold Kaufmann, *From Shakespeare to Existentialism: Studies in Poetry, Religion, and Philosophy* (Boston: Beacon Press, 1959), 326.

23. Charles Nordmann, *Notre maître le temps* (Hachette, 1924), 6.

24. Roland Barthes, "The Brain of Einstein," in *Mythologies* (New York: Hill and Wang, 1972). For Bergson's hair locks as "holy relics" see R. C. Grogin, *The Bergsonian Controversy in France: 1900–1914* (Calgary: The University of Calgary Press, 1988), ix.

25. Herbert Dingle, Introduction to *Duration and Simultaneity* (Indianapolis: Bobbs-Merrill Company, 1965), xv.

26. Gilles Deleuze, *Cinema 1: The Movement-Image*, trans. Hugh Tomlinson and Barbara Habberjam (Minneapolis: University of Minnesota Press, 1986), 60.

27. For a thorough account comparing their differences see Élie During, *Bergson et Einstein: la querelle du temps* (Paris: Presses Universitaires de France, 2013) and the dissertation by Laura Fontcuberta Famadas, "La disputa Einstein-Bergson" (Ph.D. diss, Universitat Autònoma de Barcelona, 2005).

28. André George, "Bergson et Einstein," *Les Documents de la vie intellectuelle* (January 1930): 52.

29. Pierre Lecomte du Noüy, *Biological Time* (London: Methuen and Co., 1936), 127.

30. Bergson thanked Einstein for the postcard in Bergson to Einstein, 18 June 1925, Paris. Reprinted in Angelo Genovesi, "Henri Bergson: Lettere a Einstein," *Filosofia* 49, no. 1 (1998): 31 and in Albert Einstein, *Correspondances françaises*, vol. 4, 39.

31. Einstein to Heinrich Zangger, after 27 December 1914, Berlin.

32. Jacques Chevalier, *Entretiens avec Bergson* (Paris: Plon, 1959), 34.

33. Dennis Overbye, *Einstein in Love* (New York: Penguin, 2000), 135.

34. Nordmann, *Notre maître le temps*, 156.

35. Hans Reichenbach, "The Philosophical Significance of the Theory of Relativity," in *Albert Einstein: Philosopher-Scientist*, ed. Paul Arthur Schilpp (La Salle, Ill.: Open Court, 1949), 309–310.

36. Bergson, *Durée et simultanéité*, 77.

37. Einstein's claim that the speed of light is constant in vacuo and independent of the motion of its source would be revised in his later general theory of relativity, where it varies with the strength of gravitational fields and can be significantly altered in the vicinity of black holes.

38. For the reception of Bergson's philosophy, see François Azouvi, *La Glorie de Bergson: essai sur le magistère philosophique* (Paris: Gallimard, 2007), 262.

39. The comparison of Socrates and Kant is from Édouard Le Roy. Cited in Richard A. Cohen, "Philo, Spinoza, Bergson: The Ride of an Ecological Age," in *The New Bergson*, ed. John Mullarkey (Manchester, U.K.: Manchester University Press, 1999), 18.

40. John Dewey, preface to *A Contribution to a Bibliography of Henri Bergson* (New York: Columbia University Press, 1912); also quoted in Cohen, "Philo, Spinoza, Bergson," 18.

41. William James quoted in Cohen, "Philo, Spinoza, Bergson," 18.

42. James to Bergson, 14 December 1902, in William James, *The Letters of William James*, vol. 2 (Boston: Atlantic Monthly, 1920), 179; also quoted in Grogin, *Bergsonian Controversy*, 58.

43. Jean Wahl, "At the Sorbonne," in *The Bergsonian Heritage*, ed. Thomas Hanna (New York: Columbia University Press, 1962), 153; also quoted in Cohen, "Philo, Spinoza, Bergson," 18.

44. Grogin, *Bergsonian Controversy*, 207.

45. "The greatest thinker in the world": Frédéric LeFèvre, "Une heure avec Maurice Maeterlinck," *Les Nouvelles littéraires* (7 April 1928): 8; also quoted in John Mullarkey, "Introduction: La Philosophie Nouvelle, or Change in Philosophy," in *The New Bergson*, ed. John Mullarkey (Manchester, U.K.: Manchester University Press, 1999), 1; "the most dangerous": W. Lippman, "The Most Dangerous Man in the World," *Everybody's Magazine* 27 (1912).

46. "Enchanter": Gilbert Maire, "Les années de Bergson à Clermont-Ferrand," *Glanes* 2 (March–April 1949): 20; also quoted in Grogin, *Bergsonian Controversy*, 107; "saving France": Frank Grandjean, *Une révolution dans la philosophie. La doctrine de M. Bergson* (Geneva and Paris: Atar, 1916), 211.

47. Grogin, *Bergsonian Controversy*, ix.

48. "Mr. Balfour's Objection to Bergson's Philosophy," *Current Literature* 51, no. 6 (1911).

49. Theodore Roosevelt, "The Search for Truth in a Reverent Spirit," *Outlook* 99 (1911).

50. Raïssa Maritain, *We Have Been Friends Together: Memoirs*, trans. Julie Kernan (London: Longmans, 1942), 84.

51. Morris R. Cohen, "The Insurgence Against Reason," *Journal of Philosophy* 22, no. 5 (1925).

52. Isaiah Berlin, "Impressionist Philosophy," *London Mercury* 32, no. 191 (1935); also quoted in Mullarkey, "Introduction: La Philosophie Nouvelle," 5.

53. Bertrand Russell, "The Philosophy of Bergson," *Monist* 22(1912): 323.

54. William Albert Levi, *Philosophy and the Modern World* (Bloomington: Indiana University Press, 1970), 65; also quoted in Grogin, *Bergsonian Controversy*, 29.

55. Grogin, *Bergsonian Controversy*, 39, 56.

56. P.A.Y. Gunter, ed. *Bergson and the Evolution of Physics* (Knoxville: University of Tennessee Press, 1969), 17–18. This judgment was later revoked by Father Antonin-Dalmace Sertillanges, author of *Avec Henri Bergson*.

57. Cristina Chimisso, "Introduction," in Bachelard, *The Dialectic of Duration* (Manchester: Clinamen Press, 2000).

58. Gunter, *Bergson and the Evolution of Physics*, 3–42.

59. Grogin, *Bergsonian Controversy*, ix.

60. "The Frenchman does not like the Jew, who does not like the Arab, who does not like the black man. The Arab is told: 'If you are poor, it is because the Jew has cheated you and robbed you of everything.' The Jew is told: 'You're not of the same caliber as the Arab because in fact you are white, and you have Bergson and Einstein.' The black man is told: 'You are the finest soldiers in the French empire; the Arabs think they're superior to you, but they are wrong.'" In "The So-Called Dependency Complex of the Colonized," Frantz Fanon, *Black Skin, White Masks* (New York: Grove Press, 2008), 83. In the original French: "Vous n'êtes pas sur le même pied que les Arabes parce qu'en fait vous êtes blancs et que vous, avez Bergson et Einstein."

61. From the perspective of the acclaimed Chinese writer and publisher Zhao Jiabi, their ideas of time and space shaped modern Western literature. Zhao Jiabi, "Xieshizhuyizhe de Shitaiyin [Gertrude Stein the realist]," *Wenyi fengjing* [Literary landscape] 1, no. 1 (1934).

62. Nordmann, *Notre maître le temps*, 160.

63. Ibid., 161.

64. Ibid.

65. "Double monologue": André Robinet, "Notes des éditeurs"; "the tower of Babel": Nordmann, *Notre maître le temps*, 157.

66. Dingle, Introduction, xvi.

67. Bernadette Bensaude-Vincent, "When a Physicist Turns on Philosophy: Paul Langevin (1911–39)," *Journal of the History of Ideas* 49, no. 2 (1988): 332.

68. "Locus classicus": Bruno Latour, "Some Experiments in Art and Politics," *E-Flux* 23 (2011); Michel Callon, "Whose Imposture? Physicists at War with the Third Person," *Social Studies of Science* 29, no. 2 (1999): 272.

69. Paul Valéry to Bergson, 25 June 1934, Nice, in *Mélanges*, 1511–1512.

70. Bensaude-Vincent, "When a Physicist Turns on Philosophy," 323.

71. Keith Ansell Pearson, *Germinal Life: The Difference and Repetition of Deleuze* (London: Routledge, 1999), 226n2

72. H. C. Browne, "Einstein's Paradox," *Nature* 110, no. 2768 (18 November 1922): 669.

73. Alain, *Propos* (Paris: Gallimard, 1956), 386–387; also quoted in Bensaude-Vincent, "When a Physicist Turns on Philosophy," 331.

CHAPTER 2. "MORE EINSTEINIAN THAN EINSTEIN"

1. Antonina Vallentin, *The Drama of Albert Einstein* (New York: Doubleday, 1954), 107.

2. On this topic see Michel Biezunski, *Einstein à Paris: le temps n'est plus . . .* (Vincennes: Presses Universitaires de Vincennes, 1991).

3. Johannes Stark, *Deutsche Tageszeitung*, 4 April 1922; also quoted in David E. Rowe and Robert Schulmann, eds., *Einstein on Politics: His Private Thoughts and Public Stands on Nationalism, Zionism, War, Peace and the Bomb* (Princeton, N.J.: Princeton University Press, 2007), 13–14.

4. Max Planck according to Einstein. In Einstein to Langevin, 13 March 1922. Albert Einstein, *Correspondances françaises*, vol. 4, Oeuvres Choisies (Paris: Éditions du Seuil, 1989), 168.

5. Newspaper report cited in Thomas Levenson, *Einstein in Berlin* (New York: Bantam Books, 2003), 261.

6. Xavier Léon in "La Théorie de la relativité," 92.

7. Ibid., 91.

8. Ibid.

9. Nordmann, *Notre maître le temps*, 6.

10. Einstein's first invitation came from the League of Human Rights, which was connected to the New Fatherland Association, to which he belonged.

11. For Einstein's decision to go to Paris, see Otto Nathan and Heinz Norden, eds., *Einstein on Peace* (New York: Avenel, 1960), 42–54.

12. Langevin, cited in Einstein to the Prussian Academy of Sciences, 13 March 1922.

13. When Einstein first published his theory in 1905, it received only scant attention compared to the near obsession that would surround it more than a decade and a half later.

14. *Times*, 7 November 1919; *New York Times*, 10 November 1919.

15. Paul Johnson, *Modern Times: The World from the Twenties to the Nineties* (New York: Harper Collins, 1991), 1.

16. Einstein to Marcel Grossmann, 12 September 1920 [Berlin].

17. Walter Isaacson, *Einstein: His Life and Universe* (New York Simon and Schuster, 2007), 267.

18. The meeting was recounted in "La Théorie de la relativité: séance du 6 avril 1922." It has been reprinted numerous times: Henri Bergson, *Écrits et paroles*, ed. Rose-Marie Mossé-Bastide, Bibliothèque de philosophie contemporaine (Paris: Presses Universitaires de France, 1957); Henri Bergson, "Discussion avec Einstein," in *Mélanges* (Paris: Presses Universitaires de France, 1972); and Henri Bergson, "Einstein et Bergson, 1922," in *Durée et simultanéité*. A translated extract is in Gunter, *Bergson and the Evolution of Physics*, 128–135.

19. Einstein to Paul Langevin, 6 March [1922], [Berlin].

20. Einstein to Paul Ehrenfest, 15 March [1922], [Berlin].

21. Einstein to Maurice Solovine, 14 March 1922, Berlin.

22. During his last high school year in Aarau, Einstein received his lowest grade in French. John J. Stachel, introduction to the centenary edition, *Einstein's Miraculous Year: Five Papers that Changed the Face of Physics*, ed. John J. Stachel (Princeton, N.J.: Princeton University Press, 2005), xx.

23. Maurice Montabré, "Une heure avec Einstein," *L'Intransigeant*, 10 April 1922; also quoted in Biezunski, *Einstein à Paris*, 45.

24. "La Théorie de la relativité," 91.

25. "La Théorie de la relativité," 101. Italics mine.

26. Ibid., 102.

27. Ibid., 107.

28. Ibid.

29. "Objective meaning": ibid., 99; "There are objective events": ibid., 107.

30. Ibid., 99. Italics mine.

31. Nordmann, *Notre maître le temps*, 177.

32. Einstein to Elsa Einstein, 9 [8] April 1922 [Paris].

33. Italics mine. Bergson, *Durée et simultanéité*, v.

34. Ibid., 2.

35. For the theme of unity, see Gerald Holton, *Thematic Origins of Scientific Thought; Kepler to Einstein* (Cambridge, Mass.: Harvard University Press, 1973).

36. Gerald Holton, "Einstein and the Cultural Roots of Modern Science," *Daedalus* 127, no. 1 (1998): 21.

37. On Ravaisson's influence on Bergson, see Jimena Canales, "Movement before Cinematography: The High-Speed Qualities of Sentiment," *Journal of Visual Culture* 5, no. 3 (2006): 275–294.

38. By Jacques Chevalier, Secrétaire général à l'Instruction publique from 11 September to 13 December 1940.

39. Einstein to Besso, 9 June 1937, [Princeton]. From Albert Einstein and Michele Besso, *Correspondance 1903–1955*, trans. Pierre Speziali (Paris: Hermann, 1979), 184–185.

40. Einstein to Paul Ehrenfest, 22 March 1919.

41. "Testament de Bergson," 8 February 1937, in Henri Bergson, *Correspondances* (Paris: Presses Universitaires de France, 2002), 1669–1671.

42. Einstein to Elsa Einstein, 31 August [1917], Gottmadingen, Württemberg.

43. Einstein to Elsa Einstein, [29 March 1922, Paris].

44. "Uninterrupted happiness": James Lee Christian, *Philosophy: An Introduction to the Art of Wondering*, 7th ed. (Fort Worth: Harcourt Brace College Publishers, 1998), 201.

45. Marcel Bataillon, "At the Collège de France," in *The Bergsonian Heritage*, ed. Thomas Hanna (New York: Columbia University Press, 1962), 110.

46. "Notes," *Philosophical Review* 31, no. 3 (1922).

47. Jacques Chevalier, *Henri Bergson*, trans. Lilian A. Clare (London: Rider and Co., 1928), 67–68. Other addresses include 47 boulevard Beauséjour, 32 rue Vital, 31 rue d'Erlanger: Chevalier, *Entretiens avec Bergson*, 21.

48. Quoted in Hubert Goenner, *Einstein in Berlin, 1914–1933* (Munich: C. H. Beck, 2005), 272.

49. Quoted in ibid.

50. Ilya Prigogine described their difference as "ideological." Prigogine, "The Arrow of Time: Inaugural Lecture" (Pescara, Italy: International Center for Relativistic Astrophysics, 1999); Giorgio Agamben, *Infancy and History: Essays on the Destruction of Experience* (London: Verso, 1993), 91.

51. For a note on Bergson's subtle play with capitalization see Élie During, "Introduction au dossier critique," in Bergson, *Durée et simultanéité*, 237.

52. Foreword to the second, expanded, 1923 edition (which already contained the three appendices aimed at answering some of the most pertinent objections raised by physicists), Bergson, *Durée et simultanéité*, ix. Although my reading of this sentence agrees with Élie During's ("Cet énoncé, pris à la lettre, est évidemment innacceptable pour un physicien"), I see the capitalization of "Time" as indicating a difference from the physicists' notion. During, "Dossier critique: I. Notes," 253.

53. Appendix 3 in second, expanded, 1923 edition, Bergson, *Durée et simultanéité*, 208.

54. Thomas Hanna, "Introduction," in *The Bergsonian Heritage*, ed. Thomas Hanna (New York: Columbia University Press, 1962), 23.

55. Alan D. Sokal and Jean Bricmont, *Fashionable Nonsense: Postmodern Intellectuals' Abuse of Science* (New York: Picador, 1998), 176. They also attribute this error to Merleau-Ponty's understanding of relativity.

56. Another formulation of Bergson's mistake centered on his comparison of the effects of relativity theory to the illusory perspectival shrinkage in the size of distant objects. For the relation of relativity theory to the science of perspective, see Jean-Marc Lévy-Leblond, "Le Boulet d'Einstein et les boulettes de Bergson," in *Annales Bergsoniennes 3*, ed. Frédéric Worms, *Bergson et la science* (Paris: Presses Universitaires de France, 2007).

57. Bergson, *Durée et simultanéité*, 65.

58. Henri Bergson, "Les Temps fictifs et le temps réel," *Revue de philosophie* 31, no. 3 (1924): 248. Republished as "Les Temps fictifs et le temps réel," in *Mélanges*, ed. André Robinet (Paris: Presses Universitaires de France, 1972), 1432–1449; "Les Temps fictifs et le temps réel," in *Durée et simultanéité: á propos de la théorie d'Einstein*, ed. Élie During (Paris: Presses Universitaires de France, 2010), 417–430. Translated in Gunter, *Bergson and the Evolution of Physics*, 168–186.

59. Bergson to Lorentz, 9 November 1924, Paris, in Bergson, *Correspondances*, 1119–1122, on p. 1122. Republished and annotated in Henri Bergson, *Écrits philosophiques*, ed. Arnaud Bouaniche, et al., Quadrige (Paris: Presses Universitaires de France, 2011), 556–559.

60. George, "Bergson et Einstein," 60; also quoted in André Metz, "Bergson, Einstein et les relativistes," *Archives de philosophie* 22 (1959): 378.

61. Thomas Hanna, ed. *The Bergsonian Heritage* (New York: Columbia University Press, 1962), 23. For a later text on this topic see Andrew C. Papanicolaou and P.A.Y. Gunter, *Bergson and Modern Thought: Towards a Unified Science*, Models of Scientific Thought (Chur, Switzerland: Harwood Academic Publishers, 1987).

62. Hervé Barreau, "Bergson et Einstein: à propos de Durée et simultanéité," *Les Études bergsoniennes*, no. 10 (1973): 167.

63. Auguste Champetier de Ribes, "Two Hundred and Fifteenth Day, Friday, 30 August 1946, Morning Session," in *Trial of the Major War Criminals before the International Military Tribunal* (Nuremberg, Germany: International Military Tribunal, 1948), 22: 298.

64. "'The pure present is an ungraspable advance of the past devouring the future. In truth, all sensation is already memory.' Hoshino looked up, mouth half open, and gazed at her face. 'What's that?' 'Henri Bergson,' she replied, licking the semen from the tip of his penis. [. . .] 'I can't think of anything special, but could you quote some more of that philosophy stuff? I don't know why, but it might keep me from coming so quick. Otherwise I'll lose it pretty fast.'" Haruki Murakami, *Kafka on the Shore*, trans. J. Philip Gabriel (New York: Vintage International, 2005), 273.

65. Patrick Romanell, "Bergson in México: A Tribute to José Vasconcelos," *Philosophy and Phenomenological Research* 21 (1961).

66. The group "Ateneo de la Juventud" was founded by José Vasconcelos and Antonio Caso in Mexico with the explicit purpose of reading Bergson alongside other humanists.

67. Ahmet Hamdi Tanpinar, *The Time Regulation Institute*, trans. by Alexander Dawe (New York: Penguin, 2013), 135.

68. Chevalier, *Entretiens avec Bergson*, 44.

69. Jacques Maritain, *Réflexions sur l'intelligence et sur sa vie propre* (Paris: Nouvelle Libraire Nationale, 1926), 229.

70. For some of Einstein's trips, see Siegfried Grundmann, *The Einstein Dossiers: Science and Politics* (Berlin: Springer, 2005).

71. Miguel Masriera Rubio, "La verdad sobre Einstein," *La Vanguardia*, 15 January 1925.

72. Alfonso Reyes, *Einstein: Notas de Lectura* (Mexico: Fondo de Cultura Económica, 2009), 33. Notes written in 1938 and revised in 1956.

73. Satosi Watanabe, "Le Concept de temps en physique moderne et la durée pure de Bergson," *Revue de métaphysique et de morale* 56: 128–142.

74. Yoav Di-Capua, "Arab Existentialism: An Invisible Chapter in the Intellectual History of Decolonization," *American Historical Review* 117, no. 4 (2012).

75. *Matière et mémoire* (1896), *Le Rêve* (1901), *Le Rire* (1900).

76. Ned Markosian, "Time," in *Stanford Encyclopedia of Philosophy*, ed. Edward N. Zalta (Winter 2010).

77. Irwin Edman, "Introduction" to *Creative Evolution* (New York: The Modern Library, 1944), ix, cited in P.A.Y. Gunter, "Bergsonian Method and the Evolution of Science," in *Bergson and the Evolution of Physics*, ed. P.A.Y Gunter (Knoxville: University of Tennessee Press, 1969), 20.

78. Reichenbach, "The Philosophical Significance of the Theory of Relativity," 306.

79. We know much more about the slight role experiment played in the formulation and acceptance of Einstein's first paper on relativity. The classic, original work

advancing this thesis is Gerald Holton, "Einstein, Michelson, and the 'Crucial' Experiment," *Isis* 60 (1969).

80. J. C. Hafele and Richard E. Keating, "Around-the-World Atomic Clocks: Predicted Relativistic Gains," *Science*, no. 177 (1972).

81. Arnold Sommerfeld, "To Albert Einstein's Seventieth Birthday," in *Albert Einstein: Philosopher-Scientist*, ed. Paul Arthur Schilpp (La Salle, Ill.: Open Court, 1949), 103.

82. Paul Valéry, *L'Idée fixe: Socrate et son médecin*, Oeuvres de Paul Valéry. (Paris: Éditions de la N.R.F., 1936).

83. For the comparison to Newton see Gerald Holton, "On the Origins of the Special Theory of Relativity," in *Thematic Origins of Scientific Thought* (Cambridge, Mass.: Harvard University Press, 1973).

84. Arthur I. Miller, *Einstein, Picasso: Space, Time and the Beauty That Causes Havoc* (New York: Basic Books, 2001). Miller's account has been corrected by Linda Dalrymple Henderson, "Einstein and 20th-Century Art: A Romance of Many Dimensions," in *Einstein for the 21st Century: His Legacy in Science, Art, and Modern Culture*, ed. Peter L. Galison, Gerald Holton, and Silvan S. Schweber (Princeton, N.J.: Princeton University Press, 2008).

85. On Matisse and Bergson see Mark Antliff, "The Rhythms of Duration: Bergson and the Art of Matisse," in *The New Bergson* (Manchester: Manchester University Press, 1999); Caroline A. Jones, "Rendering Time," in *Einstein for the 21st Century: His Legacy in Science, Art, and Modern Culture*, ed. Peter L. Galison, Gerald Holton, and Silvan S. Schweber (Princeton, N.J.: Princeton University Press, 2008).

86. Marcel Proust, "Letter to George de Lauris," in *Letters of Marcel Proust*, ed. and trans. Mina Curtis (New York: Random House, 1949), 197; Grogin, *Bergsonian Controversy*, 166.

87. Enid Starkie, "Bergson and Literature," in *The Bergsonian Heritage*, ed. Thomas Hanna (New York: Columbia University Press, 1962), 98. Proust himself minimized the influence of Bergson on his work.

88. Grogin, *Bergsonian Controversy*, 192.

89. Ibid., 119, 185, 186.

90. Criton, pseudonym for Charles Maurras, *Action française* 3 (10 April 1913).

91. Charles Maurras, "À propos de Bergson," *Action française* 1, no. 42 (11 February 1914); also quoted in François Azouvi, *La Glorie de Bergson*, 168.

92. Léon Daudet, "L'Assaut à l'Académie-Le Juif Bergson et F. Bourgeois," *Action française* 1, no. 43 (12 February 1914). A previous article on this same theme is "Un Juif à l'Académie française-L'Intrigue Bergson," *Action française* 1, no. 27 (27 January 1914); Daudet, "Le Juif Bergson à l'Académie. Le suprême effort de l'intrigue," *Action française* 1, no. 39 (8 February 1914).

93. Julien Benda, *Belphégor: essai sur l'esthétique de la présente société française* (Paris: Émile-Paul Frères, 1918). 157–158; Grogin, *Bergsonian Controversy,* 181. Charles Péguy similarly described the Bergson-Benda affair as "the old quarrel between Alexandrian and Rabbinical Jews," Charles Péguy, "Lettres et entretiens," *Cahiers de la quinzaine* 1, no. 154 (1927); also quoted in Grogin, *Bergsonian Controversy,* 180.

94. Grogin, *Bergsonian Controversy*, ix.

95. "Petit tarabiscoté" in Léon Daudet, "Le Stupide XIXe siècle," *Souvenirs et polémiques* (Paris: R. Laffont, 1992), 1254.

96. "Testament de Bergson," 8 February 1937, in Bergson, *Correspondances*, 1669–1671.

97. Wahl, "At the Sorbonne," 154.

98. Bergson, *Durée et simultanéité*, vi.
99. Ibid., 1.
100. Ibid., 208.
101. Ibid., vi. Italics mine.
102. Ibid., 162.
103. Hans Vaihinger to Einstein, 27 April 1919, Halle; Einstein to Hans Vaihinger, 3 May 1919, Berlin.
104. For this reason, I disagree with the assertion that "the whole thread running through Bergson's work and through the revolt against mechanism generally is that reality is fundamentally a spiritual activity." Grogin, *Bergsonian Controversy*, 56.
105. Quoted in ibid., 54.
106. Albert Einstein, "Autobiographical Notes," in *Albert Einstein: Philosopher-Scientist*, ed. Paul Arthur Schilpp (La Salle, Ill.: Open Court, 1949), 7, 13.
107. "La Théorie de la relativité," 103.

CHAPTER 3. SCIENCE OR PHILOSOPHY?

1. Benjamin Franklin, "Advice to a Young Tradesman," in *Franklin: The Autobiography and Other Writings on Politics, Economics, and Virtue* (Cambridge: Cambridge University Press, 2004). Originally published in 1748.
2. Joseph Delboeuf, "Déterminisme et liberté," *Revue philosophique de la France et de l'étranger* 13 (1882): 622.
3. Watanabe, "Le Concept de temps," 128.
4. Robert Boyle, "An Examen of Mr. T. Hobbes His Dialogus Physicus de Natura Aëris," in *The Works of the Honourable Robert Boyle*, ed. Thomas Birch (London: J. & F. Rivington, 1772), 197; also quoted in Steven Shapin, *A Social History of Truth: Civility and Science in Seventeenth-Century England* (Chicago: University of Chicago Press, 1994), 292.
5. For an account of the general philosophical impact of relativity in a broader context, see Klaus Hentschel, *Interpretationen und Fehlinterpretationen der speziellen und der allgemeinen Relativitätstheorie durch Zeitgenossen Albert Einsteins*, Science Networks Historical Studies (Basel: Birkhäuser, 1990).
6. Ken Alder, *The Measure of All Things: The Seven-Year Odyssey and Hidden Error That Transformed the World* (New York: Free Press, 2002), 308.
7. P. M. Harman, *The Natural Philosophy of James Clerk Maxwell* (Cambridge: Cambridge University Press, 1998), 10.
8. Pierre Wagner, "Introduction," in *Les Philosophes et la science*, ed. Pierre Wagner, *Collection folio/essais* (Gallimard, 2002), 17.
9. Diderot, quoted in ibid., 15, 23.
10. Henri Bergson, *Bulletin de la société française de philosophie* (28 November 1907) 21: 128; also quoted in Grogin, *Bergsonian Controversy*, 127.
11. André Lalande, "Philosophy in France, 1922–1923," *Philosophical Review* 33, no. 6 (1924): 543.
12. Rose-Marie Mossé-Bastide, *Bergson éducateur* (Paris: Presses Universitaires de France, 1955), 126.
13. Bergson, *Durée et simultanéité*, 65.
14. The question of just *how real* was the time shown in the relativity equations was amply debated at the time. An author of a book on relativity simply said that Einstein

treated the "discord between clocks as real." But that it would "no doubt" be better to qualify this by saying that these cases should be treated "as if" they were real. Even Max Born, a friend and strong supporter of Einstein, considered relativity theory more as a particular way of understanding time than as a reflection of its "physical" nature. "Thus the contraction is merely a consequence of the way we regard the things; it is not a real physical change," he explained. G. Fontené, *La Relativité restreinte* (Paris: Vuibert, 1922), 132; Max Born, *Die Relativitätstheorie Einsteins und ihre physikalische Grundlagen* (Berlin: Springer, 1921), 189.

15. Bergson, *Durée et simultanéité*, 206.

16. Henri Bergson, *Essai sur les données immédiates de la conscience*, ed. Arnaud Bouaniche, 9th ed. (Paris: Presses Universitaires de France, 2011), 80.

17. Bergson, *Durée et simultanéité*, 96.

18. Bergson, *Essai sur les données immédiates de la conscience*, 80.

19. Henri Bergson, *La Pensée et le mouvant: essais et conférences* (Paris: Presses Universitaires de France, 2009), 13.

20. Bergson, *Durée et simultanéité*, 180.

21. Henri Bergson, *Creative Evolution* (Mineola, N.Y.: Dover, 1998), 5–6.

22. "Mais dès lors toute différence est abolie entre le perception et le souvenir, puisque le passé est par essence *ce qui n'agit plus*, et qu'en méconnaissant ce caractère du passé on devient incapable de le distinguer réellement du présent, c'est-à-dire de l'*agissant*." Henri Bergson, *Matière et mémoire: Essai sur la relation du corps à l'esprit*, ed. Camille Riquier, 8th ed. (Paris: Presses Universitaires de France, 2009), 71.

23. Bergson, *Durée et simultanéité*, 180.

24. Ibid., 162–163.

25. Einstein to Richard B. Haldane, 11 September 1922, Berlin.

26. Einstein, travel diary to Japan, Palestine, and Spain [6 October 1922 to 12 March 1923]. See the entry for 9 October 1922: "Yesterday I looked into Bergson's book on relativity and time. Strange that time alone is problematic to him but not space. He strikes me as having more linguistic skill than psychological depth. He is not very scrupulous about the objective treatment of psychic factors. But he does seem to grasp the substance of relativity theory and doesn't set himself in opposition to it. The philosophers constantly dance around the dichotomy: the psychologically real and physically real, and differ only in evaluations in this regard. Either the former appears as a 'mere individual experience' or the second as 'mere construct of thought.' Bergson belongs to the latter kind but objectifies in *his* way without noticing." A different translation is given by P.A.Y. Gunter: "Yesterday I immersed myself in Bergson's book on relativity and time. Amazingly, he considers time, but not space, to be problematic. He seems to me to possess more linguistic skill than psychological depth. He does not hesitate to objectivize the psychologically-given. But he seems to really (sachlich) understand relativity theory and not to put himself in contradiction to it. The philosophers deal (tanzen) constantly here with the contrast between psychologically real and physically real and in this respect distinguish themselves only through what they value. Either the first appears as 'mere individual experience' or the second as 'mere thought-construction.' Bergson's views are of the latter sort, he objectivizes in his own way without realizing it." No. 29–131 in the Control Index of *The Collected Papers of Albert Einstein*. For another account of Einstein's impressions of Bergson from his travel diary, see Armin Hermann, *Einstein: Der Weltweise und sein Jahrhundert; eine Biographie*

(München: Piper, 1996), 283: "Einstein hatte sich mit Reiselektüre versorgt. Zuerst nahm er sich Bergson's Buch über Relativität und Zeit . . . Der französische Philosoph, den er in Paris kennengelernt hatte, schien ihm 'mehr sprachliches Geschick als psychologische Tiefe' zu besitzen."

27. Chevalier, *Entretiens avec Bergson*, 69.

28. Ibid.

29. Diary entry for 16 January 1922 in Franz Kafka, *Gesammelte Werke: Taschenbuchausgabe in sieben Bänden*, ed. Max Brod, vol. 7 (Frankfurt am Main: Fischer Taschenbuch Verlag, 1976), 405.

30. "La Théorie de la relativité," 107.

31. Ibid.

32. "Il n'y a donc pas un temps des philosophes ; il n'y a qu'un temps psychologique différent du temps du physicien." Ibid.

33. Hilary Putnam, "Time and Physical Geometry," *Journal of Philosophy* 64, no. 8 (1967): 247.

34. Maurice Merleau-Ponty, *Éloge de la philosophie et autres essais* (Paris: Gallimard, 1960), 309–320. Reprinted in Maurice Merleau-Ponty, "Einstein and the Crisis of Reason," in *Signs*, ed. John Wild (Evanston, Ill.: Northwestern University Press, 1964), 197.

35. Merleau-Ponty, "Einstein and the Crisis of Reason," 195.

36. Maurice Merleau-Ponty, *La Nature: notes, cours du Collège de France*, ed. Dominique Séglard, Traces écrites (Paris: Seuil, 1995); Maurice Merleau-Ponty, *Nature: Course Notes from the Collège de France*, ed. Dominique Séglard, trans. Robert Vallier, Northwestern University Studies in Phenomenology and Existential Philosophy (Evanston, Ill.: Northwestern University Press, 2003).

37. Merleau-Ponty, "Einstein and the Crisis of Reason," 197.

38. *Bergson et nous, Congrès Bergson: Paris, 17–20 mai 1959*, vol. 10, Bulletin de la Société française de philosophie: Actes du Xe Congrès des sociétés de philosophie de langue française (Paris: A. Colin, 1959).

39. Merleau-Ponty, "Einstein and the Crisis of Reason," 197.

40. Ibid., 194.

41. Maurice Merleau-Ponty, "At the Sorbonne," in *The Bergsonian Heritage*, ed. Thomas Hanna (New York: Columbia University Press, 1962), 139.

42. Maurice Merleau-Ponty, *Phenomenology of Perception*, International Library of Philosophy and Scientific Method (New York: Humanities Press, 1962), 375.

43. Merleau-Ponty, "Bergson in the Making," in *Signs*, edited by John Wild (Evanston, Ill.: Northwestern University Press, 1964), 184. Paper read at the conclusion of the Congrès Bergson (May 17–20, 1959) and published in *Bergson et nous, Congrès Bergson: Paris, 17–20 mai 1959* (Paris: Colin, 1959), 10.

44. Merleau-Ponty, "Einstein and the Crisis of Reason," 197.

CHAPTER 4. THE TWIN PARADOX

1. Giulio Cesare Ferrari, "IVe Congrès international de philosophie (Bologne, mars-avril 1911)," *Revue de métaphysique et de morale* 18, no. 3 (1910): 35–36.

2. Bataillon, "At the Collège de France," 108.

3. "Paradoxical": Paul Langevin, "L'Évolution de l'espace et du temps," *Scientia* 10 (1911), 41; also published as Paul Langevin, "L'Évolution de l'espace et du temps," *Revue*

de métaphysique et morale 10 (July 1911). For Langevin, see Eva Telkes-Klein, eds., *Paul Langevin: propos d'un physicien engagé* (Paris: Vuibert et Société française d'histoire des sciences et des techniques, 2007) and the work of Bernadette Bensaude-Vincent.

4. Langevin, "L'Évolution de l'espace et du temps," *Scientia* 10 (1911): 50.

5. Ibid.

6. Albrecht Fölsing, *Albert Einstein: A Biography* (New York: Viking, 1997), 192.

7. Albert Einstein, "Relativitätstheorie," *Vierteljahrsschrift Naturforschende Gesellschaft Zürich* 56 (1911): 6; also quoted in Peter Galison, *Einstein's Clocks, Poincaré's Maps: Empires of Time* (New York: W. W. Norton and Company, 2003), 266.

8. Poincaré delivered a talk titled "L'Évolution des lois."

9. Wildon Carr, "The IVth International Congress of Philosophy, Bologna, April 6th–11th, 1911," *Proceedings of the Aristotelian Society* 11 (1910–11): 225.

10. Ibid., 226.

11. Ibid.

12. Einstein to Langevin, 13 March 1922.

13. Xavier Léon in "La Théorie de la relativité," 92.

14. Albert Einstein, "Paul Langevin," *La Pensée* 12 (May–June 1947).

15. Maritain, *Réflexions sur l'intelligence et sur sa vie propre*, 214.

16. Bataillon, "At the Collège de France," 107. He referred to the years 1911–1912.

17. Mary Jo Nye, "Science and Socialism: The Case of Jean Perrin in the Third Republic," *French Historical Studies* 9, no. 1 (1975).

18. Georges Politzer, *La Fin d'une parade philosophique: le bergsonisme*, vol. 3, Libertés nouvelles (Paris: J. J. Pauvert, 1967), 12.

19. Peter Pesic, "Einstein and the Twin Paradox," *European Journal of Physics* 24 (2003).

20. Michel Paty, *Einstein philosophe: la physique comme practique philosophique*, Philosophie d'aujourd'hui (Paris: Presses Universitaires de France, 1993), 159.

21. Paul Langevin, "Le temps, l'espace et la causalité dans la physique moderne," *Bulletin de la Société française de philosophie* 12(1912): 32–33.

22. Langevin, "Le temps, l'espace et la causalité," 38.

23. Ibid., 24.

24. Ibid., 42.

25. For Bergson's view of the relation of science to philosophy in comparison with Léon Brunschvicg see Frédéric Worms, "Entre critique et métaphysique: la science chez Bergson et Brunschvicg," in *Les Philosophes et la science*, ed. Pierre Wagner, *Folio essais* (Paris: Gallimard, 2002).

26. Langevin, "Le Temps, l'espace et la causalité," 42.

27. Brunschvicg, in ibid., 43.

28. Brunschvicg, in ibid.

29. Bensaude-Vincent, "When a Physicist Turns on Philosophy," 330. He dedicated a chapter to Einstein's theory in his monumental *L'expérience humaine et la causalité physique* (1922).

30. Langevin, "Le Temps, l'espace et la causalité," 46.

31. The label of Bergsonian is from Abel Rey in "La Théorie de la physique chez les physiciens contemporains," *Bulletin de la Société française de philosophie* 9 (1909): 165.

32. Langevin, "Le Temps, l'espace et la causalité," 46.

33. Metz, "Bergson, Einstein et les relativistes," 378.

34. Édouard Le Roy, "Science et philosophie," *Revue de métaphyisique et morale* 7 (1899).

35. Some physicists would find Bergson's criticisms and his suggestions productive and adopt them in their own work. When the Japanese physicist Satosi Watanabe, who was inspired by Bergson, used the term "time," he was very clear to stress that it simply referred to a variable in an equation: "We understand by 'physical time' the variable that intervenes in the equations of physics under the name of 'time.'" He meant by that word nothing more and nothing less. Watanabe, "Le Concept de temps," 129.

36. Bergson, *Durée et simultanéité*, 96–97.

37. Ibid.

38. Le Roy explained his decision not to republish *Duration and Simultaneity* in Le Roy, Lettre-Préface, 29 September 1953, in Bergson, *Écrits et paroles*, vii–viii.

CHAPTER 5. BERGSON'S ACHILLES' HEEL

1. For a clear exposition of the twin paradox using only the special theory of relativity see Tim Maudlin, *Philosophy of Physics: Space and Time*, Princeton Foundations of Contemporary Philosophy (Princeton, N.J.: Princeton University Press, 2012).

2. Albert Einstein, "Zur Elektrodynamik bewegter Körper," *Annalen der Physik* 17(1905): 904.

3. Jürgen Renn, M. Janssen, and Matthias Schemmel, *The Genesis of General Relativity*, 4 vols., Boston Studies in the Philosophy of Science (Dordrecht: Springer, 2007).

4. "We remain": Bergson, *Durée et simultanéité*, vii; "straight": ibid., 77.

5. Ibid., 33.

6. Ibid., 206.

7. Ibid., 79.

8. Henri Bergson, "Les Temps fictifs et le temps réel," 254.

9. Bergson, *Durée et simultanéité*, 210.

10. Henri Bergson, "Les Temps fictifs et le temps réel," 254.

11. For debates about the "three clocks paradox," see Élie During's notes on *Duration and Simultaneity* in During, "Dossier critique: I. Notes," 292–293.

12. Bergson, *Durée et simultanéité*, 65.

13. André Lalande, "Philosophy in France, 1921," *Philosophical Review* 31, no. 6 (November 1922): 559.

14. "Notes," 319.

15. W. T. Bush, "The Paris Philosophical Congress," *Journal of Philosophy* 19, no. 9 (27 April 1922).

16. Harry Graf Kessler, *Tagebücher, 1918–1937* (Frankfurt am Main: Insel-Verlag, 1961); also quoted in Grundmann, *The Einstein Dossiers*, 134.

17. Prime Minister of France from 12 September 1917 to 16 November 1917.

18. "Notes," 319.

19. Solovine to Einstein, 27 April 1922.

20. "Notes," 317.

21. Ibid., 319.

22. Ibid., 318.

23. Ibid.

24. H.T.H. Piaggio, "Geometry and Relativity," *The Mathematical Gazette* 11, no. 159 (July 1922).

25. For biographical details, see Élie During's note in Bergson, *Durée et simultanéité*, 402n1.

26. Nordmann, "Einstein à Paris."

27. Ibid., 937.

28. "Whispering": Figure caption in Lucien Jonas, *L'Illustration* 80 (8 April 1922), 304.

29. *L'Humanité*, 4 April 1922; also quoted in Biezunski, *Einstein à Paris*, 21.

30. "La Théorie de la relativité," 98.

31. Ibid.

32. Bergson, *Durée et simultanéité*, 199.

33. Jeffrey Crelinsten, *Einstein's Jury: The Race to Test Relativity* (Princeton, N.J.: Princeton University Press, 2006).

34. Lisa Crystal, "Quantum Times: Physics, Philosophy, and Time in the Postwar United States" (Ph.D. diss., Harvard University, 2013).

CHAPTER 6. WORTH MENTIONING?

1. Henri Poincaré, "L'Espace et le temps," in *Dernières pensées* (Paris: Flammarion, 1917), 42.

2. Gerald Holton, "On the Thematic Analysis of Science: The Case of Poincaré and Relativity," in *Mélanges Alexandre Koyré, à l'occasion de son soixante-dixième anniversaire* (Paris: Hermann, 1964); Stanley Goldberg, "Poincare's Silence and Einstein's Relativity: The Role of Theory and Experiment in Poincaré's Physics," *British Journal for the History of Science* 5 (1970); Olivier Darrigol, "The Mystery of the Einstein-Poincaré Connection," *Isis* 95, no. 4 (2004); Yves Gingras, "The Collective Construction of Scientific Memory: The Einstein-Poincaré Connection and its Discontents, 1905–2005," *History of Science* 46 (2008); Galison, *Einstein's Clocks*.

3. One paper in particular, Poincaré's "Sur la dynamique de l'électron," has been the subject of intense attention in comparison to Einstein's "Zur Elektrodynamik bewegter Körper." Abraham Pais explained that Einstein could not have read the full version before completing his own paper in 30 June 1905, since Poincaré delivered it as a lecture on 23 July 1905 and published it in 1906. However, a short abstract was sent earlier, on 5 June 1905, to the Académie des sciences. In it, Poincaré referred to the changes that would occur in science if one rejected Lorentz's contraction hypothesis and stuck solely to his definition of lengths in terms of the time taken by light to transverse them. For Pais' account see Pais, *"Subtle is the Lord . . . ,"* 165.

4. Referring to Lucien Fabre. Solovine to Einstein, 16 March 1921, Paris.

5. Poincaré, "L'Espace et le temps," 109.

6. Bergson, "Conférence pour l'Academy of Science," 1917, in Bergson, *Correspondances*: 726.

7. "La Théorie de la relativité," 92.

8. Xavier Léon, quoted in Bensaude-Vincent, "When a Physicist Turns on Philosophy," 322.

9. Henri Bergson, "La Philosophie," in *La Science française* (Paris: 1915), 13; also quoted in ibid., 321–322.

10. "Poincaré never understood the basis of special relativity" and later "why did Poincaré never understand special relativity?" in Pais, *"Subtle is the Lord . . . ,"* 21, 164; "For his part, Poincaré seems never to have fully understood Einstein's breakthrough." Isaacson, *Einstein*, 135; "When he [Poincaré] finally addressed the issue of relativity squarely, in 1912, it was clear that he didn't understand it." Overbye, *Einstein in Love*, 145.

11. Henri Poincaré, "La Mesure du temps," *Revue de metaphisique et de morale* 6 (1898): 6.

12. Henri Poincaré, "La Théorie de Lorentz et le principe de réaction," *Archives néerlandaises des sciences exactes et naturelles* 5 (1900) .

13. Henri Poincaré, *La Valeur de la science*, Bibliothèque de philosophie scientifique (Paris: Ernest Flammarion, 1908), 210. Lorentz and Gustav Mie also worked on the relation of gravitational forces and electromagnetic ones.

14. Galison, *Einstein's Clocks*.

15. Henri Poincaré, "Sur la dynamique de l'électron," *Rendiconti del circolo matematico di Palermo* 21 (1906): 4.

16. Paul Lévy in "La Théorie de la relativité," 98.

17. "Objective meaning": ibid., 99; "Objective events": "La Théorie de la relativité," 107.

18. Galison, *Einstein's Clocks*, 297.

19. Quoted in ibid., 299. Einstein described Poincaré's position in a letter to Zangger, 16 November 1911, Prague. The encounter was described in Théo Kahan, "Sur les origines de la théorie de la relativité restreinte," *Revue d'Histoire des Sciences* 12 (1959).

20. Galison, *Einstein's Clocks*, 294.

21. Ibid., 295.

22. Overbye, *Einstein in Love*.

23. Henri Poincaré, "Rapport sur les travaux de H.A. Lorentz, ca. 31 January 1910," in Scott Walter, Étienne Bolmont, André Coret, eds. *La Correspondance entre Henri Poincaré et les physiciens, chimistes et ingénieurs* (Basel: Birkhäuser, 2007), 438.

24. Einstein's Kyoto talk is cited in Pais, *"Subtle is the Lord . . . ,"* 212.

25. Albert Einstein, "Geometry and Experience," in *Ideas and Opinions*, ed. Carl Seelig (New York: Three Rivers, 1954), 235.

26. Ibid., 238.

27. Henri Poincaré, *Science and Hypothesis*, 50.

28. Ibid.

29. See Peter Galison, "Einstein's Clock," *Critical Inquiry* 26 (Winter 2000); Galison, *Einstein's Clocks*.

30. Cited in Philipp Frank, *Einstein: His Life and Times* (New York: Alfred Knopf, 1947), 314.

31. For criticisms of Einstein in Germany in the 1920s, see Milena Wazeck, *Einsteins Gegner: die öffentliche Kontroverse um die Relativitätstheorie in den 1920er Jahren* (Frankfurt: Campus Verlag, 2009).

32. Albert Einstein, "Meine Antwort. Ueber die anti-relativitätstheoretische G.m.b.H.," *Berliner Tageblatt* (27 August 1920).

33. Albert Einstein, "Zionists Greet Einstein, Here for Palestine," *New York Tribune*, 3 April 1921, 3.

34. Bertrand Russell, *An Essay on the Foundations of Geometry*. Reviewed in Louis Couturat, "Compte-rendu critique de B. Russell, 'Essai sur les fondements de la géométrie,'" *Revue de métaphysique et morale* 6 (1898).

35. Poincaré, *La Valeur de la science*, 214. In 1894 Bergson congratulated Xavier Léon about the latest volume of the *Revue de métaphysique et de morale*, seeing it as "one of the most interesting ones that has to date appeared." This praise was due to an article therein by Poincaré, which Bergson claimed would be "deeply admired." The article was "Sur la nature du raisonnement mathématique," 371–384. Bergson to Xavier Léon, 20 August 1894, in Bergson, *Correspondances*, 23–24, on p. 23.

36. On Émile Boutroux and Poincaré, see Mary Jo Nye, "The Boutroux Circle and Poincaré's Conventionalism," *Journal of the History of Ideas* 40, no. 1 (1979).

37. Le Roy, "Science et philosophie," 381.

38. Poincaré, *La Valeur de la science*, 214. Poincaré responded to an article by Le Roy titled "Sur la logique de l'invention."

39. Bergson to V. Norström, 12 April 1910 in Bergson, *Correspondances*, 346–350, on p. 347.

40. Bergson to V. Norström, ibid., 348.

41. Ibid., 350.

42. Ibid.

43. "Par une voie toute différente, plus directe, par l'analyse des conditions auxquelles est soumise la construction des concepts scientifiques, le grand mathématicien Henri Poincaré est arrivé à des conclusions du même genre : il montre ce qu'il y a de relatif à l'homme, de relatif aux exigences et aux préférences de notre science, dans le réseau de lois que notre pensée étend sur l'univers." Henri Bergson, "La Philosophie française," *Revue de Paris* (15 May 1915). Reprinted in Bergson, *Écrits philosophiques*, 452–479, on p. 469.

44. Bergson, *Durée et simultanéité*, 166.

45. Michel Souriau, *Le Temps*, Nouvelle Encyclopédie philosophique (Paris: Félix Alcan, 1937), 9n1, 10.

46. Paul Doumergue, "Préface" to *Le Matérialisme actuel* (Paris: Ernest Flammarion, 1913).

47. Poincaré, "La Théorie de Lorentz et le principe de réaction," and Henri Poincaré, *Science and Method* (New York: Dover Publications, 1952).

48. Poincaré, "La Mesure du temps."

49. Bergson to V. Norström, 12 April 1910, in Bergson, *Correspondances*, 349.

50. Poincaré could be seen as having "blasted the popular view, espoused by the influential French philosopher Henri Bergson, that we have an intuitive understanding of time, simultaneity and duration." Galison, *Einstein's Clocks*, 32–33.

51. Poincaré, "La Mesure du temps," 1.

52. Bergson to V. Norström, 12 April 1910, in Bergson, *Correspondances*, 349.

53. W. Berteval, "Bergson et Einstein," *Revue philosophique de la France et l'étranger*, 132 (1943).

54. Paty, *Einstein philosophe: La physique comme practique philosophique*; Arthur Fine, *The Shaky Game: Einstein, Realism, and the Quantum Theory*, 2nd ed., Science and Its Conceptual Foundations (Chicago: University of Chicago Press, 1996), 86–111; Michael Friedman, *Foundations of Space-Time Theories: Relativistic Physics and Philosophy of Science* (Princeton, N.J.: Princeton University Press, 1983); Holton, "On the Origins of the Special Theory of Relativity," 219–259; Don Howard, "Realism and Conventionalism in Einstein's Philosophy of Science: The Einstein-Schlick Correspondence," *Philosophia Naturalis* 21, no. 1984 (1993); Don Howard, "Was Einstein Really a Realist?" *Perspectives on Science: Historical, Philosophical, Social* 1, no. 2 (1993).

55. Quoted in Holton, "On the Origins of the Special Theory of Relativity," 195. He expressed a similar view in Einstein, "Zum Relativitätsproblem," *Scientia* 15 (1914).

56. Albert Einstein, "Die Grundlage der allgemeinen Relativitätstheorie," *Annalen der Physik* 49 (1916): 777.

57. Ibid., 775.

58. Albert Einstein, "Dialog über Einwände gegen die Relativitätstheorie," *Die Naturwissenschaften* 6, no. 48 (29 November 1918).

59. Albert Einstein, "Induktion und Deduktion in der Physik," *Berliner Tageblatt*, 25 December 1919.

60. Oliver L. Reiser, "The Problem of Time in Science and Philosophy," *Philosophical Review* 35, no. 3 (1926).

CHAPTER 7. BERGSON WRITES TO LORENTZ

1. O. W. Richardson, "Hendrik Antoon Lorentz," *Journal of the London Mathematical Society* 4, no. 1 (1929): 192.

2. Einstein to Lorentz, 23 November 1911.

3. Pais, *"Subtle is the Lord . . . ,"* 164.

4. Bergson to Lorentz, 9 November 1924, Paris, in Bergson, *Correspondances*, 1119–1122, on p. 1122. Republished and annotated in Bergson, *Écrits philosophiques*, 556–559.

5. Bergson to Lorentz, 28 November 1924, in Bergson, *Correspondances*, 1126.

6. Einstein to Lorentz, 23 November 1911.

7. Rowe and Schulmann, eds., *Einstein on Politics*, 69.

8. Nordmann, "Einstein à Paris," 926.

9. Poincaré, "L'Hypothèse des quanta," 166. "In all instances in which it differs from that of Newton, the mechanics of Lorentz endures. We continue to believe that no body in motion will ever be able to exceed the speed of light; that the mass of a body is not a constant, but depends on its speed and the angle formed by this speed with the force which acts upon the body; that no experiment will ever be able to determine whether a body is at rest or in absolute motion either in relation to absolute space or even in relation to the ether." Henri Poincaré, "The Quantum Theory," in *Last Essays* (New York: Dover, 1963), 75.

10. Poincaré et al. to the Nobel Prize Committee, ca. 28 January 1902, in Scott Walter, Étienne Bolmont, André Coret, eds. *La Correspondance entre Henri Poincaré et les physiciens, chimistes et ingénieurs* (Basel: Birkhäuser, 2007), 399.

11. Hendrik A. Lorentz, *The Theory of Electrons and Its Applications to the Phenomena of Light and Radiant Heat: A Course of Lectures Delivered in Columbia University, New York, in March and April, 1906*, vol. 29, Sammlung von Lehrbüchern auf dem Gebiete der mathematischen Wissenschaften mit Einschluss ihrer Anwendungen (Leipzig: B.G. Teubner, 1909), 230.

12. Walter Kaufmann, *Annalen* (30 November 1905); quoted in Arthur I. Miller, *Albert Einstein's Special Theory of Relativity: Emergence (1905) and Early Interpretation (1905–1911)* (New York: Springer, 1998), 214.

13. Albert Einstein, "Bestimmung des Verhältnisses der transversalen und longitudinalen Masse des Elektrons" (1906): 586; also quoted in Richard Staley, *Einstein's Generation: The Origins of the Relativity Revolution* (Chicago: University of Chicago Press, 2008), 308.

14. Albert Einstein, "Über das Relativitätsprinzip und die aus demselben gezogenen Folgerungen," *Jahrbuch der Radioaktivität und Elektronik* 4 (1907).

15. Ibid.; also quoted in Staley, *Einstein's Generation*, 311.

16. Minkowski, "Raum und Zeit."

17. Hermann Minkowski, *Zwei Abhandlungen über die Grundgleichungen der Elektrodynamik, mit einem Einführungswort von Otto Blumenthal*, vol. 1, Fortschritte der mathematischen Wissenschaften in Monographien hrsg. von O. Blumenthal. (Leipzig: B.G. Teubner, 1910).

18. On the omission of Poincaré, see Scott Walter, "Minkowski, Mathematicians, and the Mathematical Theory of Relativity," in *The Expanding worlds of General Relativity*, ed. Hubert Goenner, et al. (Boston: Birkäuser, 1999), cited in Staley, *Einstein's Generation*, 313n30.

19. Minkowski, "Raum und Zeit."

20. The footnote was added to the republished article that appeared in Hendrik A. Lorentz et al., *The Principle of Relativity: A Collection of Original Memoirs on the Special and General Theory of Relativity*, trans. W. Perrett and G.B. Jeffery (London: Methuen & Co., 1923).

21. Einstein, "Relativitätstheorie," 6.

22. He referred to Hendrik A. Lorentz, "Electromagnetic Phenomena in a System Moving With Any Velocity Less Than That of Light," *Proceedings of the Royal Netherlands Academy of Arts and Science, Amsterdam* 6 (1904). Republished in Hendrik A. Lorentz, *Collected Papers*, 9 vols. (The Hague: M. Nijhoff, 1934), 172–197.

23. Lorentz, *Collected Papers*, 173.

24. Ibid.

25. Poincaré, *La Valeur de la science*, 187.

26. Poincaré, *Science and Hypothesis*, 172.

27. Einstein, "Über das Relativitätsprinzip und die aus demselben gezogenen Folgerungen." For the role of these historical accounts of relativity, see Staley, *Einstein's Generation*.

28. Minkowski, "Raum und Zeit," 58.

29. In 1909 the already eighteen-year-old Michelson-Morley experiment gained a second life when the research of Alfred Bucherer showed that electrons behaved in a similar way as the light waves described by Einstein. These experiments on electrons facilitated Einstein's point, that his theory was not simply ad hoc.

30. Lorentz, *The Theory of Electrons*, 230.

31. Hendrik A. Lorentz, "Alte und neue Fragen der Physik," *Physikalische Zeitschrift* 11 (1910): 1236.

32. "Welcher der beiden Denkweisen man sich anschliessen mag, bleibt wohl dem einzelnen überlassen." Ibid.

33. On the place of epistemology in Lorentz's attempts to distinguish his work from Einstein's, see Staley, *Einstein's Generation*, 329. Although he traces the origins of Lorentz's view about the role of epistemology to 1910 (by reference to "Alte un Neue Fragen"), I see it arising more strongly in 1913; see Hendrik A. Lorentz, *Das Relativitätsprinzip: Drei Vorlesungen gehalten in Teylers Stiftung zu Haarlem*, Beihefte zur Zeitschrift für mathematischen und naturwissenschaftlichen Unterricht (Leipzig: B.G. Teubner, 1914).

34. Einstein to Julius, 16 November [1911], Prague.

35. Einstein to Grossmann, 10 December 1911, Prague.

36. "Einstein sagt kurz und gut, daß alle soeben genannten Fragen keinen Sinn haben." Lorentz, *Das Relativitätsprinzip*, 23.

37. Ibid. "Die Bewertung dieser Begriffe gehört größtenteils zur Erkenntnislehre, und man kann denn auch das Urteil ihr überlassen, im Vertrauen, daß sie die besprochenen Fragen mit der benötigten Gründlichkeit betrachten wird. Sicher ist es aber, daß es für einen großen Teil von der Denkweise abhängen wird, an die man gewöhnt ist, ob man sich am meisten zur einen oder zur andern Auffassung angezogen fühlt."

38. Svante Arrhenius, 10 December 1922, in *Nobel Lectures in Physics (1901–1921)*, 479.

39. Albert Einstein, "Die Relativitätstheorie," in *Die Kultur der Gegenwart. Ihre Entwicklung und ihre Ziele*, ed. Emil Warburg (Leipzig: Teubner, 1915), 706.

40. Lorentz to Einstein, [between 1 and 23 January 1915, Haarlem].

41. Ibid.

42. Ibid.

43. Ibid.

44. Einstein to Lorentz, 23 January 1915, Berlin.

45. For an analysis of the political correspondence between Einstein and Lorentz, see Françoise Balibar and Jean-Philippe Mathieu, "Einstein-Lorentz: une correspondance scientifique et politique," *Mil neuf cent*, no. 8 (1990).

46. Hendrik A. Lorentz, *The Einstein Theory of Relativity: A Concise Statement* (New York: Brentano's, 1920), 62.

47. Ibid., 61–62.

48. Hendrik A. Lorentz, "The Principle of Relativity for Uniform Translations," in *Lectures on Theoretical Physics* ed. A. D. Fokker (London: 1931); also quoted in Holton, "On the Origins of the Special Theory of Relativity," 200.

49. Lorentz to Einstein, 6 June 1916, Haarlem.

50. Hendrik A. Lorentz, *Problems of Modern Physics: A Course of Lectures Delivered in the California Institute of Technology* (Boston: Ginn and Company, 1927), 221.

51. Arthur D'Abro, *Bergson ou Einstein* (Paris: H. Goulon, 1927), 75.

52. Hendrik A. Lorentz, "Report," *The Astrophysical Journal* 68, no. 5 (1928): 350.

53. Ibid.

54. Ibid., 349.

55. Ibid., 351.

56. Ibid., 349.

CHAPTER 8. BERGSON MEETS MICHELSON

1. Historians have argued that Einstein showed a surprisingly cavalier unconcern for the experimental results of the Michelson-Morley experiment. The classic work advancing this thesis is Holton, "Einstein, Michelson, and the 'Crucial' Experiment." Revised editions of Holton's work have updated the growing literature on this topic.

2. Chevalier, *Entretiens avec Bergson*, 46.

3. "The most recent applications of interferometry." Albert A. Michelson, "Les plus récentes applications de la méthode interférentielle," *Bulletin de la Société française de philosophie* 21 (12 May 1921).

4. For his meeting and impression of Michelson see Bergson to Xavier Léon, 8 May 1921, Paris in Bergson, *Correspondances*, 934.

5. Albert A. Michelson, "The Relative Motion of the Earth and the Luminiferous Ether," *American Journal of Science* 22 (1881); Albert A. Michelson and Edward Williams

Morley, "On the Relative Motion of the Earth and the Luminiferous Ether," *American Journal of Science*, no. 34 (1887).

6. Albert A. Michelson, "Report," *Astrophysical Journal* 68, no. 5 (1928): 344.

7. Albert Einstein, "Über das Relativitätsprinzip und die aus demselben gezogenen Folgerungen."

8. Henri Bergson, "Durée et simultanéité: a propos de la théorie d'Einstein," in *Mélanges* (Paris: Presses Universitaires de France, 1972), 63.

9. Chevalier, *Entretiens avec Bergson*, 50.

10. Einstein to Wilhelm Ostwald, 19 March 1901, Zurich.

11. Thomas Nagel, *The View from Nowhere* (New York: Oxford University Press, 1986); Robert P. Crease, *World in the Balance: The Historic Quest for an Absolute System of Measurement* (New York: W.W. Norton, 2011).

12. The budget for establishing a natural system of weights and measures in the post-revolutionary period was roughly three times as much as the entire operating costs of the pre–Revolutionary Académie des Sciences. Alder, *The Measure of All Things*. Bruno Latour mentions that, according to the National Bureau of Standards, the United States spends 6% of its gross national product in maintaining physical constants, that is, three times what is spent on research and development. Bruno Latour, *Science in Action: How to Follow Scientists and Engineers through Society* (Cambridge, Mass.: Harvard University Press, 1987), 251.

13. Albert A. Michelson, "Experimental Determination of the Velocity of Light," *U.S. Nautical Almanac Office Astronomical Papers* 1, Part 3 (1879).

14. For the history of the expedition to find the length of the meter see Alder, *The Measure of All Things*.

15. Antoine Lavoisier to Méchain, quoted in ibid., 67.

16. An alternative to basing the meter on a portion of the Earth's circumference developed in England. In 1824, an ordnance survey determined that the length of a pendulum beating seconds in London was 39.13929 inches long. A standard of length, some argued, could be defined by reference to the second of time. This standard, however, would be directly affected by changes in the Earth's rotation. If sidereal seconds were getting longer, then these errors would certainly spread to those of the yard. The pendulum standard, which never caught on, was soon abandoned. Simon Schaffer, "Metrology, Metrification and Victorian Values," in *Victorian Science in Context*, ed. Bernard Lightman (Chicago: Chicago University Press, 1997), 445.

17. James Clerk Maxwell, *Scientific Papers* (New York: Dover Publications, 1965), 225; also quoted in Schaffer, "Metrology, Metrification and Victorian Values," 461.

18. Thomas Henry Huxley, *Method and Results* (New York: D. Appleton, 1897), 79; also quoted in Schaffer, "Metrology, Metrification and Victorian Values," 464.

19. Albert A. Michelson and Edward Williams Morley, "On the Feasibility of Establishing a Light-Wave as the Ultimate Standard of Length," *American Journal of Science* 38, no. 225 (1889). Another publication on this topic is Albert A. Michelson, "Comparison of the International Meter with the Wave Length of the Light of Cadmium," *Astronomy and Astrophysics* 12 (1893).

20. For a description of the work of Michelson with Benoît see Jules Andrade, *Le Mouvement: mesures de l'étendue et mesures du temps* (Paris: Félix Alcan), 264–266.

21. J.-René Benoît, "De la précision dans la détermination des longueurs en métrologie," in *Rapports présentés au Congrès International de Physique réuni à Paris en 1900*, ed.

Ch.-Éd. Guillaume and L. Poincaré (Paris: Gauthier-Villars, 1900), 69, 70. In 1967 the meter was redefined as 1650763.73 times the wavelength of the orange-red light emitted by ^{86}Kr, a natural krypton isotope.

22. Charles-Édouard Guillaume, "L'Oeuvre du Bureau International des poids et mesures," in *La Création du Bureau International des poids et mesures et son oeuvre*, ed. Ch.-Éd. Guillaume (Paris: Gauthier-Villars, 1927), 198.

23. K. B. Hasselberg, "Presentation Speech," in *Nobel Lectures in Physics (1901–1921)*, 162.

24. Schaffer, "Metrology, Metrification and Victorian Values," 457.

25. For Michelson's work and its relation to the solar parallax and the transits of Venus, see Octave Callandreau, "Histoire abrégée des déterminations de la parallaxe solaire," *Revue scientifique* 2 (1881). For a history of these connections see Jimena Canales, *A Tenth of a Second: A History* (Chicago: Chicago University Press, 2009); Jimena Canales, "Photogenic Venus: The 'Cinematographic Turn' and Its Alternatives in Nineteenth-Century France," *Isis* 93 (2002).

26. When one of Jupiter's satellites passed into the shadow of the planet, it was seen approximately 480 seconds earlier when Earth was on the same side of the Sun as Jupiter than when Earth was on the opposite side. Since the distance between Jupiter and the Earth was shorter by a whole diameter of the Earth's orbit, when the Earth was on the same side of the Sun as Jupiter, then, the speed of light could be calculated.

27. James Clerk Maxwell, "On a Possible Mode of Detecting a Motion of the Solar System through the Luminiferous Ether," *Nature* 21 (1880): 314–315.

28. Michelson, "Report."

29. Michelson to Rayleigh, 17 May 1893, quoted in Staley, *Einstein's Generation*, 116.

30. Albert A. Michelson, "On the Relative Motion of the Earth and the Ether," *American Journal of Science* 3 (1897): 478; also quoted in Staley, *Einstein's Generation*, 117.

31. Problems with the sidereal clock and in determinations of the Earth-Sun distance were described in detail in Simon Newcomb, *Popular Astronomy* (New York: Harper & Bros., 1887); Simon Newcomb, *The Elements of the Four Inner Planets and the Fundamental Constants of Astronomy*, Supplement to the American Ephemeris and Nautical Almanac for 1897 (Washington, D.C.: U.S. Government Printing Office, 1895); Simon Newcomb, *Astronomy for Everybody: A Popular Exposition of the Wonders of the Heavens* (London: Isbiter and Company, 1903). Through the aberration constant, scientists determined that light from the Sun took about 499.6 seconds to reach the Earth. If light is found to travel at 299,860 km/s, then the Earth-Sun distance is the simple multiplication of these numbers: 93,075,480 miles.

32. Poincaré, "Sur la dynamique de l'électron."

33. Later in 1907, when he referred to the Michelson-Morley experiment, he took the experiment as providing experimental proof for the constancy of light.

34. Bergson, *Durée et simultanéité*, 134.

35. Ibid., 119–120.

36. Nordmann, *Notre maître le temps*, 223.

37. Léon Brillouin, *Relativity Reexamined* (New York: Academic Press, 1970), 5.

38. Ibid.

39. For Brillouin's recollection of Einstein's visit to Paris, see his interview at the Rockefeller Institute, New York, on 29 March 1962. Paul Peter Ewald, George

Uhlenbeck, Thomas S. Kuhn, Ella Ewald, "Interview with Dr. Léon Brillouin," in *Oral History Transcript* (American Institute of Physics, 29 March 1962).

CHAPTER 9. THE DEBATE SPREADS

1. Léon Brunschvicg in "La Théorie de la relativité," 99.

2. Entire books have been written about the political involvement of both men. For primary sources see Rowe and Schulmann, *Einstein on Politics.* For Einstein's early political views see Levenson, *Einstein in Berlin*. For later ones see Fred Jerome, *The Einstein File: J. Edgar Hoover's Secret War Against the World's Most Famous Scientist* (New York: St. Martin's Press, 2002). For the politics of Bergson, the classic account is Philippe Soulez, *Bergson politique* (Paris: Presses Universitaires de France, 1989).

3. Rowe and Schulmann, *Einstein on Politics,* 66–67.

4. Ibid., 137. Official quota limits for non-Aryans in Germany were set after 1933.

5. Albert Einstein, "Zur Errichtung der hebräischen Universität in Jerusalem," *Jüdische Pressezentrale Zürich* (26 August 1921).

6. Einstein to Langevin, [23 March 1922].

7. Einstein to Langevin, [23 March 1922].

8. Bergson cited in Jean-Jacques Renoliet, *L'UNESCO oubliée: La Société des Nations et la coopération intellectuelle (1919–1946)* (Paris: Publications de la Sorbonne, 1999), 31.

9. Nordmann, *Notre maître le temps*, 93.

10. Ibid., 82.

11. Ibid., 90.

12. Ibid., 99.

13. Ibid., 76.

14. Ibid., 91.

15. Ibid., 115.

16. Ibid., 52–53.

17. William Wallace Campbell and Joel Stebbins, "Report on the Organization of the International Astronomical Union," *Proceedings of the National Academy of Sciences of the United States of America* 6, no. 6 (1920): 394–395.

18. Ibid., 394.

19. Ibid., 393–394.

20. According to Brigitte Schroeder-Gudehus, competition was particularly evident in both institutions' bibliographic efforts. Brigitte Schroeder-Gudehus, "Challenge to Transnational Loyalties: International Scientific Organizations after the First World War," *Science Studies* 3, no. 2 (1973): 98.

21. Joel Stebbins, "The International Astronomical Union," *Popular Astronomy* 27 (1919): 609.

22. "Metrology by Light Waves at the National Physical Laboratory," *Nature* 131, no. 3302 (11 February 1933): 193.

23. Vallentin, *Drama of Albert Einstein*, 129–130.

24. The place of German scientists within this system was particularly contentious. The International Research Council (IRC) initially excluded Germans after the Great War and only eventually accepted neutral countries. Schroeder-Gudehus, "Challenge to Transnational Loyalties: International Scientific Organizations after the First World War."; A. G. Cock, "Chauvinism and Internationalism in Science: The International

Research Council, 1919–1926," *Notes and Records of the Royal Society of London* 37, no. 2 (1983). On the IRC, see Daniel J. Kevles, "'Into Hostile Political Camps': The Reorganization of International Science in World War I," *Isis* 62 (1971). The International Council for Science (ICSU) was founded in 1931. On the ICSU, see H. Spencer Jones, "The Early History of the ICSU 1919–1946," *ICSU Review* 2 (1960). In 1926 the IRC decided to stop the exclusion and invited German scientists to join as members. The Germans, however, declined. Germany was entirely left out of the most pertinent forums for discussing standards of time.

25. Campbell and Stebbins, "Report on the Organization of the International Astronomical Union," 358.

26. Ibid.

27. Ibid., 368.

28. Albert Einstein, "Die hauptsächlichen Gedanken der Relativitätstheorie," (after December 1916) published as "The Principal Ideas of the Theory of Relativity" in *The Collected Papers of Albert Einstein*, vol. 7 (Princeton, N.J.: Princeton University Press, 2002), 5.

29. On Einstein's vacillations about his membership, see Nathan and Norden, *Einstein on Peace*, 59, 64.

30. "13 septembre 1922: Commission Internationale de Coopération Intellectuelle," in Bergson, *Mélanges* (Paris: Presses Universitaires de France, 1972), 1352–1363.

31. Lord Robert Cecil of South Africa argued for declassification as a form of determent while Gilbert Murray argued for the opposite.

32. "13 septembre 1922," 1362.

33. Gilbert Murray to Einstein, 10 July 1922, Oxford.

34. Quoted in Renoliet, *L'UNESCO oubliée*, 27.

35. Albert Einstein, *Die Friedens-Warte*, 23 (June 1923), 186. Republished in Rowe and Schulmann, eds., *Einstein on Politics*, 194–195.

36. Born to Einstein, 7 April 1923, Göttingen, in *The Born-Einstein Letters: Friendship, Politics and Physics in Uncertain Times* (New York: Macmillan, 2005), 74.

37. Einstein to Solovine, 20 May 1923, in Albert Einstein, *Letters to Solovine* (New York: Philosophical Library, 1987), 58–59, on p. 59. The original reads: "Bergson hat in seinem Buch über Rel. Theorie schwere Böcke geschossen." The last sentence of the letter was cited as "God forgive him" in Pais, *Einstein Lived Here*, 75.

38. Einstein to Marie Curie, 25 December 1923. The letter appears in Carl Seelig, ed. *Albert Einstein: eine dokumentarische Biographie* (Zürich: Europa Verlag, 1954), 209–210, on p. 210. It is also in Nathan and Norden, *Einstein on Peace*, 64–65.

39. Bergson, *Mélanges*, 1104; Genovesi, "Henri Bergson: Lettere a Einstein," 12n2.

40. Germany joined in 1926 after Locarno (1925).

41. Quoted in Frank, *Einstein: His Life and Times*, 197.

42. Einstein to Marie Curie, 25 December 1923.

43. Einstein to Besso, 5 January 1924, [Berlin].

44. Mossé-Bastide, *Bergson éducateur*, 123.

45. Ibid., 125.

46. Renoliet, *L'UNESCO oubliée*, 37n79.

47. Einstein was "named by the Commission, as everyone else, without having asked for it." "25 julliet 1924: Commission Internationale de Coopération Intellectuelle," in *Mélanges* (Paris: Presses Universitaires de France, 1972), 1455.

48. Benrubi, *Souvenirs sur Henri Bergson*.

49. "Es wird Gras darüber wachsen, und dann wird man mit mehr Objectivität darüber urteilen." Ibid., 107–108; Genovesi, "Henri Bergson: Lettere a Einstein," 8–9; Mossé-Bastide, *Bergson éducateur*, 126.

50. Albert Einstein, "Von der Tagung der Völkerbundkommission für intellektuelle Zusammenarbeit aus Genf zuruckgekehrt, habe ich das Bedurfnis," *Frankfurter Zeitung*, 29 August 1924. Republished in Rowe and Schulmann, *Einstein on Politics,* 196–198.

51. Mossé-Bastide, *Bergson éducateur*, 145.

52. Bergson to Einstein, 5 February 1925, in Bergson, *Correspondances*, 1147. The original letter is Document no. 342600, Box 14, Einstein Archives, Hebrew University of Jerusalem.

53. Albert Einstein, "Wie ich Zionist wurde," *Jüdische Rundschau*, 21 June 1921, 352. This article was not written by Einstein but resulted from an interview approved by him.

54. Fritz Haber to Einstein, 9 March 1921, Berlin.

55. Einstein to Besso, 28 July 1925, Geneva.

56. Renoliet, *L'UNESCO oubliée*, 72.

57. Ibid., 93. For requesting to use Langevin as a substitute, see Einstein to Painlevé, 6 February 1926.

58. Einstein, "Report on Meeting of the Committee on Intellectual Cooperation," Document no. 28–037 in Einstein Archives, Hebrew University of Jerusalem. Reprinted in Nathan and Norden, *Einstein on Peace*, 77–78. To strengthen Germany's international standing among scientists and intellectuals, the Weimar Republic created the National Committee on Intellectual Cooperation (in 1928) and named Einstein as one of its members.

59. Ibid., 98.

60. "Keen disappointment": Albert Einstein, "Statement to *Berliner Tageblatt* (1927)," in Rowe and Schulmann, eds., *Einstein on Politics*, 206–207; "weak" and "by no means": Albert Einstein, "Interview in *Neue Züricher Zeitung* 20 November 1927, in ibid., 207–209.

61. Einstein to Jacques Hadamard, 24 September 1929.

62. Albert Einstein, "Einstein Considers Seeking New Home," *New York Times*, 22 December 1930.

63. Einstein to Painlevé, 9 April 1930.

64. Ibid.

65. Murray to Einstein, 21 June 1932, on p. 2 of 2. Document no. 34 892 in Einstein Archives, Hebrew University of Jerusalem.

66. Quoted in Niels Bohr, "Discussion with Einstein on Epistemological Problems in Atomic Physics," in *Albert Einstein: Philosopher-Scientist*, ed. Paul Arthur Schilpp (La Salle, Ill.: Open Court, 1949), 224.

67. Ibid.

68. Ibid., 228.

69. Ibid., 236.

70. Ibid.

71. Henri Bergson, *Les Deux sources de la morale et de la religion*, 10th ed. (Paris: Presses Universitaires de France, 2008), 306; also quoted in Mossé-Bastide, *Bergson éducateur*, 143.

72. Renoliet, *L'UNESCO oubliée*, 317.

73. Einstein to Willy Hellpach, Summer 1929.
74. Einstein, "Wie ich Zionist wurde," 352.
75. Albert Einstein, "A Re-examination of Pacifism," *Polity* 3, no. 1 (January 1935).
76. Renoliet, *L'UNESCO oubliée*, 7.

CHAPTER 10. BACK FROM PARIS

1. Einstein to Maurice Solovine, 20 April 1922.
2. Walter Rathenau to Einstein, 10–11 May 1917.
3. Charles Nordmann, "Avec Einstein dans les régions dévastées," *L'Illustration* 80, no. 4128 (15 April 1922).
4. Einstein to Paul Ehrenfest, 22 March 1919.
5. Cassirer to Einstein, 10 May 1920, Hamburg.
6. Max Wertheimer to Einstein, 15 May 1920, Berlin.
7. Alf Nyman, "Einstein-Bergson-Vaihinger," *Annalen der Philosophie und philosophischen Kritik* 6 (1927).
8. Max Wertheimer to Einstein, 15 May 1920, Berlin.
9. Einstein to Elsa Einstein, [19 May 1919], [Leyden].
10. Einstein to Elsa Einstein, [20 May 1919], [Leyden].
11. Einstein to Max Wertheimer, [21 May 1919], [Leyden].
12. Elsa Einstein to Einstein, [24 May 1920], [Berlin].
13. Einstein to Cassirer, 5 June 1920, Berlin.
14. Cassirer to Einstein, 16 June 1920, Hamburg.
15. Cassirer to Einstein, 28 August 1920, Hamburg.
16. Einstein, "Meine Antwort."
17. Ibid.
18. Cassirer to Einstein, 28 August 1920, Hamburg.
19. For the English translation, see Ernst Cassirer, *Substance and Function and Einstein's Theory of Relativity*, trans. William Curtis Swabey and Marie Collins Swabey (New York: Dover, 1953).
20. For Cassirer and Bergson, see Hisashi Fujita, "Cassirer, lecteur de Bergson," *Annales Bergsoniennes* 3 (2007).
21. Bergson, *Essai sur les données immédiates de la conscience*, 174.
22. Toni Cassirer, *Mein Leben mit Ernst Cassirer* (Hildesheim: Gerstenberg, 1981); also quoted in Fujita, "Cassirer, lecteur de Bergson," 69n61.
23. For Cassirer "immediate perception" escaped mediation. Cassirer, *Substance and Function*, 357.
24. Ibid.
25. Ibid., 375.
26. Einstein, "Zur Elektrodynamik bewegter Körper," 891–892.
27. Ernst Cassirer, *An Essay on Man: An Introduction to a Philosophy of Human Culture* (New Haven, Conn.: Yale University Press, 1944), 207.
28. Ibid., 220.
29. Ibid., 161.
30. For the themes of spontaneity and receptivity in Cassirer, see Peter E. Gordon, *Continental Divide: Heidegger, Cassirer, Davos* (Cambridge, Mass.: Harvard University Press, 2010).

31. Cassirer, *An Essay on Man*, 88.

32. Ernst Cassirer, *The Philosophy of Symbolic Forms*, vol. 4 (New Haven, Conn.: Yale University Press, 1996), 209–211.

CHAPTER 11. TWO MONTHS LATER

1. "Unendlich bedeutsam, viel wichtiger als der Besuch Einstein's." Koyré to Husserl, May 21, 1922, in Edmund Husserl, *Briefwechsel*, vol. 3, Husserliana Dokumente (Dordrecht: Kluwer Academic Publishers, 1994), 357.

2. Ethan Kleinberg, *Generation Existential: Heidegger's Philosophy in France, 1927–1961* (Ithaca, N.Y.: Cornell University Press, 2005), 59.

3. For Alexandre Koyré in Egypt, see Di-Capua, "Arab Existentialism."

4. For how phenomenology was driven by a desire to go "a step beyond Bergson," see Stefanos Geroulanos, *An Atheism That Is Not Humanist Emerges in French Thought* (Stanford, Calif.: Stanford University Press, 2010), 67.

5. Roman Ingarden, *On the Motives Which Led Husserl to Transcendental Idealism*, vol. 64, Phaenomenologica (Den Haag: Martinus Nijhoff, 1975). It was extremely critical of the French philosopher. The *Gutachten* praised Ingarden for showing how Bergson's philosophy was nothing more than a thorough skepticism and for revealing "die Wiedersinngkeit dieses Skepticismus."

6. Also see ibid., 12.

7. Martin Heidegger, *Parmenides*, trans. André Schuwer and Richard Rojcewicz (Bloomington: University of Indiana Press, 1992), 77.

8. Ibid.

9. Martin Heidegger, "The Concept of Time in the Science of History," in *Supplements: From the Earliest Essays to Being and Time and Beyond*, ed. John van Buren (Albany, N.Y.: State University Press of New York, 2002), 55.

10. Ibid.

11. Martin Heidegger, *Being and Time* (New York: Harper-Collins, 1962), 499n.iv.

12. Ibid.

13. Martin Heidegger, *The Concept of Time* (Oxford: Blackwell, 1992), 3, 21. Originally "Der Begriff der Zeit" lecture delivered at Marburg Theological Society in July 1924.

14. Martin Heidegger, *The Concept of Time* (London: Continuum, 2011), 59n. Originally review article "Der Begriff der Zeit" written in 1924 for *Deutsche Vierteljahresschrift fur Literaturwissenschaft und Geistesgeschichte* but not published at the time.

15. Martin Heidegger, *History of the Concept of Time: Prolegomena* (Bloomington: Indiana University Press, 1979), 2–3, 9.

16. Ibid., 9.

17. Martin Heidegger, "Wilhelm Dilthey's Research and the Struggle for a Historical Worldview (1925)," in *Supplements*, 169.

18. Editor's prologue in Heidegger, *History of the Concept of Time*, 321.

19. Heidegger, *Being and Time*, 49.

20. Ibid., 39.

21. Gordon, *Continental Divide*, 93.

22. Edmund Husserl, *The Phenomenology of Internal Time-Consciousness* (Bloomington: Indiana University Press, 1964), 67.

23. Ibid.

24. Peter E. Gordon labels Heidegger a "convinced Nazi" in "Heidegger in Black," *New York Review of Books* 61, no. 15 (2014). For François Fédier's argument that Heidegger is the "wrong suspect" in these accusations, see Georg Blume, "Der Philosoph François Fédier über Martin Heidegger," *Die Zeit*, no. 3, 2014.

25. Translated and reprinted in Edmund Husserl, "The Vienna Lecture," in *The Crisis of European Sciences and Transcendental Phenomenology: An Introduction to Phenomenological Philosophy*, ed. John Wild (Evanston, Ill.: Northwestern University Press, 1970).

26. Ibid., 125.

27. Ibid.

28. Ibid., 126.

29. The comparison of Bergson to Heraclitus appears in Russell, "The Philosophy of Bergson," 323, 339 and in Hans Reichenbach, *The Direction of Time* (Berkeley: University of California Press, 1956), 16–17. The comparison of Einstein to Parmenides is from Émile Meyerson, *La Déduction relativiste* (Paris: Payot, 1925), 108; Karl R. Popper, *Conjectures and Refutations: The Growth of Scientific Knowledge* (New York: Basic Books, 1962), 38.

30. Ebenezer Cunningham, *The Principle of Relativity* (Cambridge: Cambridge University Press, 1914), 191.

31. Heidegger, *Being and Time*, 141.

32. Ibid.

33. Ibid.

34. Ibid., 463.

35. Martin Heidegger, *The Metaphysical Foundations of Logic*, trans. Michael Heim, Studies in Phenomenology and Existential Philosophy (Bloomington: Indiana University Press, 1984), 149. Originally Martin Heidegger, *Metaphysische Anfangsgrunde der Logik im Ausgang von Leibniz*, the last Marburg course given in the summer of 1928.

36. Ibid., 203.

37. Martin Heidegger, *The Fundamental Concepts of Metaphysics: World, Finitude, Solitude*, trans. William McNiell and Nicholas Walker, Studies in Continental Thought (Bloomington: Indiana University Press, 1995), 141.

38. Heidegger, *Parmenides*, 77.

39. Ibid., 86.

40. Ibid., 85.

41. Parmenides's poem, according to Heidegger, was about the "unconcealedness of Being"; that is, it was about the relation of Being to his essence, which Heidegger considered to be the word.

42. Martin Heidegger, "The Question Concerning Technology," in *The Question Concerning Technology and Other Essays* (New York: Harper and Row, 1977), 18.

43. In this respect he differs most notably from the position developed later by the philosopher Jürgen Habermas in *The Structural Transformation of the Public Sphere: An Inquiry Into a Category of Bourgeois Society*, Studies in Contemporary German Social Thought (Cambridge, Mass.: MIT Press, 1989).

44. The course delivered in the winter of 1913–1914 was taught by Heinrich Rickert. Peter Fenves, *The Messianic Reduction: Walter Benjamin and the Shape of Time* (Stanford, Calif.: Stanford University Press, 2011), 6, 15, 118.

45. Ibid., 58.

46. Ibid., 177.

47. "A critique of the concept of such a progression must be the basis of any criticism of the concept of progress itself." Walter Benjamin, "Theses on the Philosophy of History," in *Illuminations*, ed. Hannah Arendt (New York: Schocken Books, 1969), 261.

48. Walter Benjamin, "Über einige Motive bei Baudelaire," *Zeitschrift für Sozialforschung* 8 (1939): 51.

49. Fenves, *The Messianic Reduction: Walter Benjamin and the Shape of Time*, 32.

50. Max Horkheimer, "On Bergson's Metaphysics of Time," *Radical Philosophy* 131 (May–June 2005).

51. Max Horkheimer, "Traditionelle und kritische Theorie," *Zeitschrift für Sozialforschung* 6 (1937).

52. Max Horkheimer, "Society and Reason. Five Public Lectures Delivered at the Department of Philosophy of Columbia University (February/March 1944)," in Max Horkheimer Archives, Frankfurt, LIX, 36, 6a, lecture 2: 5–12; also quoted in James Schmidt, "The Eclipse of Reason and the End of the Frankfurt School in America," *New German Critique* 100 (2007): 65.

CHAPTER 12. LOGICAL POSITIVISM

1. Rudolf Carnap, *Der logische Aufbau der Welt* (Berlin-Schlachtensee: Weltkreis, 1928), 258–259. Section "182. Intuitive Metaphysics."

2. Remarks by Carnap written in 1957 appended to Rudolf Carnap, "Überwindung der Metaphysik durch Logische Analyse der Sprache," *Erkenntnis* 2 (1932). Translated by Arthur Pap in A.J. Ayer, *Logical Positivism* (New York: The Free Press, 1959), 80.

3. Reichenbach, "The Philosophical Significance of the Theory of Relativity," 310.

4. Rudolf Carnap, *Der Raum. Ein Beitrag zur Wissenschaftslehre* (Berlin: Reuther and Reichard, 1922), 64.

5. Reichenbach, "The Philosophical Significance of the Theory of Relativity," 310.

6. Marco Giovanelli, "Talking at Cross-Purposes: How Einstein and the Logical Empiricists Never Agreed on What They Were Disagreeing About," *Synthese* 190, no. 17 (2013): 3819–3863.

7. David Hilbert quoted in ibid., 3826.

8. Einstein to Walter Dällenbach, [after 15 February 1917], [Berlin].

9. Hermann Weyl, "Erwiderung auf Einsteins Nachtrag zu H. Weyl, Gravitation und Elektrizität," *Sitzungsberichte der Königlich Preussischen Akademie der Wissenschaften* (1918): 479; also quoted in Giovanelli, "Talking at Cross-Purposes," 3831.

10. Einstein to Walter Dällenbach, [after 15 June 1918], [Berlin].

11. Hermann Weyl, *Philosophy of Mathematics and Natural Science*, trans. Olaf Helmer (Princeton, N.J.: Princeton University Press, 1949), 288.

12. Einstein's reply to Laue in Max von Laue, "Theoretisches über neuere optische Beobachtungen zur Relativitätstheorie," *Physikalische Zeitschrift* 21, no. 659–662 (1920); also quoted in Thomas Ryckman, *The Reign of Relativity: Philosophy in Physics 1915–1925*, Oxford Studies in Philosophy of Science (Oxford: Oxford University Press, 2005), 90 and in Thomas Ryckman, "Einstein Agonists: Weyl and Reichenbach on Geometry and the General Theory of Relativity," in *Origins of Logical Empiricism*, ed. Ronald N. Giere and Alan W. Richardson (Minneapolis: University of Minnesota Press, 1996), 179.

13. Moritz Schlick to Einstein, 9 October 1920, Rostock.

14. Michael Friedman, "Geometry as a Branch of Physics: Background and Context for Einstein's 'Geometry and Experience,'" in *Reading Natural Philosophy*, ed. David Malament (Chicago: Open Court, 2002).

15. See Einstein's late-in-life reformulation of the questions in "Geometry and Experience" in Albert Einstein, "Reply to Criticisms," in *Albert Einstein: Philosopher-Scientist*, ed. Paul Arthur Schilpp (La Salle, Ill.: Open Court, 1970), 676–681.

16. Albert Einstein, "Geometry and Experience," 238.

17. Hans Reichenbach, "Der gegenwärtige Stand der Relativitätsdiskussion. Eine kritische Untersuchung," *Logos* 10 (1921): 365–366.

18. Ibid.

19. Hans Reichenbach, "La Signification philosophique de la théorie de la relativité," *Revue philosophique de la France et de l'étranger* 93 (1922): 35. In technical terms, he argued that "the world" admitted a "univocal" definition of measurement.

20. "Matter of fact": Hans Reichenbach, *Philosophie der Raum-Zeit-Lehre* (Berlin: W. de Gruyter & Co., 1928), 25.

21. Einstein, "Geometry and Experience," 236.

22. Hans Reichenbach, *The Rise of Scientific Philosophy* (Berkeley: University of California Press, 1951), 133.

23. Einstein to Reichenbach, 30 June 1920, [Berlin].

24. "Der Urmensch, der seinen Arm hoch hob, um seinem Genossen zu winken, hat telegraphiert." Hans Reichenbach, *Was ist Radio? Mit 27 Abbildungen und einer Tafel der Sendestationen*, vol. 1, Das Radio-Reihe (Berlin: Richard Carl Schmidt & Co., 1924), 5.

25. Reichenbach, "The Philosophical Significance of the Theory of Relativity," 310.

26. Reichenbach, *Rise of Scientific Philosophy*, 121, 123.

27. Ibid., 145.

28. Ibid., 121.

29. Ibid., 132, 146, 131, 147.

30. Ibid., 153.

31. Sandra G. Harding, ed. *Can Theories Be Refuted? Essays on the Duhem-Quine Thesis* (Dordrecht, Netherlands: Reidel, 1976).

CHAPTER 13. THE IMMEDIATE AFTERMATH

1. Einstein to Elsa Einstein, [31 March 1922, Paris].

2. "There were well around one hundred people at the Becquerel's." Montabré, "Une heure avec Einstein."

3. Ibid.

4. Ibid.

5. Metz, "Bergson, Einstein et les relativistes."

6. Ibid., 373.

7. Many authors have explored Einstein's reception in France. In most of these accounts, the role of Bergson in physics plays a minor or nonexistent role. See Michel Biezunski, "La Diffusion de la théorie de la relativité en France" (Ph.D. diss., Université Paris Diderot, 1981); Dominique Pestre, *Physique et physiciens en France, 1918–1940*, Histoire des sciences et des techniques (Paris: Éditions des archives contemporaines, 1984);

Michel Paty, "The Scientific Reception of Relativity in France," in *The Comparative Reception of Relativity, Boston Studies in the Philosophy of Science* (Dordrecht: D. Reidel, 1987); Thomas F. Glick, ed. *The Comparative Reception of Relativity*, vol. 103, Boston Studies in the Philosophy of Science (Dordrecht: D. Reidel, 1987); Alexandre Moatti, *Einstein, un siècle contre lui*, Sciences (Paris: Odile Jacob, 2007); Michel Paty, "Einstein et la philosophie en France: à propos du sejour de 1922," *La Pensée* 210 (1980); Michel Biezunski, "Einstein's Reception in Paris in 1922," in *The Comparative Reception of Relativity*, ed. Thomas F. Glick, *Boston Studies in the Philosophy of Science* (Dordrecht: D. Reidel, 1987); Vincent Borella, "L'Introduction de la relativité en France, 1905–1922" (Ph.D. diss., Université Nancy 2, 1998). Max Jammer considered the relation of Einstein's work to Bergson and Whitehead too technical for inclusion. Max Jammer, *Concepts of Simultaneity: From Antiquity to Einstein and Beyond* (Baltimore, Md.: Johns Hopkins University Press, 2006), 162.

8. The president of the Société française de physique attended all of Einstein's conferences at the Collège de France. On those who welcomed Einstein, see Nordmann, "Einstein à Paris." Nordmann's claim that the Société française de physique rejected Einstein (pp. 935–936) is incorrect. Einstein refused their invitation to speak there. See Einstein to Thomas Barclay, 14 March 1922, [Berlin].

9. Jean Becquerel, "Débats sur la relativité: I. critique par M. J. Becquerel de l'ouvrage 'Durée et simultanéité' de M. Bergson," *Bulletin scientifique des étudiants de Paris* (March 1923).

10. Chevalier, *Entretiens avec Bergson*, 45–46.

11. Jean Becquerel, Préface to *La Relativité: exposé dans formules des théories d'Einstein et réfutation des erreurs contenues dans les ouvrages les plus notoires* (Paris: Étienne Chiron, 1923), vi. For criticisms see Henri Bouasse, *La Question préable contre la théorie d'Einstein* (Paris: Blanchard, 1923).

12. Bergson, *Durée et simultanéité*, 193.

13. Becquerel, "Débats sur la relativité: I. critique par M. J. Becquerel de l'ouvrage 'Durée et simultanéité' de M. Bergson," 28.

14. Bergson, *Durée et simultanéité*, 185.

15. Einstein, "Die Relativitätstheorie," 708–709.

16. André Metz, "Le Temps d'Einstein et la philosophie: à propos de l'ouvrage de M. Bergson, Durée et simultanéité," *Revue de philosophie* 31 (1924); André Metz, *La Relativité: Exposé dans formules des théories d'Einstein et réfutation des erreurs contenues dans les ouvrages les plus notoires* (Paris: Étienne Chiron, 1923); Metz, "Bergson, Einstein et les relativistes"; Eva Telkes-Klein, "Meyerson dans les milieux intellectuels français dans les années 1920," *Archives de Philosophie* 70, no. 3 (2007): 370.

17. Einstein's comments were included in the second edition.

18. Maritain, *Réflexions sur l'intelligence et sur sa vie propre*, 225n221.

19. Metz to Émile Meyerson, 12 January 1925, 408/70, folder Metz, Archives Centre Sèvres; also quoted in Telkes-Klein, "Meyerson dans les milieux intellectuels français dans les années 1920," 371.

20. Translated in Gunter, *Bergson and the Evolution of Physics*, 135–190.

21. Bergson, *Durée et simultanéité*, 202.

22. Henri Bergson, "Les Temps fictifs et le temps réel," 248.

23. Ibid.

24. Einstein to Metz, 2 July 1924. Published after Bergson's reply, Henri Bergson, "Réponse de M. Henri Bergson," *Revue de philosophie* 31 (1924): 440. Republished in Gunter, *Bergson and the Evolution of Physics*, 189–190.

25. Einstein, travel diary to Japan, Palestine, and Spain, 9 October 1922.

26. Miguel Masriera Rubio (1901–1981) was a professor of physical chemistry in Barcelona. He became Einstein's defender and Bergson's attacker in the Spanish-speaking world. In articles published in the prestigious *Vanguardia* newspaper, he brought the debate to the public. Einstein to Masriera Rubio: "Kurz: Bergson vergass, dass raumzeitlische Koinzidenz auch nach der Relativitätstheorie absoluten Character hat," p. 1 of 2. During the Spanish Civil War, Masriera Rubio went into exile. Einstein to Masriera Rubio, 7 October 1925, Berlin, Scientific Correspondence, Folders M–Misc, 1, Box 6, Einstein Archives, Hebrew University of Jerusalem.

27. "Il est bien entendu que la lettre que je sollicite de vous est destinée à être insérée (sauf avis contraire de votre part) dans la Revue de Philosophie.- J'ai d'ailleurs fait bien souvent usage de vos lettres (qui m'ont servi à dissiper bien d'erreurs) et j'espère que vous n'y voyez pas d'inconvénient." See Metz to Einstein, 23 November 1924, Bonn, on p. 2 of 5, Document no. 18 253, Folder 18.8. A. Metz, Paris(5), Box 7, Einstein Archives, Hebrew University of Jerusalem.

28. Henri Bergson, "Réponse de M. Henri Bergson," *Revue de philosophie* 31 (1924): 440. Reprinted in Bergson, *Mélanges*, 1450. This phrase was translated as "Metz believes that registering instruments are sufficient, with no need for an observer to observe what they indicate," in Gunter, *Bergson and the Evolution of Physics*, 189–190.

29. Bergson, *Durée et simultanéité*, 207–208n1.

30. Ibid.

31. Bergson, "Réponse de M. Henri Bergson," 440.

32. For Meyerson and Metz, see Élie During, "Entre Bergson et Saint Thomas: La Correspondance Metz-Meyerson," *Corpus, revue de philosophie* 58 (2010). For Meyerson, see Telkes-Klein, "Meyerson."

33. For his assessment of Meyerson's work see, Bergson, "Identité et réalité. Par Émile Meyerson," *Séances et travaux de l'Académie des sciences morales et politiques* 171 (23 January 1909).

34. Mario Biagioli, "Meyerson: Science and the 'Irrational,'" *Studies in History and Philosophy of Science* 19, no. 1 (1988): 35.

35. Einstein, "À propos de la deduction relativiste de M. Émile Meyerson," *Revue philosophique 105* (1928).

36. Meyerson to Einstein, 28 May 1927.

37. Metz to Einstein, 8 January 1926.

38. Déclaration de succession, Archives de l'enregistrement, Paris, quoted in Telkes-Klein, "Meyerson," 370.

39. References to this meeting in Paris appear in Metz to Einstein, 14 April 1927 and Metz to Einstein, Paris, 20 January 1927.

40. Albert Einstein, *Correspondances françaises*, vol. 4, Oeuvres Choisies (Paris: Éditions du Seuil, 1989). 219n5.

41. Metz to Einstein, 20 January 1927, Paris.

42. Bensaude-Vincent, "When a Physicist Turns on Philosophy," 330.

43. Émile Meyerson, *Identité et réalité* (Paris: Vrin, 1951). 438.

44. Émile Meyerson, *Du cheminement de la pensée* (Paris: Alcan, 1931), 599–600.

45. Metz to Helene Dukas, Antony (Seine) to Princeton, 10 January 1960, Document no. 18 269, Folder 18.8. A. Metz, Box 7, Einstein Archives. For her response and his interest in recovering Einstein's letters with "his opinion about Bergson," see their correspondence from Antony (Seine) to Princeton, Documents no. 18 270 and 18 271, Folder 18.8. A. Metz, Box 7, Einstein Archives, Hebrew University of Jerusalem.

46. Metz, "Bergson, Einstein et les relativistes," 374.

47. Ibid., 376.

48. Henri Arzeliès, *La Cinématique relativiste*, vol. 1, Études relativistes (Paris: Gauthier-Villars, 1955), 149.

49. His accomplishments included five military mentions: Fourragère aux couleurs de la medaille militaire; Chevalier de la légion d'honneur; Croix de guerre with four palms and four stars; Titulaire de la medaille de l'Yser; and Titulaire de la medaille de la resistance.

50. Metz to Meyerson, 12 January 1925, Document 408/70, Folder Metz, Archives Centre Sèvres; also quoted in Telkes-Klein, "Meyerson," 371.

51. Comment by André Metz in Jean-Louis Destouches "La théorie physique et ses principes fondamentaux," *Bulletin de la Société française de philosophie* 42, no. 1 (1947): 446.

52. Comment by Étienne Wolff, in ibid., 448.

CHAPTER 14. AN IMAGINARY DIALOGUE

1. Arthur Lynch, *The Case Against Einstein* (London: P. Allan, 1932); also quoted in Allan Ferguson, "Reviews: The Case against Einstein by Arthur Lynch," *The Mathematical Gazette* 17, no. 225 (1933): 272.

2. Wildon Carr, *A Theory of Monads: Outlines of the Philosophy of the Theory of Relativity* (1922), 339.

3. Lynch, *The Case Against Einstein*; also quoted in Ferguson, "Reviews: The Case against Einstein by Arthur Lynch," 272.

4. Einstein to Arnold Sommerfeld, 13 July 1921.

5. Matthew Stanley, "An Expedition to Heal the Wounds of War: The 1919 Eclipse and Eddington as Quaker Adventurer," *Isis* 94 (2003); Matthew Stanley, *Practical Mystic: Religion, Science, and A. S. Eddington* (Chicago: University of Chicago Press, 2007).

6. Arthur Stanley Eddington, "Einstein's Theory of Space and Time," *Contemporary Review* 116 (1919): 639; also quoted in Stanley, "Expedition to Heal," 84.

7. Eddington, "Einstein's Theory of Space and Time," 639; also quoted in Stanley, "Expedition to Heal," 84.

8. Historians and scientists alike often claim that the 1919 expedition was "one of the three 'classic' relativistic effects predicted by Einstein." Stanley, "Expedition to Heal," 70. See the entry "Crucial Experiment" in Nicholas Bunnin and Jiyuan Yu, *The Blackwell Dictionary of Western Philosophy* (Malden, Mass.: Blackwell, 2004).

9. Newcomb, *Astronomy for Everybody*, 166; Newcomb, *Elements of the Four Inner Planets* (Washington, D.C.: U.S. Government Printing Office, 1895).

10. See "Einstein-Besso Manuscript on the Motion of the Perihelion of Mercury" in vol. 4 of the documentary edition of *The Collected Papers of Albert Einstein*.

11. Einstein, "Meine Antwort."

12. Some authors see in these speculations the first descriptions of a black hole. "The possibility of Newtonian gravity affecting light had been investigated a handful of

times (notably by Henry Cavendish) over the previous two centuries, but the idea was never taken particularly seriously." Stanley, "Expedition to Heal," 71n37.

13. Norman Robert Campbell, *Théorie quantique des spectres. La Relativité* (1924), 216; Esclangon, *Les Preuves astronomiques de la relativité* (Paris: Gauthier Villars, 1922), 11; Chazy, *La Théorie de la relativité de la mécanique céleste*, vol. 1 (Paris: Gauthier-Villars,1928), 180.

14. Ernest Esclangon, *Les Preuves astronomiques de la relativité*, 11.

15. Einstein to David Hilbert, 18 November 1915.

16. Arthur Stanley Eddington, *Space, Time, and Gravitation: An Outline of the General Relativity Theory*, Cambridge Science Classics (Cambridge: Cambridge University Press, 1987), 51.

17. "past and future": Arthur Stanley Eddington, *The Nature of the Physical World.* Everyman's Library: Science (London: J. M. Dent & Sons, 1935), 89; "fancy": ibid., 95.

18. Ibid., 46

19. Ibid.

20. Ibid. 47.

21. Ibid.

22. Ibid., 79–80.

23. Ibid., 47–48.

24. Einstein to Besso, 29 July 1953, Princeton.

CHAPTER 15. "FULL-BLOODED" TIME

1. Whitehead, *Science and the Modern World*, 10; also quoted in Frank, *Einstein: His Life and Times*, 140. For Whitehead and Einstein, see Élie During, "Philosophical Twins? Bergson and Whitehead on Langevin's Paradox and the Meaning of 'Space-Time,'" in *Alfred North Whitehead's Principles of Natural Knowledge*, ed. G. Durand and M. Weber (Frankfurt: Ontos, 2007).

2. Alfred North Whitehead, *Science and the Modern World* (1925), 10; also quoted in Frank, *Einstein: His Life and Times*, 140.

3. Ronny Desmet, "Did Whitehead and Einstein Actually Meet?" in *Researching with Whitehead: System and Adventure*, ed. Franz Riffert and Hans-Joachim Sander (Freiburg: Alber Verlag, 2008).

4. Quoted in G. J. Whitrow, ed. *Einstein: The Man and His Achievement* (New York: Dover Publications, 1963), 44–45.

5. For the reception of Einstein's work in England, see Andrew Warwick, *Masters of Theory: Cambridge and the Rise of Mathematical Physics* (Chicago: University of Chicago Press, 2003).

6. Jacques Chevalier cited Lancelot Law Whyte. Chevalier, *Entretiens avec Bergson*, 96.

7. Chevalier, *Entretiens avec Bergson*, 96.

8. John Tyler Bonner, *Life Cycles: Reflections of an Evolutionary Biologist* (Princeton, N.J.: Princeton University Press, 1993), 67.

9. Alfred North Whitehead, *The Principle of Relativity with Applications to Physical Science* (Cambridge: Cambridge University Press, 1922), 88.

10. Filmer S. C. Northrop, "Whitehead's Philosophy of Science," in *The Philosophy of Alfred North Whitehead*, ed. Paul Arthur Schilpp (Evanston,Ill.: Northwestern University Press, 1941), 185.

11. "But it is purely a matter for experiment to decide which formula gives the small corrections which are observed in nature. So far as matters stand at present both formulae give the motion of Mercury's perihelion, my formula gives a possible shift of the spectral lines dependent upon the structure of the molecule and on the interplay of the gravitational and electromagnetic fields, and lastly, assuming a well-known modification of Maxwell's equations giving such an interplay, the famous eclipse results follow." Whitehead, *Principle of Relativity*, 84. For recent assessments of Whitehead's gravitational theory, see Jonathan Bain, "Whitehead's Theory of Gravity," *Studies in History and Philosophy of Modern Physics* 29 (1998); Gary Gibbons and Will M. Clifford, "On the Multiple Deaths of Whitehead's Theory of Gravity," *Studies in History and Philosophy of Modern Physics* 39 (2008).

12. Northrop, "Whitehead's Philosophy of Science," 184.

13. Ibid., 204.

14. Bergson, *Durée et simultanéité*, 62 and 62n61.

15. Paul Arthur Schilpp, preface to *The Philosophy of Alfred North Whitehead*, ed. Paul Arthur Schilpp (Evanston, Ill.: Northwestern University Press, 1941), xv.

16. Alfred North Whitehead, *Process and Reality* (New York: Macmillan, 1929), vii.

17. Alfred North Whitehead, "Autobiographical Notes," in *The Philosophy of Alfred North Whitehead*, ed. Paul Arthur Schilpp (Evanston, Ill.: Northwestern University Press, 1941), 13. Whitehead joined in 1915, and from 1922 to 1923 he was the society's president.

18. Richard B. Haldane, *The Reign of Relativity*, 3rd ed. (London: J. Murray, 1921), 121.

19. Ibid., 63.

20. Victor Lowe, *Alfred North Whitehead: The Man and His Work*, vol. 2 (Baltimore: Johns Hopkins University Press, 1985), 82.

21. Bergson to Haldane, 14 July 1921, Paris.

22. Evander Bradley McGilvary, "Space-Time, Simple Location, and Prehension," in *The Philosophy of Alfred North Whitehead*, ed. Paul Arthur Schilpp (Evanston, Ill.: Northwestern University Press, 1941), 234.

23. For Russell and Whitehead's different positions with regard to Einstein, see Ushenko, "Einstein's Influence on Contemporary Philosophy."

24. Russell to Karl Popper, 6 May 1960, Plas Penrhyn [Penrhyndeudraeth, Merioneth]. Published in I. Grattan-Guinness, "Russell and Karl Popper: Their Personal Contacts," *Russell: The Journal of the Bertrand Russell Archives*, no. 12 (1992): 18.

25. Milič Čapek, "La Pensée de Bergson en Amérique," *Revue international de philosophie* 31 (1977): 349.

26. For Bergson's opinion, see Chevalier, *Entretiens avec Bergson*, 197. The talk Bergson criticized was Bertrand Russell, "On the Relations of Universals and Particulars," *Proceedings of the Aristotelian Society, New Series* 12 (1911–1912).

27. Russell's book was positively reviewed by Louis Couturat, who was highly critical of Bergson. For Couturat on Bergson, see Louis Couturat, *De l'infini mathématique* (Paris: Félix Alcan, 1896). His review of Russell's *An Essay on the Foundations of Geometry* appeared in the *Revue de métaphysique et morale* (1898).

28. Poincaré, *Revue de métaphysique et morale* (1900).

29. Russell, "The Philosophy of Bergson."

30. "Difficult": ibid., 327; "not always easy to follow": ibid., 324.

31. For an Oxo advertisement, see *Graphic*, 30 November 1912.

32. Russell, "The Philosophy of Bergson," 332.

33. Ibid., 341.

34. Ibid., 334.

35. Ibid., 346.

36. Bertrand Russell, *The Philosophy of Bergson* (Cambridge: Bowes and Bowes, 1914).

37. Russell, "The Philosophy of Bergson," 337.

38. Russell, *The Philosophy of Bergson*, back of title page.

39. Bertrand Russell, *A History of Western Philosophy* (New York: Simon and Schuster, 1945), 793.

40. Alfred North Whitehead, "Symposium: The Problem of Simultaneity: Is There a Paradox in the Principle of Relativity in Regard to the Relation of Time Measured to Time Lived?" in *Relativity, Logic, and Mysticism. Proceedings of the Aristotelian Society, Supplementary Volumes* 3 (1923), 38.

41. Wildon Carr, "On Mr. Russell's Reasons for Supposing that Bergson's Philosophy Is Not True," in Bertrand Russell, *The Philosophy of Bergson*. Originally published in *Cambridge Magazine* 2 (1913).

42. Bergson in Russell, *The Philosophy of Bergson*, back of title page.

43. Carr delivered a lecture, "Bergson's Theory of Knowledge and Einstein's Theory of Relativity," at the Lyceum Club of London in November 1924. Mentioned in Maren Tova Linett, *The Cambridge Companion to Modernist Women Writers* (Cambridge: Cambridge University Press, 2010), 8–9.

44. Carr, *Theory of Monads*, 339.

45. Bertrand Russell, "The Ultimate Constituents of Matter," address delivered to the Philosophical Society of Manchester in February 1915. Reprinted in Bertrand Russell, *Mysticism and Logic* (New York: Longmans, Green and Co., 1918), 403.

46. Russell, *History of Western Philosophy*, 791.

47. Bergson's will, in Bergson, *Correspondances*, 1670.

48. Bergson's will, in ibid.

49. Alfred North Whitehead, *The Concept of Nature* (1920).

50. Whitehead, *Principle of Relativity*, 7.

51. Ibid., 66.

52. "Historical route": ibid., 68; "full-bloodedness": ibid., 69.

53. Ibid., 49.

54. Ibid., 65.

55. Whitehead, "Symposium."

56. Ibid., 37.

57. Whitehead confronted Carr earlier in Alfred North Whitehead, "Discussion: The Idealistic Interpretations of Einstein's Theory," *Proceedings of the Aristotelian Society* 22 (1922): 130–134.

58. Whitehead, "Symposium," 37.

59. Ibid., 40.

60. Ibid.

61. Whitehead, *Principle of Relativity*, 62.

62. Whitehead, *Science and the Modern World*, 145.

63. Alfred North Whitehead, *Adventures of Ideas* (Cambridge: Cambridge University Press, 1933), 237.

64. Alfred North Whitehead, *Modes of Thought* (New York: Macmillan Company, 1956), 370.

65. McGilvary, "Space-Time, Simple Location, and Prehension," 211.

66. Whitehead, "Symposium," 20.

67. Herbert Dingle, "History and Philosophy of Science," *Higher Education Quarterly* 6 (1952): 344.

68. Lowe, *Alfred North Whitehead*, 2: 64–65.

69. See Dingle's description of his experience that day in Whitrow, *Einstein*, 44–45.

70. G. P. Thomson, *The Foreseeable Future*, (Cambridge: Cambridge University Press, 1955), 89n81.

71. Dingle, introduction, xxxviii.

72. "One of the chief objectors to the view that asymmetrical aging is compatible with the relativity postulate was the philosopher Bergson, who wrote a book on the subject, *Duration and Simultaneity*, in 1922." Herbert Dingle, *Science at the Crossroads* (London: Martin, Brian and O'Keefe, 1972), 131.

73. Dingle, introduction, xliii.

74. Ibid., xli.

75. Hasok Chang, "A Misunderstood Rebellion: The Twin-Paradox Controversy and Herbert Dingle's Vision of Science," *Studies in the History and Philosophy of Science* 24 (1994): 781.

76. Dingle, "History and Philosophy of Science," 339.

77. Dingle, introduction, xv.

78. The offprint-collection of Percy W. Bridgman included numerous titles by Dingle.

79. Dingle, introduction, xxii.

80. Ibid.

81. Ibid.

82. Albert Einstein, *The Meaning of Relativity: Four Lectures Delivered at Princeton University, May, 1921* (London: Methuen & Co., 1922), 31.

83. Ibid.

84. Dingle, "History and Philosophy of Science," 337.

85. Ibid.

CHAPTER 16. THE PREVIOUS SPRING

1. For Bergson in the United States, see W. I. Riley, "La philosophie française en Amérique III. Le bergsonisme," *Revue philosophique de la France et de l'étranger* 91 (1921); Čapek, "La Pensée de Bergson en Amérique."

2. Gary A. Cook, *George Herbert Mead: The Making of a Social Pragmatist* (Urbana: University of Illinois Press, 1993), 210n216.

3. Grogin, *Bergsonian Controversy*, ix.

4. Cook, *George Herbert Mead*, 210n216.

5. Mead Papers, Special Collections Library, University of Chicago, Regenstein Library: "On relativity," 34 pages, in Box A3.5; "Relativity," 46 pages, in Box 9.13;1923, Philosophy 46, Relativity from the Standpoint of Pragmatism, notes by Van Meter Ames, 45 pages, in Box A4.10; George Herbert Mead, "Relative Space-Time and Simultaneity" was published posthumously in *Review of Metaphysics* 17 (1963–1964), edited and with

an introduction by David L. Miller. Hans Joas asserts that these are student notes and not original Mead texts.

6. Fritz Haber to Einstein, 9 March 1921, Berlin.

7. For the influence of Bergson on Mead, see Jon S. Moran, "Bergsonian Sources of Mead's Philosophy," *Transactions of the Charles S. Pierce Society* 32 (1996).

8. Mead, "Fragments on Relativity," Essay 30 in George Herbert Mead, *The Philosophy of the Act* (Chicago, Ill.: University of Chicago Press, 1938).

9. Ibid., 605.

10. Mead to Irene Tufts Mead, 14 September 1926, Mead Papers, Special Collections Library, University of Chicago, Box 1a, Folder 14; also quoted in Cook, *George Herbert Mead*, 138.

11. Mead to Irene Tufts Meas, 2 August 1920, Mead Papers, Special Collections Library, University of Chicago, Box 1a, Folder 4; also quoted in Moran, "Bergsonian Sources of Mead's Philosophy," 46.

12. Mead, "The Experiential Basis of Natural Science," Essay 15 in Mead, *The Philosophy of the Act*, 232.

13. George Herbert Mead, *The Philosophy of the Present*, Lectures upon the Paul Carus Foundation (Chicago: Open Court, 1932), 14.

14. Cook, *George Herbert Mead*, xvii.

15. Mead, *The Philosophy of the Present*, 32.

16. Arthur E. Murphy, preface to ibid., vii.

17. The program for the congress was published in *Journal of Philosophy* 23 (16 September 1926).

18. Mead to Irene Tufts Mead, 14 September 1926, Mead Papers, Special Collections Library, University of Chicago, Box 1a, Folder 14; also quoted in Cook, *George Herbert Mead*, 138.

19. Mead, *The Philosophy of the Present*, 43.

20. Ibid., 20–21.

21. Moran, "Bergsonian Sources of Mead's Philosophy," 58.

22. Mead, *The Philosophy of the Act*, xv.

23. Isaacson, *Einstein*, 298.

24. Arthur O. Lovejoy, "The Problem of Time in Recent French Philosophy," *Philosophical Review* 21 (1912). Today Arthur O. Lovejoy is remembered mostly as the founder of the history of ideas and is famous for his book *The Great Chain of Being*. Lovejoy's first engagement with Bergson was critical. In 1930 he published *The Revolt against Dualism*, claiming that the theory of relativity was "simply a vindication of dualism" and stating on various occasions that he did not accept whole cloth many of its conclusions. Lovejoy's book was an extended argument against all kinds of monists, including Bergson (because he allegedly denied a distinction between mind and matter), Whitehead, and Russell.

25. Arthur O. Lovejoy, *The Revolt Against Dualism: An Inquiry Concerning the Existence of Ideas* (W. W. Norton and Company, 1930), 147.

26. Arthur O. Lovejoy, "The Dialectical Argument Against Absolute Simultaneity. II," *Journal of Philosophy* 27, no. 24 (20 November 1930): 654.

27. Ibid., 651–652.

28. Lovejoy continued this argument in "The Travels of Peter, Paul, and Zebedee," *Philosophical Review* 41, no. 5 (1932).

29. Semester paper in Contemporary Philosophy by W.V.O. Quine, "On Bergson's notion of an 'élan vital.'" Typescript, 11 June 1930. 1 folder in Houghton Library, Harvard University.

30. One of Bridgman's first papers on time was "The Concept of Time," *Scientific Monthly* 35, no. 2 (1932).

31. William Marias Malisoff, "The Universe of Operations," *Philosophy of Science* 3, no. 3 (1936): 362.

32. Percy W. Bridgman, *Einstein's Theories and the Operational Point of View*, Vol. 7 of the Library of Living Philosophers (La Salle, Ill.: Open Court, 1949), 343.

33. Ibid., 354.

34. Ibid., 335.

35. Bridgman, *Einstein's Theories*, 354.

36. Ibid., 341.

37. Ibid., 341.

38. Ibid.

39. Ibid., 342.

40. Ibid.

CHAPTER 17. THE CHURCH

1. 8 February 1937, in Bergson, *Correspondances*, 1669–1671.

2. The radio address was given by Jacques Chevalier, who was then secrétaire d'État à l'Éducation nationale.

3. Some of the details pertaining to Bergson's death were described in the article published anonymously by Georges Politzer, "Après la mort de Bergson," *La Pensée libre*, no. 1 (February 1941).

4. Paul Bourget, a student of the philosopher Boutroux urged readers to acknowledge "the final bankruptcy of scientific knowledge." Paul Bourget, *Essais de psychologie contemporaine* (Paris: Gallimard, 1883); also quoted in Gerald Holton, "Ernst Mach and the Fortunes of Positivism in America," *Isis* 83(1992): 42. The theme of "crisis" appeared prominently in an influential book by Abel Rey, *La Théorie de la physique chez les physiciens contemporains: exposé des théories* (Paris: Félix Alcan, 1923). It announced a crisis in contemporary physics due to the failure of the mechanistic view that had dominated research during the nineteenth century. Poincaré initially considered that this "crisis" talk was exaggerated. As he remarked on its influence, he tried to combat it, initially blaming Édouard Le Roy as the main propagator of the bankruptcy-of-science view. Poincaré subtitled chapter eight in the *Value of Science* "The Crisis of Mathematical Physics."

5. For some of the debates within Catholicism involving Bergson, see chap. 7, "The Catholic Revival," in Grogin, *Bergsonian Controversy*, 24.

6. Maritain's "De la métaphysique des physiciens ou de la simutanéité selon Einstein" was composed between 1922 and 1924. It was published in Jacques Maritain, *Réflexions sur l'intelligence et sur sa vie propre*, Bibliothèque française de philosophie (Paris: Nouvelle Librairie Nationale, 1924). See also *Revue Universelle* (1 April 1924).

7. Jacques Maritain, *La Philosophie bergsonienne* (Paris: Marcel Rivière, 1913). Translated as Part 1 of Jacques Maritain, *Bergsonian Philosophy and Thomism* (New York: Philosophical Library, 1955).

8. Ibid., 306, 308. These quotations were eliminated from the second edition of the book.

9. Nordmann, *Notre maître le temps*, 160.

10. Ibid.

11. Azouvi, *La Glorie de Bergson*, 262.

12. Maritain, *Réflexions sur l'intelligence et sur sa vie propre*, 202.

13. "Naïve," "parade," "buffoonery": Maritain, *Réflexions sur l'intelligence et sur sa vie propre*, 248; "monstrous": ibid., 253; "physical dogma": ibid., 224; "metaphysical misery": ibid, 253.

14. Ibid., 202.

15. Ibid., 208.

16. Ibid., 203.

17. Ibid., 204.

18. Ibid., 261.

19. Ibid., 228.

20. Ibid., 207.

21. Ibid., 208.

22. Ibid., 260.

23. Ibid., 213.

24. Ibid., 208.

25. Ibid., 215.

26. Ibid., 258.

27. Ibid., 259.

28. Ibid., 239.

29. Ibid., 243.

30. Ibid., 247.

31. Ibid., 245.

32. Ibid., 247.

33. Ibid., 248.

34. "Internal logic": ibid., 346; "indemnity," ibid., 364.

35. Bergson, *Durée et simultanéité*, 96–97; also quoted in Maritain, *Réflexions sur l'intelligence et sur sa vie propre*, 369.

36. Ibid.

37. Maritain, *Bergsonian Philosophy and Thomism*, 155.

38. Maritain, *Réflexions sur l'intelligence et sur sa vie propre*, 216.

39. Einstein to Brecht, 4 May 1939.

40. "Et le physicien Duhem les avait tous deux précédés dans cette voie critique." Bergson, "La Philosophie française," in *Écrits philosophiques*, 470.

41. Maritain, *Réflexions sur l'intelligence et sur sa vie propre*, 254.

42. For Einstein and Duhem, see Don Howard, "Einstein and Duhem," *Synthese*, no. 83 (1990).

43. Le Roy in "La Théorie de la physique chez les physiciens contemporains," 183.

44. Gaston Bachelard, *Épistémologie*, SUP (Paris: Presses Universitaires de France, 1971), 137–138.

45. Thomas S. Kuhn, *The Structure of Scientific Revolutions*, 2nd ed. (Chicago: University of Chicago Press, 1970), 76.

46. Poincaré, "Sur la dynamique de l'électron," 4. He referred to Lorentz's contributions.

47. Max Planck, *Die Stellung der neueren Physik zur mechanischen Naturanschauung. Vortrag gehalten am 23 September 1910*, Versammlung Deutscher Naturforscher und Ärzte in Königsberg (Leipzig: Verlag von S. Hirzel, 1910). He did not refer exclusively to Einstein.

48. Paul Langevin, "L'Aspect général de la theorie de la relativité," *Bulletin scientifique des étudiants de Paris* 2 (1922): 2; also quoted in Bensaude-Vincent, "When a Physicist Turns on Philosophy," 326.

49. Jeremy Gray, *Henri Poincaré: A Scientific Biography*.

50. Quoted in ibid., 73, 424.

51. Poincaré, *Science and Hypothesis*, 117.

52. Ibid.

53. Bergson, *Durée et simultanéité*, vi.

54. Ibid., 123.

55. Nordmann, *Notre maître le temps*, 6.

56. Ibid., 195.

57. William Pepperell Montague, "The Einstein Theory and a Possible Alternative," *Philosophical Review* 33, no. 2 (1924): 146.

58. Eddington, *The Nature of the Physical World*, 62.

59. Chevalier, *Entretiens avec Bergson*, 90.

60. Ibid., 56.

61. Although Sesmat considered Einstein's theory perfectly "objective," he still felt that "one could, in conformity with the ideas of Lorentz, return to the absolutist conception of relativity theory." Quoted in André Lalande, "Philosophy in France, 1936–37," *Philosophical Review* 47, no. 1 (1938): 8.

62. Antonin-Dalmace Sertillanges, *Henri Bergson et le catholicisme* (Paris: Ernest Flammarion, 1941).

63. Ibid., 6,7.

64. Ibid., 5.

65. Yaron Ezrahi, "Einstein and the Light of Reason," in *Albert Einstein: Historical and Cultural Perspectives*, ed. Gerald Holton and Yehuda Elkana (Princeton, N.J.: Princeton University Press, 1986).

66. Bergson, *Durée et simultanéité*, 95.

67. This conventional reality was different from the "really real," which he defined as "that which is perceived or can be." Ibid.

68. Charles Péguy, "Note sur M. Bergson et la philosophie bergsonienne," *Cahiers de la quinzaine* (26 April 1914).

CHAPTER 18. THE END OF UNIVERSAL TIME

1. *L'Oeuvre*, 7 April 1922; also quoted in Biezunski, *Einstein à Paris*, 21.

2. Charles Nordmann, "Einstein expose et discute sa théorie," *Revue des deux mondes* 9 (1 May 1922).

3. *Le Producteur*, June 1922; also quoted in Biezunski, *Einstein à Paris*, 21.

4. *L'Oeuvre*, 6 April 1922; also quoted in Biezunski, *Einstein à Paris*, 47–48.

5. Chevalier, *Entretiens avec Bergson*, 41. Communicated to Chevalier on 7 February 1922.

6. Albert Einstein, "Le Principe de relativité et ses conséquences dans la physique moderne," *Archives des sciences physiques et naturelles (Genève)* 29, no. 1–2 (1910).

7. Einstein to Guillaume, 24 September 1917, Berlin. For the Einstein-Guillaume correspondence see Angelo Genovesi, *Il Carteggio tra Albert Einstein ed Édouard Guillaume*, Filosofia (Milano: Franco Angeli, 2000) in addition to the letters in the *Collected Papers of Albert Einstein*

8. Guillaume to Einstein, 3 October 1917, Berne.

9. *L'Oeuvre*, 6 April 1922; also quoted in Biezunski, *Einstein à Paris*, 47–48.

10. Bergson, *Creative Evolution*, 9.

11. Ibid., 21.

12. Ibid., 22.

13. Édouard Guillaume, "La Question du temps, d'après M. Bergson," *Revue générale des sciences pures et appliquées* 33(1922): 573.

14. Ibid.

15. Bergson to Guillaume, 21 August 1922, Saint Cèrgue to Berne.

16. Lorentz, *Das Relativitätsprinzip*, 23.

17. Guillaume to Einstein, 3 October 1917, Berne.

18. Ibid.

19. Einstein to Guillaume, 9 October 1917, Berlin.

20. Einstein to Guillaume, 24 October 1917, Berlin.

21. From February and April 1917. See Guillaume, "La Question du temps," 21.

22. Édouard Guillaume, "La Théorie de la relativité et le temps universel," *Revue de Métaphysique et morale* 25, no. 3 (May–June 1918); Édouard Guillaume, "Sur la théorie de la relativité, compte rendu de la séance de la Societé suisse de physique tenue à Berthond le 10 mai 1919," *Archives des sciences physiques et naturelles* 124, no. 3 (1919); Édouard Guillaume, "Représentation et mesure de temps," *Archives des sciences physiques et naturelles* 125, no. 2 (1920); Édouard Guillaume, "La Théorie de la relativité et sa signification," *Revue de Métaphysique et morale* 27, no. 4 (October–December 1920); Édouard Guillaume, *La Théorie de la relativité: résumé des conférences faites à l'Université de Lausanne au semestre d'été* (Lausanne, France: F. Rouge, 1921).

23. Ch. Taillens, "À propos des conférences du Dr. Guillaume," *Gazette de Lausanne* (10 June 1920).

24. Einstein to Grossmann, 27 February 1920, Berlin. Édouard Guillaume, "Les Bases de la théorie de la relativité," *Revue générale des sciences* 31, no. 7 (1920).

25. Guillaume to Grossmann, 19 June 1920, Bern.

26. Grossmann to Guillaume, 20 June 1920, Zurich.

27. Einstein to Grossmann, 12 September 1920, [Berlin].

28. Marcel Grossmann, "Mise à point mathématique," *Archives des sciences physiques et naturelles* 125, no. 6 (1920).

29. Marcel Grossmann, "Physikprofessoren," *Neue Schweizer Zeitung* (15 June 1920).

30. Guillaume to Einstein, 20 June 1920.

31. Einstein to Guillaume, 4 July 1920.

32. Grossmann, "Mise à point mathématique."

33. Einstein to Guillaume, 15 December 1920.

34. Fabre to Einstein, 17 May 1920.

35. Guillaume, "Temps relatif et temps universel," in Lucien Fabré and Albert Einstein, *Une Nouvelle figure du monde: les théories d'Einstein* (Paris: Payot, 1922), 245–249.

36. Chevalier, *Entretiens avec Bergson*, 90; Leslie Frewin, *La Notion du temps d'apres Einstein* (Paris: F. Alcan, 1921).

37. Paul Dupont, "Trois conceptions du temps physique," *Revue philosophique de la France et l'étranger* 96 (1923): 100n1.

38. Einstein to Guillaume, 20 January 1921, [Berlin].

39. Guillaume, "La Question du temps," 581.

40. Ibid.

41. A. I. Miller erroneously attributed this text to Charles-Édouard Guillaume.

42. Guillaume, "La Question du temps," 582.

43. Henri Poincaré, "Sur la dynamique de l'électron."

44. Bergson, *Durée et simultanéité*, 133.

45. Bergson quoted in Guillaume, "La Question du temps," 580.

46. He became associate director of the bureau in 1902 and from 1915 until his retirement in 1936, he was director.

47. *Paris-Midi*, 5 April 1922; also quoted in Biezunski, *Einstein à Paris*, 90.

48. Maritain, *Réflexions sur l'intelligence et sur sa vie propre*, 216–217n1.

49. Lorraine Daston, "Intelligences: Angelic, Animal, Human," in *Thinking with Animals: New Perspectives on Anthropomorphism*, ed. Lorraine Daston and Gregg Mitman (Chicago: University of Chicago Press, 2005).

50. Nordmann, *Notre maître le temps*, 15.

51. Maritain, *Réflexions sur l'intelligence et sur sa vie propre*, 216–217n 1.

52. [Felix Eberty], *The Stars and the Earth, or Thoughts Upon Space, Time and Eternity* (Boston: Lee and Shepard, 1882), 22.

53. Charles Hartshorne, "Whitehead's Idea of God," in *The Philosophy of Alfred North Whitehead*, ed. Paul Arthur Schilpp (Evanston, Ill.: Northwestern University Press, 1941), 545.

54. Newton, "General Scholium," added as appendix to the second edition of the *Principia* in 1713.

55. François Arago, "Notice scientifique sur la tonnerre," *Annuaire du bureau des longitudes* (1838): 475.

56. Pierre-Simon Laplace, *Essai philosophique sur les probabilités* (Paris: Courcier, 1814), 4.

57. Bruno Latour refers to the persistent denials of God's place in our understanding of the cosmos as "the crossed-out God" in Bruno Latour, *We Have Never Been Modern* (Cambridge, Mass.: Harvard University Press, 1991).

58. Poincaré quoted in Émile Richard-Foy, "Le Temps et l'espace du sens commun et les théories d'Einstein," *Revue philosophique de la France et l'étranger* 94 (1922): 160.

59. Léon Brunschvicg, *L'Expérience humaine et la causalité physique* (Paris: Félix Alcan, 1922), 411–412.

60. Lorentz to Einstein, [between 1 and 23 January 1915], [Haarlem].

61. Hartshorne, "Whitehead's Idea of God," 545.

62. "La Théorie de la relativité," 105.

63. Guillaume to Einstein, 23 February 1948, Neuchâtel.

CHAPTER 19. QUANTUM MECHANICS

1. Karl Marx, *The Eighteenth Brumaire of Louis Bonaparte* (New York: International Publishers, 1963), 15.

2. "Le 12 novembre 1929, Valéry est allé voir Bergson en compagnie d'Einstein, qui avait été invité à exposer sa théorie du champ unitaire à l'Institut Poincaré." Robinson

Judith, "Valéry, critique de Bergson," *Cahiers de l'Association internationale des études francaises*, no. 17 (1965): 206.

3. Einstein to Philipp Frank, cited in Isaacson, *Einstein*, 332.

4. Watanabe, "Le Concept de temps," 129.

5. Review by George Boas in B[oas], *Journal of Philosophy* 31, no. 25 (1934).

6. Jacques Chevalier, "Le Continu et le Discontinu," *Proceedings of the Aristotelian Society, Supplementary Volumes* 4 (1924).

7. Ibid., 185.

8. Ibid., 176.

9. George F. Barker, "Some Modern Aspects of the Life Question: Address of Professor George F. Barker, the Retiring President of the Association," *Science: A Weekly Record of Scientific Progress* 1(4 September 1880): 366.

10. A special number of the *Cahiers de la nouvelle journée*.

11. Review of *Continu et Discontinu* in *Revue de métaphysique et de morale* 37, no. 4 (1930).

12. Nye, "Science and Socialism."

13. John Hellman, "Jacques Chevalier, Bergsonism, and Modern French Catholic Intellectuals," *Biography* 4, no. 2 (1981).

14. Arrhenius, *Nobel Lectures in Physics (1901–1921)*.

15. Bergson, *Creative Evolution*, 320.

16. J.W.A. Hickson, "Causality and Recent Physics," *Philosophical Review* 44, no. 6 (1935): 542–543.

17. J.W.A. Hickson, "Review of *Der Kampf um das Kausalgesetz in der jüngsten Physik*. Hugo Bergmann. Braunschweig, Germany: Vieweg & Sohn. 1929. Pp. 78," *Journal of Philosophy* 26, no. 24 (1929); J.W.A. Hickson, "Recent Attacks on Causal Knowledge," *Philosophical Review* 47, no. 6 (1938); Hickson, "Causality and Recent Physics."

18. Du Noüy, *Biological Time*, 28n21.

19. Valéry to Bergson, 25 June 1934, Nice, in Bergson, *Mélanges*, 1511–1512.

20. Ibid.

21. Henri Mondor, *Propos familiers de Paul Valéry* (Paris: Grasset, 1957), 69.

22. André Rousseaux, "De Bergson à Louis de Broglie," in *Henri Bergson: Essais et témoignages inédits*, ed. Albert Béguin and Pierre Thévenaz (Neuchatel: La Baconnière, 1941), 280.

23. Ibid.

24. Ibid., 281.

25. Ibid., 280.

26. Chevalier, *Entretiens avec Bergson*, 255; Satosi Watanabe, "Contribution du point de vue de la mécanique ondulatoire, à l'étude du deuxième théorème de la thermodynamique" (Université de Paris, 1935). Another article in the *Revue de métaphysique et de morale* also contained a prefactory paragraph by de Broglie.

27. "Si donc on se décide à concevoir sous cette seconde forme la relation causale, on peut affirmer *a priori* qu'il n'y aura plus entre la cause et l'effet un rapport de détermination nécessaire, car l'effet ne sera plus donné dans la cause. Il n'y résidera qu'à l'état de pur possible, et comme une représentation confuse qui ne sera peut-être pas suivie de l'action correspondante." Bergson, *Essai sur les données immédiates de la conscience*, 159.

28. Louis de Broglie, "Les conceptions de la physique contemporaine et les idées de Bergson sur le temps et sur le mouvement," *Revue de métaphyisique et morale* (1941). In quantum mechanics "at each instant nature is described as hesitating between a multiplicity of possibilities."

29. Louis de Broglie, *Physique et microphysique*, Sciences d'aujourd'hui (Paris: A. Michel, 1947).

30. Quoted in Lewis S. Feuer, *Einstein and the Generations of Science* (New York: Basic Books, 1974), 221.

31. Watanabe, "Le Concept de temps."

32. Ibid., 135.

33. Olivier Costa de Beauregard, "Le Principe de relativité et spatialisation du temps," *Revue de questions scientifiques* (1947): 63.

34. "Illustrious": Olivier Costa de Beauregard, *Le Second principe de la science du temps* (Paris: Le Seuil, 1963), 126. "Irreparable": ibid., 7.

35. David Kaiser, *How the Hippies Saved Physics: Science, Counterculture, and the Quantum Revival* (New York: Norton, 2011), 171–175, 271.

36. Ibid., 99.

37. On lived time, see Bergson to Minkowski, 25 August 1937; "cited more and more" in Bergson to Minkowski, 24 September 1939.

38. Norbert Wiener, *The Human Use of Human Beings: Cybernetics and Society* (New York: Avon Books, 1967), 20–21.

39. Norbert Wiener, *Cybernetics: Or, Control and Communication in the Animal and the Machine*, 2nd ed. (New York: MIT Press, 1961), 44.

40. Gunter, *Bergson and the Evolution of Physics*, v.

CHAPTER 20. THINGS

1. Nordmann, *Notre maître le temps*.

2. "The initial day of a calendar serves as a historical time-lapse camera," explained Walter Benjamin at the very same moment when he noted the predicament of historians charged with selecting particular salient moments in history hiding among a mass of homogenous, empty time. Benjamin, "Theses on the Philosophy of History," 261.

3. Albert Einstein, "Die Freie Vereinigung für technische Volksbildung," *Neue Freie Presse* (1920).

4. Bergson, *Les Deux sources de la morale et de la religion*, 310.

5. Bergson to the president of the Nobel Committee, 10 November 1928, in Bergson, *Mélanges*, 1488–1490, on p. 1489.

6. Piéron was the expert in studying subjective perception of both time and simultaneity. In 1915 he reviewed the existing literature on the shortest perceptible time interval between two flashes of light, finding that it lay somewhere between 12 and 44 thousandths of a second. Henri Piéron, "2. Rhythm: sens du temps," *L'Année psychologique* 21, no. 1 (1914). Reviewed by Knight Dunlap, "The Shortest Perceptible Time-Interval Between Two Flashes of Light," *Psychology Review* 22 (1915).

7. "La Théorie de la relativité," 112.

8. Ibid.

9. Ibid., 113.

10. Nordmann, *Notre maître le temps*, 163.

11. Hervé Faye, "Sur les erreurs d'origine physiologique," *Comptes rendus de l'Académie des sciences* 59 (12 September 1864).

12. Poincaré commented on Auguste Calinon, "Étude critique sur la mécanique," *Bulletin de la société des sciences de Nancy* 7 (1885). See Calinon to Poincaré, 15 August 1886, Pompey in Poincaré, *La Correspondance*, 102.

13. "La Théorie de la relativité."

14. Canales, *A Tenth of a Second*.

15. Bergson, *Durée et simultanéité*, vi.

16. For an investigation into the "newness" of media technologies, see Lisa Gitelman, *Always Already New: Media, History, and the Data of Culture* (Cambridge, Mass.: MIT Press, 2006).

17. See discussion following Einstein's lecture "Relativitätstheorie" published in *Vierteljahrsschrift Naturforschenden Gesellschaft Zürich* 56 (1911).

18. Ibid.; Einstein, "The Theory of Relativity."

19. Langevin, "Le Temps, l'espace et la causalité," 339.

20. Ibid., 340.

21. Arthur Stanley Eddington, "The Relativity of Time," *Nature* 106, no. 2677 (1921): 804.

22. Arthur Stanley Eddington, *Espace, temps et gravitation*, trans. J. Rossignol (Paris: J. Hermann, 1921), 10; also quoted in Gaston Bachelard, *La Valeur inductive de la relativité* (Paris: J. Vrin, 1929), 149.

23. Eddington, *Nature of the Physical World*, 64–66.

24. Albert Einstein, *The Meaning of Relativity*, 31.

25. Charles Nordmann, *Einstein et l'univers*, 1922 ed., Le Roman de la science (Paris: Hachette, 1921), 241–242.

26. Whitehead, "Symposium," 39.

27. Whitehead, *Principle of Relativity*, 9, 36.

28. Ibid., 10.

29. Alexandre Koyré, "Compte rendu de H. Bergson, Le temps et la durée, 2e éd., 1925," *Versty* 1 (1926). Extract reprinted in Bergson, *Durée et simultanéité*, 434.

30. Bergson, *Creative Evolution*, 197.

31. Bachelard, *La Valeur inductive de la relativité*, 145.

32. Chimisso, "Introduction," 1.

33. Because Bergson had focused on continuity, Bachelard would focus on *dis*continuity: "From Bergsonism we accept almost all, except continuity," in Bachelard, *The Dialectic of Duration* (Manchester: Clinamen Press, 2000), 28–29. He wholly accepted the value of Einstein's work, and of new scientific discoveries to come from it.

34. Gaston Bachelard, "The Philosophic Dialectic of the Concepts of Relativity," in *The Library of Living Philosophers*, ed. Paul Arthur Schilpp (La Salle, Ill.: Open Court, 1949), 572.

35. Dominique Lecourt, *La Philosophie des sciences*, 4th. ed., Que sais-je? (Paris: Presses Universitaires de France, 2008).

36. Cariou, *Bergson et Bachelard*, 105.

37. For Bachelard and Bergson see Frédéric Worms and Jean-Jacques Wunenburger, *Bachelard et Bergson, continuité et discontinuité?: actes du Colloque international de Lyon,*

28–30 septembre 2006 (Paris: Presses Universitaires de France, 2008); Cariou, *Bergson et Bachelard*.

CHAPTER 21. CLOCKS AND WRISTWATCHES

1. Nordmann, *Notre maître le temps*, 136.
2. *La Presse* (10 April 1922); also quoted in Biezunski, *Einstein à Paris*, 73.
3. "l'horologe optique—je veux dire la propagation de la lumière": Bergson, *Durée et simultanéité*, 134.
4. Hermann von Helmholtz, "On the Interaction of Natural Forces," in *Science and Culture: Popular and Philosophical Essays* (Chicago: University of Chicago Press, 1995), 40.
5. Nordmann, *Notre maître le temps*, 109.
6. Canales, *A Tenth of a Second*.
7. Nordmann, *Notre maître le temps*, 137.
8. Ibid., 114.
9. Einstein, "Zur Elektrodynamik bewegter Körper," 893.
10. Einstein to Hans Albert and Edouard Einstein, [after 1 September 1921], [Berlin].
11. Heidegger, *Being and Time*, 499n. iv.
12. Kafka, diary entry for 16 January 1922, in Kafka, *Gesammelte Werke: Taschenbuchausgabe in sieben Bänden* 7: 405.
13. Einstein to Pauline Winteler, Thursday [May 1897], Zurich.
14. Einstein, 29 April 1899.
15. Einstein, 28 July 1899.
16. E. P. Thomson, "Time, Work-Discipline, and Industrial Capitalism," *Past and Present* 38 (1967).
17. The example of the lump of melting sugar is one of Bergson's most famous. "If I want to mix," "coincides," "is no longer": Bergson, *Creative Evolution*, 9–10; "bite": ibid., 8.
18. Bergson, *Creative Evolution*, 336.
19. Descartes to the Marquis of Newcastle, 23 November 1646.
20. Marx to Engels, 28 January 1863, [London].
21. Leibniz in "Note L to Bayle's article 'Rorarius' (1702) and Leibniz's private comments on it."
22. Bergson, *Durée et simultanéité*, 8.
23. Einstein, "Zur Elektrodynamik bewegter Körper," 898.
24. "Thus these clocks are set one over the other by an exchange of optical signals, typically electromagnetic, on the hypothesis that the signal covers the same trajectory going as coming back." "La Théorie de la relativité," 105.
25. Mary Frances Cleugh, *Time and Its Importance in Modern Thought* (London: Methuen & Co., 1937), 57.
26. Galison, "Einstein's Clock"; Galison, *Einstein's Clocks*.
27. Isaacson, *Einstein*, 25.
28. William Crookes, "Some Possibilities of Electricity," *Fortnightly Review* 102, no. 173–181 (1 February 1892).
29. Élie Mascart to Poincaré, 29 October 1903, Paris. Poincaré, *La Correspondance*, 262.

30. Max Planck, *A Survey of Physical Theory* (New York: Dover Publications, 1993). 49. The problem would be even more complicated if latitudinal points were not fixed on Earth. Some geological evidence suggested that the Earth's crust moved around significantly, and that even the place of the poles was found to shift in the span of a few years. This was evidence for the "curious theory" of Alfred Wegener, a geologist known for having found the continental drift. Nordmann, *Notre maître le temps*, 152.

31. "Darum braucht man diese Zeit gar nicht zu beachten, und kann mit gutem Gewissen sagen, dass die Wellen zur gleichen Zeit in Neuyork eintreffen, wie sie in Nauen abgesandt werden." Reichenbach, *Was ist Radio? Mit 27 Abbildungen und einer Tafel der Sendestationen*, 1: 23–24.

32. "Nur die Geschwindigkeit des Lichtes ist ebenso gross; und Lichtwellen sind ja auch von derselben Natur wie Radio-Wellen. Beide sind elektrische Wellen, nur von sehr verschiedener Länge." Ibid.

CHAPTER 22. TELEGRAPH, TELEPHONE, AND RADIO

1. Bergson, *Durée et simultanéité*, 65.
2. Ibid., 9–10.
3. Ibid.
4. Ibid., 166.
5. Langevin, "L'Évolution de l'espace et du temps."
6. Becquerel, "Débats sur la relativité: I. critique par M. J. Becquerel de l'ouvrage 'Durée et simultanéité' de M. Bergson," 26.
7. An important publication on this topic is Henri Poincaré, "Étude de la propagation du courant en période variable sur une ligne munie de récepteur," *Éclairage électrique* 40 (1904). See Poincaré's summary of his work in this area "Mes principaux ouvrages relatives à la physique," ca. June 1908 and Gaston Darboux et al. to the Nobel Prize Committee, ca. 1 January 1910 in Poincaré, *La Correspondance*: 415, 433.
8. Wildon Carr, "Symposium: The Problem of Simultaneity: Is There a Paradox in the Principle of Relativity in Regard to the Relation of Time Measured to Time Lived?" in *Relativity, Logic and Mysticism. Proceedings of the Aristotelian Society, Supplementary Volumes 3* (1923): 15.
9. Reichenbach, "The Philosophical Significance of the Theory of Relativity," 308; "Radio-telephone": Reichenbach, *The Rise of Scientific Philosophy*, 155.
10. "Mr. Whitehead is thus obviously in error when he says of Einstein's definition of simultaneity: 'The very meaning of simultaneity is made to depend on light signals.' He is also in error when he says: 'In fact the determination of simultaneity in this way [i.e., by electromagnetic signals] is never made.'" McGilvary, "Space-Time," 216.
11. For Bergson's numerous references to light, optical and electromagnetic signals see Bergson, *Durée et simultanéité*, 2–4, 6–7, 9–11, 14, 22, 55, 68, 70, 72, 86, 88–89, 92, 94, 97, 110, 120, 125–126, 128, 132, 186.
12. Henri Bergson, "Presidential Address," *Proceedings of the Society for Psychical Research* 26 (1913): 462. Later republished as "Fantômes et vivants."
13. Bergson, *Durée et simultanéité*, 91, 108.
14. Ibid., v.
15. Ibid., 26–27.
16. Ibid., 171.

17. Franz Kafka to Milena Jesenská, March 1922. In Franz Kafka, *Letters to Milena* (New York: Schocken, 1953), 229.

CHAPTER 23. ATOMS AND MOLECULES

1. Max Born, "Einstein's Statistical Theories," in *Albert Einstein: Philosopher-Scientist*, ed. Paul Arthur Schilpp (La Salle, Ill.: Open Court, 1949).
2. "La Théorie de la relativité," 109.
3. Einstein to Paul Bernays, [13 October 1916 or after 1920].
4. Weyl, *Philosophy of Mathematics*, 221; Hermann Weyl, *Philosophie der Mathematik und Naturwissenschaft* (München: R. Oldenbourg, 1966), 268. The reference to Bergson does not appear in the original 1927 essay published in the *Handbuch der Philosophie*, edited and compiled by A. Bäumler und M. Schröter.
5. Lewis, *Time and Western Man*, xviii.
6. James Joule, "On the Calorific Effects of Magneto-Electricity, and On the Mechanical Value of Heat," *Philosophical Magazine* 23, no. 152 (1843): 442.
7. Ludwig Boltzmann, *Vorlesungen über Gastheorie* (Leipzig: Johann Ambrosius Barth, 1898), 253.
8. Wilhelm Ostwald, *Die Überwindung des wissenschaftlichen Materialismus. Vortrag gehalten in der 3. Allgemeinen Sitzung der Versammlung der Gesellschaft deutscher Naturforscher und Ärzte zu Lübeck am 20. September 1895*, vol. 1, Versammlung Deutscher Naturforscher und Ärzte zu Lübeck (Leipzig: Von Veit, 1895), 21.
9. William Thomson, "The Sorting Demon of Maxwell," *Proceedings of the Royal Institution* 9 (1879).
10. Giovanni Cantoni and Eusebio Oehl, "Esperienze sulla produzione dei vibrioni in liquidi bolliti," 1865.
11. Jean Perrin, *Atoms* (Woodbridge, Conn.: Ox Bow Press, 1990), 85.
12. Lucretius, *On the Nature of Things*, trans. H.A.J. Munro (London: Bell and Sons, 1919). Originally c. 60 BCE.
13. Albert Einstein, "Über die von der molekularkinetischen Theorie der Wärme geforderte Bewegung von in ruhenden Flüssigkeiten suspendierten Teilchen," *Annalen der Physik* 17 (1905), 549.
14. Einstein to Maric, September 1900.
15. Albert Einstein, "Kinetische Theorie des Wärmegleichgewichtes und des zweiten Hauptsatzes der Thermodynamik, *Annalen der Physik* 9 (1902).
16. Einstein to Conrad Habicht, 18 or 25 May 1905, Bern.
17. Einstein, "Über die von der molekularkinetischen Theorie der Wärme geforderte Bewegung von in ruhenden Flüssigkeiten suspendierten Teilchen," 549.
18. Weyl, *Philosophy of Mathematics*, 208. "Der Gegensatz von *links und rechts*, der das mythische Denken zwar viel beschäftigt, stellt dem wissenschaftichen Denken kein so tiefgründiges Problem dar wie Vergangenheit und Zukunft. Es kann kein Zweifel bestehen, dass all naturegesetze invariant sind in bezug auf eine Vertauschung von links und rechts." From Weyl, *Philosophie der Mathematik und Naturwissenschaft*, 263–264.
19. Bergson, *Creative Evolution*, 241, 244.
20. Ibid., 246.
21. Ibid., 247.
22. Martin Heidegger, *The Concept of Time* (London: Continuum, 2011), 68.

CHAPTER 24. EINSTEIN'S FILMS: REVERSIBLE

1. Bergson, *Durée et simultanéité*, 157.

2. For Bergson on cinematography see Élie During, "Vie et mort du cinématographe: de L'Évolution créatrice à Durée et Simultanéité," in *Bergson*, ed. Camille Riquier (Paris: Cerf, 2012); Élie During, "Notes on the Bergsonian Cinematograph," in *Cine-dispositives. Essays in Epistemology across Media*, ed. F. Albera and M. Tortajada (Amsterdam: Amsterdam University Press, 2013).

3. Bergson, *Creative Evolution*, 330.

4. Ibid.

5. Ibid., 305.

6. His criticisms of cinematography appeared in print in 1907 although they were delivered as well in his lectures at the Collège de France. Paul Fontana, "Cours de Bergson au Collège de France: Théories de la volonté," in *Mélanges*, ed. André Robinet (Paris: Presses Universitaires de France, 1972). Bergson reinterpreted his earlier work in terms of the development of cinematography. Reflecting on his first important book, *Essai sur les données immediate de la conscience* (1889), he claimed that—although he had not yet used the term, it already criticized the cinematographic method: "I did not characterize that process, since then, as cinematographic. But the cinematographic camera had not yet been invented."

7. Bergson, *Creative Evolution*, 342.

8. Henri Bergson, "La Perception du changement," in *Mélanges*, ed. André Robinet (Paris: Presses Universitaire de France, 1972), 911.

9. Charlotte Bigg, "Evident Atoms: Visuality in Jean Perrin's Brownian Motion Research," *Studies In History and Philosophy of Science* 39 (2008); Charlotte Bigg, "A Visual History of Jean Perrin's Brownian Motion Curves," in *Histories of Scientific Observation*, ed. Lorraine Daston and Elizabeth Lunbeck (Chicago: University of Chicago Press, 2011).

10. Einstein to Besso, 29 July 1953, Princeton.

11. Albert Einstein, "Über die von der molekularkinetischen Theorie der Wärme geforderte Bewegung von in ruhenden Flüssigkeiten suspendierten Tielchen," *Annalen der Physik* 17 (1905).

12. Albert Einstein, "Theoretische Bemerkung über die Brownsche Bewegung," *Zeitschrift fur Elektrochemie und angewandte physikalische Chemie* 13 (1907).

13. Victor Henri, "Études cinematographique du mouvement brownien," *Comptes rendus de l'Académie des sciences* 146 (1908).

14. Victor Henri, "Influence du milieu sur les mouvements browniens," *Comptes rendus de l'Académie des sciences* 147 (1908).

15. Henri, "Études cinématographique."

16. Jean Perrin, "Mouvement brownien et réalité moléculaire," *Annales de chimie et de physique* 8, no. 18 (1909).

17. Jean Perrin, "La Réalité des molécules," *Revue scientifique* 49 (1911); Bigg, "Evident Atoms."

18. Watanabe, "Le Concept de temps," 130.

19. Reichenbach, *Rise of Scientific Philosophy*, 149.

20. Lecture 5 in Richard Phillips Feynman, *The Character of Physical Law* (London: British Broadcasting Corp., 1965).

21. Reiser, "The Problem of Time in Science and Philosophy," 241.

CHAPTER 25. BERGSON'S MOVIES: OUT OF CONTROL

1. Russell, "The Ultimate Constituents of Matter," address delivered to the Philosophical Society of Manchester in February 1915. Reprinted in Russell, *Mysticism and Logic*: 402.

2. Russell, "The Philosophy of Bergson," 339.

3. Russell, *Mysticism and Logic*, 402.

4. Ibid.

5. Ibid.

6. Ibid.

7. Ibid., 402–403.

8. Bergson, *Creative Evolution*, 342.

9. Two versions of Bergson's "Le Rêve" article appeared in 1901: Henri Bergson, "Le Rêve," *Bulletin de l'Institut général psychologique* 1, no. 3 (1901) and Henri Bergson, "Le Rêve," *Revue scientifique* 15, no. 23 (1901).

10. Bergson, "Le Rêve," *Revue scientifique* 15, no. 23 (1901): 713.

11. Bergson, *Essai sur les données immédiates de la conscience*, 171–172.

12. Bergson, "Le Rêve," 713.

13. Henri Bergson, *L'Énergie spirituelle*, ed. Frédéric Worms, Élie During, and Arnaud François, 9th ed. (Paris: Presses Universitaires de France, 2009).

14. Einstein to Aurel Stodola, 31 March 1919, Berlin.

15. Béla Balázs, "Béla Balázs, Visible Man, or the Culture of Film," *Screen* 48, no. 1 (2007): 107.

16. "La cinégraphie sera bergsonienne ou ne sera pas!" in Émile Vuillermoz, "Devant l'écran," *Le Temps* (10 October 1917): 3.

17. Georges-Michel Michel was the pseudonym of Georges Dreyfus.

18. Eddington, *The Nature of the Physical World*, 95.

19. Du Noüy, *Biological Time*, 102.

20. Chevalier, *Entretiens avec Bergson*, 256.

21. André George, "Le Temps, la vie, et la mort," *Les Documents de la vie intellectuelle* 43, no. 1 (1946).

22. Hannah Landecker, "Cellular Features: Microcinematography and Film Theory," *Critical Inquiry* 31 (2005); Hannah Landecker, *Culturing Life: How Cells Became Technologies* (Cambridge, Mass.: Harvard University Press, 2007).

23. Alexis Carrel, "Physiological Time," *Science* 74 (1931): 620; also quoted in Landecker, "Cellular Features," 922; Landecker, *Culturing Life*, 87.

24. Du Noüy, *Biological Time*, 103.

25. Ibid.

26. Ibid., 160–161.

27. Ibid., 172.

28. Ibid., 127.

29. Ibid., 172.

30. Watanabe, "Sur le sens unique dans le courant du temps," quoted in Pierre Lecomte du Noüy, *L'Homme devant la science*, Bibliothèque de philosophie scientifique (Paris: Flammarion, 1939), 120.

31. Bergson, *Les Deux sources de la morale et de la religion*, 286.

32. Ibid., 288.

33. Ibid., 28.

34. Ibid., 34.
35. Ibid., 17.
36. Ibid., 5.

CHAPTER 26. MICROBES AND GHOSTS

1. Nordmann, "Einstein à Paris," 931.
2. Dunoyer, "Einstein et la relativité (I)," 182.
3. Jacques Morland, "Une heure chez Monsieur Bergson," in Bergson, *Écrits philosophiques*, 406.
4. Félicien Challaye, *Bergson*, Les Maitres de la pensée moderne (Paris: Librairie Mellottée, 1929), 12.
5. "Bald fetus": Léon Daudet, *"Verts d'académie et vers de presse"* (Paris: Éditions du Capitole, 1930), 40; "black hair": Nordmann, "Einstein à Paris," 932.
6. Einstein to Heinrich Zangger, 16 February 1917, [Berlin].
7. Ibid.
8. For secondary material on Bergson's example of the microbe, see Christina Vagt, "Im Äther. Einstein, Bergson und die Uhren der Mikrobe," in *Übertragsungsräume*, ed. Eva Johach and Diethart Sawicki (Weisbaden: Ludwig Reichert, 2013).
9. "La Théorie de la relativité," 106; Bergson, *Durée et simultanéité*, 55.
10. Ibid.
11. "La Théorie de la relativité," 105.
12. "Discussion," 9–10.
13. "La Théorie de la relativité," 98.
14. Edwin Abbott Abbott, *Flatland: A Romance of Many Dimensions* (London: Seeley, 1884).
15. Poincaré, *Value of Science*.
16. Quoted in Isaacson, *Einstein*, 196.
17. Albert Einstein, *Relativity: The Special and the General Theory* (New York: Crown, 1961).
18. Charles MacArthur, "Einstein Baffled in Chicago: Seeks Pants in Only Three Dimensions, Faces Relativity of Trousers," *Chicago Herald and Examiner*, 3 May 1921.
19. Einstein, "The Principal Ideas of the Theory of Relativity," 3.
20. Felix Eberty, *Die Gestirne und die Weltgeschichte; Gedanken über Raum, Zeit und Ewigkeit*, ed. Gregory Itelson (Berlin: Gregor Rogoff Verlag, 1923). See Jimena Canales, "Las ficciones de Einstein," *Investigación y ciencia*, no. 453 (June 2014).
21. The eleven-year-old Einstein received Aaron Bernstein's book as a gift from Max Talmud in 1889. In his autobiography, he claimed to have read it "with breathless attention." Einstein, "Autobiographical Notes," 13–15.
22. Frederick Gregory, "The Mysteries and Wonders of Natural Science: Aaron Bernstein's Naturwissenschaftliche Volksbücher and the Adolescent Einstein," in *Einstein: The Formative Years, 1879–1909*, ed. Don Howard and John J. Stachel (Boston: Birkhauser, 2000).
23. Aaron Bernstein, *Naturwissenschaftliche Volksbücher*, vol. 5, part 21 (Berlin: Dümmler, 1873–1874), 100–101.
24. Langevin, "L'Évolution de l'espace et du temps," 47.
25. Eddington, *Space, Time, and Gravitation*.

26. Bertrand Russell, *The ABC of Relativity*, Harpers Modern Science Series (New York: Harper, 1925), 78.

27. Bergson, *Durée et simultanéité*, 166.

28. Leo Gilbert, "A Satire on the Principle of Relativity," *Monist* 24, no. 2 (April 1914): 302.

29. The secondary literature on thought experiments is now vast. A paradigmatic text is Alexandre Koyré, "Le 'De motu gravium' de Galilée: De l'expérience imaginaire et de son abus," *Revue d'Histoire des Sciences* 13 (1960). The term was coined by Hans Christian Oersted in *Die Naturwissenschaft und die Geistesbildung* (Leipzig: Carl B. Lorch, 1851).

30. Bergson, *Durée et simultanéité*, 77.

31. "La Théorie de la relativité," 105.

32. Maritain, *Réflexions sur l'intelligence et sur sa vie propre*, 217.

33. Whitehead, *Principle of Relativity*.

34. Eddington, *Nature of the Physical World*, 42.

35. Poincaré, *Science and Method*, 108–109.

36. Ibid., 109.

CHAPTER 27. ONE NEW POINT: RECORDING DEVICES

1. Bergson, "Réponse de M. Henri Bergson," 440.

2. André Metz, "Réplique de M. André Metz," *Revue de philosophie* 31 (1924): 438.

3. Russell, *ABC of Relativity*, 219. An earlier text explaining how the senses could be replaced by machines is "The Relation of Sense-Data to Physics" *Scientia* 16 (1914): 1–27. Reprinted in Russell, *Mysticism and Logic*.

4. Henri Bergson, "Les Temps fictifs et le temps réel," 241.

5. Bergson, *Durée et simultanéité*, 210.

6. Jimena Canales, "Recording Devices." In *Companion to the History of Science* (Malden, Mass.: Wiley-Blackwell, forthcoming).

7. Einstein, "Relativitätstheorie," 6.

8. Einstein, *Relativity: The Special and the General Theory*, 75.

9. [Eberty], *Stars and the Earth*, 68. On Eberty, see Karl Clausberg, "A Microscope for Time: What Benjamin and Klages, Einstein and the Movies Owe to Distant Stars," in *Given World and Time*, ed. Tyrus Miller (Budapest: Central European University Press, 2008).

10. [Eberty], *Stars and the Earth*, 79.

11. Bergson, *Durée et simultanéité*, 57.

12. "Close friend": J. Thiebault, "Les Fantômes du professeur Charles Richet devant Einstein," *Revue contemporaine* (August–September 1922); also quoted in Grogin, *Bergsonian Controversy*, 24.

13. Charles Richet, "L'Unité psychologique du temps," *Comptes rendus de l'Académie des sciences* 173, no. 25 (1921): 1316.

14. Ibid., 1317.

15. Ibid.

16. Montluys, *L'Écho de Paris*, 29 March 1922; also quoted in Biezunski, *Einstein à Paris*, 105–106. The director of the Strasbourg Observatory, Ernest Esclangon, dedicated a couple of articles to the topic. Ernest Esclangon, *Le Temps*, 8 January 1922, 27 January

1922, and 22 March 1922, and Ernest Esclangon, "Sur la relativité du temps," *Comptes rendus de l'Académie des sciences* (1921). Also see Nordmann, "Avec Einstein dans les régions dévastées."

17. The newspaper article, published in the *L'Écho de Paris*, referred to Charles Richet's article sent to the Académie des sciences on 24 December 1921. See Biezunski, *Einstein à Paris*, 106.

18. Charles Richet, *Our Sixth Sense*, trans. Fred Rothwell (London: Rider, 1928), 5. Cited in Grogin, *Bergsonian Controversy*, 58.

19. Graeme J. N. Gooday, *The Morals of Measurement: Accuracy, Irony and Trust in Late Victorian Electrical Practice* (Cambridge: Cambridge University Press, 2004).

20. For how the "burden of interpretation" of scientific images has shifted in different historical periods, see Peter Galison, "Judgment against Objectivity," in *Picturing Science, Producing Art*, ed. Caroline A. Jones and Peter Galison (New York: Routledge, 1998); Lorraine Daston and Peter Galison, *Objectivity* (New York: Zone Books, 2007).

21. Vilém Flusser, "Die Auswanderung der Zahlen aus dem alphanumerischen Code," in *Literatur im Informationszeitalter*, ed. Dirk Matejovski and Friedrich Kittler (Frankfurt: Campus Verlag, 1996).

22. Lissa Roberts, "The Death of the Sensuous Chemist: The 'New' Chemistry and the Transformation of Sensuous Technology," in *The Empire of the Senses*, ed. David Howes (Oxford: Berg, 2005).

23. Eddington, *Nature of the Physical World*, 252.

24. Husserl, *Crisis of European Sciences*, 126, Section 134b.

25. Susanne K. Langer, *Philosophy in a New Key* (Cambridge, Mass.: Harvard University Press, 1951), 20. Because of changes in how scientists "read" nature, Langer urged philosophers to develop a new philosophical understanding of knowledge practices, which she called "philosophy in a new key."

26. Canales, *A Tenth of a Second*.

27. Jimena Canales, "Einstein's Discourse Networks," *Zeitschrift für Medien- und Kulturforschung* 5, no. 1 (2014): 11–39.

28. Most readers of Einstein have argued that his science was radical in that reality was defined by measurement: only that which could be measured was real. What I want to add is that Einstein's work was concerned not with measurement in general but with a particular form of measurement: detection of light signals.

29. Lorraine Daston and Elizabeth Lunbeck, *Histories of Observation* (Chicago: University of Chicago Press, 2010).

30. Metz, "Le Temps d'Einstein et la philosophie."

31. Mario Biagioli, "Stress in the Book of Nature: The Supplemental Logic of Galileo's Realism," *MLN* 118, no. 3 (2003).

32. Bergson, *Durée et simultanéité*, 87n81; "La Théorie de la relativité," 113.

CHAPTER 28. BERGSON'S LAST COMMENTS

1. Bergson to Eugène Minkowski, 6 August 1936, Vevey, Switzerland.

2. Henri Bergson, "Message au Congrès Descartes," in *Écrits philosophiques*, ed. Frédéric Worms (Paris: Presses Universitaires de France, 2011).

3. "Mais je le vois aussi dans le barque où des bateliers complotant de voler et de le jeter par-dessus bord, les devinant, tirant l'epée et tenant en respect les bandits." Ibid., 700.

4. Ibid.

5. Ibid.

6. Albert Einstein, "Speech in Royal Albert Hall," 3 October 1933.

7. The highest-grossing feature-length film of the silent film era (*The Birth of a Nation* by D. W. Griffith) had premiered in 1915.

8. Pais, *"Subtle is the Lord . . . ,"* 185–186.

9. Einstein to Ehrenfest, [before 9 September 1920], [Berlin].

10. Einstein to Solovine, 8 March 1921, [Berlin]; Einstein to Besso [before 30 May 1921], [New York]

11. Einstein to Max and Hedwig Born, 9 September 1920, [Berlin].

12. Max Born to Einstein, 13 October 1920, Frankfurt-am-M[ain].

13. Milena Wazeck, "The 1922 Einstein Film: Cinematic Innovation and Public Controversy," *Physics in Perspective* 12 (2010). The film is described in Ronald William Clark, *Einstein: The Life and Times* (New York: World Pub. Co., 1971), 290.

14. "Einstein und kein Ende!" *Casseler Allgemeine Zeitung*, 18 May 1922; quoted in Wazeck, "The 1922 Einstein Film: Cinematic Innovation and Public Controversy," 175.

15. Ibid.

16. The Irvine Ranch in Orange County was one of the many sites where Michelson's experiment was performed, as well as being the place used for filming the battle scenes of *All Quiet on the Western Front*.

17. Document no. 28–817 in Einstein Archives, Hebrew University of Jerusalem.

18. Einstein to Hans Albert Einstein, 18 June [1921], [Berlin].

19. Vallentin, *Drama of Albert Einstein*.

20. Einstein to Heinrich Zangger, [before 4 December 1915], [Berlin].

21. Einstein to Elsa Einstein, [after 26 July 1914], Berlin.

22. Einstein to Elisabeth Ney, [30 September 1920], [Stuttgart].

23. Elsa Einstein to Einstein, 24 May 1920, Berlin.

24. Fritz Haber to Einstein, 30 August 1920, Gastein.

25. Lorentz to Einstein, 3 September 1920, Haarlem.

26. Maja Winteler-Einstein to Einstein, 1 September 1920, Lucerne.

27. Paul Ehrenfest to Einstein, 11 September 1920.

28. Einstein to Max and Hedwig Born, 9 September 1920, [Berlin].

29. Einstein, "Opinion on the War," written in late October or early November 1915. These lines were deleted from the published version.

30. He used the term "public life" in Einstein to Willem and Betsy Julius, 11 September 1920, [Berlin]. About the difference between his writing intended for the press and that which was not, see Albert Einstein, "Response to the Declaration of the Prussian Academy," 5 April 1933.

31. Einstein to Marcel Grossman, 12 September 1920, [Berlin].

32. The term "life in the raw" appeared in Einstein to Freud, 30 July 1932.

33. Einstein, "Speech in Royal Albert Hall."

34. Albert Einstein, "Einsteins amerikanische Eindrüche. Was er wirklich sah," *Vossische Zeitung*, 10 July 1921.

35. "Een interview met prof. Albert Einstein," *Nieuwe Rotterdamsche Courant*, 4 July 1921.

36. Einstein, "Einsteins amerikanische Eindrüche."

37. Einstein, "Autobiographical Notes," 33.

38. Otto Neurath to Einstein, 12 January 1921, Vienna; Paul M. Warburg to Einstein, 12 April 1921, New York. Starting in the 1920s Einstein was frequently requested to allow his lectures to be "stenographically recorded" and they were then slightly "modified for publication."

39. Henri Bergson, "Conférence de Madrid: La Personnalité," in *Écrits philosophiques*, ed. Frédéric Worms (Paris: Presses Universitaires de France, 2011), 513.

40. "À l'Académie française: réception de Monsieur Henri Bergson, l'éminent philosophe qui prend possession du fauteuil de Monsieur Émile Ollivier," Reference: 1805GJ 00003, Subject: Journal Actualité Gaumont (Journal Gaumont), 00:00:16 Black and White, Silent, 1918, Gaumont Pathé Archives.

41. *T.S.F Tribune*, 3 April 1934. Broadcasted on 10 April 1934.

42. Jimena Canales, "The Media of Relativity: Einstein and Communications Technologies," *Technology and Culture* (forthcoming).

43. Arnaud François, "Notices et notes: introduction ii," in *La Pensée et le mouvant: Essais et conférences* (Paris: Presses Universitaires de France, 2009), 335.

44. Henri Bergson, *La Pensée et le mouvant,* 37n31. Italics mine.

45. Ibid.

46. Bergson, *Durée et simultanéité*, 55.

47. Raïssa Maritain, "Henri Bergson," *La Relève* (1941).

48. Paul Valéry, "Allocution de Paul Valéry devant l'Académie française," (9 January 1941). Reprinted in Bergson, *Correspondances*, 1671.

CHAPTER 29. EINSTEIN'S LAST THOUGHTS

1. Besso to Einstein, 24 December 1951, Geneva. The Besso-Einstein letters are published in Albert Einstein and Michele Besso, *Correspondance 1903–1955*, trans. Pierre Speziali (Paris: Hermann, 1979).

2. Einstein, "Autobiographical Notes," 3.

3. Einstein to Besso, 13 July 1952, [Princeton].

4. Palle Yourgrau, *A World Without Time: The Forgotten Legacy of Gödel and Einstein* (New York: Basic Books, 2005).

5. Besso to Einstein, 13 June 1952, Geneva.

6. Einstein to Besso, 13 July 1952, [Princeton].

7. Einstein to Besso, 29 July [1953].

8. Einstein to Besso's family, 21 March 1955, Princeton.

9. Northrop, "Whitehead's Philosophy of Science," 204.

10. Ibid.

11. Ibid.

12. Ibid., 194.

13. On second thought, Einstein did not think the objections to his concept of locality were valid. Einstein articulated his reasons clearly: "On that theory there would be no meaning to two observers speaking about the same event." Ibid., 204. Einstein's reply was similar to Russell's view on this same topic, in which science was seen as depending on a difference between the local and the distant. For him the "fallacy of simple location, when avoided," led to the demise of science: "such a view, if taken seriously, is incompatible with science." Even more damaging, Russell added that this denial "involves a mystic pantheism." McGilvary, "Space-Time," 211, 234.

14. Einstein, "Zur Elektrodynamik bewegter Körper," 893.
15. "La Théorie de la relativité," 105.
16. Einstein, "Zur Elektrodynamik bewegter Körper," 893.
17. Ibid., 893n1.
18. Ibid.
19. Bergson, *Durée et simultanéité*, 63–64.
20. "(There is a somewhat similar problem about the relation of the human consciousness to brain-cell events.)" Hartshorne, "Whitehead's Idea of God," 545. Criticisms of this kind reached its apogee in the work of J. J. Gibson and his "little man in the brain" theory.
21. For Heidegger's position on measurement in relation to the "now" see Canales, *A Tenth of a Second*, 14n28.
22. Moritz Schlick, "On the Relation Between Psychological and Physical Concepts," in *Philosophical Papers*, ed. Henk L. Mulder and Barbara F. B. van de Velde-Schlick (Dordrecht: Reidel, 1979), 424; also quoted in Daston and Galison, *Objectivity*.
23. Karl Marx, *Economic and Philosophic Manuscripts of 1844*, trans. Martin Milligan (New York: International Publishers, 1968), 139–141.
24. Einstein, "Autobiographical Notes," 59–60.
25. Ibid., 61.
26. "If, however, one introduces as unit of time instead of the second the time in which light travels 1 cm, *c* no longer occurs in the equations." Ibid.
27. Ibid.
28. Ibid.
29. Einstein, "Reply to Criticisms," 673.
30. Albert Einstein, "Remarks on Bertrand Russell's Theory of Knowledge," in *The Philosophy of Bertrand Russell*, ed. Paul Arthur Schilpp (La Salle, Ill.: Open Court, 1944).
31. Einstein, "Reply to Criticisms," 673. For the distinction between the subjective and objective in Einstein, see Daston and Galison, *Objectivity*, 302–305.
32. Bergson, *Durée et simultanéité*, 63–64.
33. A. J. Ayer, "Editor's Introduction," in *Logical Positivism*, ed. A.J. Ayer (New York: The Free Press, 1959), 8.
34. Bergsonian philosophy and Bergson himself was feminized by his critics, especially by Julien Benda. The newspaper *Bonsoir* indicated that Bergson's popularity among women hurt his scientific standing: if "Einstein accepted them" he "would be doing him a great favor." But Einstein did not. Charles Nordmann described that the difference between going to a lecture by Bergson and to one of Einstein's was that women would attend Bergson's. In Einstein's lectures, there "were few known actresses or *dames du monde*." *Bonsoir* (3 April 1922); Nordmann, "Einstein expose et discute sa théorie," 131–132; also quoted in Biezunski, *Einstein à Paris*, 53–54. For the role of women as Bergson's followers, see Gunter, *Bergson and the Evolution of Physics*, 16.
35. Eddington, *Nature of the Physical World*
36. Einstein to Besso, 9 June 1937, Princeton.
37. The association of Bergson's philosophy with instinct as opposed to intellect runs throughout Russell's work from Bertrand Russell, "The Philosophy of Bergson," *Monist* 22 (1912) to Russell, *History of Western Philosophy*, 756–765.
38. Cassirer, *An Essay on Man*, 220. Cassirer used these categories to describe the difference between philosophies associated with Einstein's work (such as his own) and

those that sided with Bergson (such as Heidegger's). He used these terms to differentiate his approach (focusing on "spontaneity") to Bergson's and Heidegger's, which he argued were best characterized by "receptivity."

39. Martin Heidegger, *Parmenides*, trans. André Schuwer and Richard Rojcewicz (Bloomington: University of Indiana Press, 1992), 77.

40. Bridgman, *Einstein's Theories*, 7: 354.

41. The comparison of Bergson to Heraclitus appeared in Russell, "The Philosophy of Bergson," 323; Hans Reichenbach and Maria Reichenbach, *The Direction of Time* (Mineola, N.Y.: Dover Publications, 1999), 16–17. The comparison of Einstein to Parmenides is from Russell, *A History of Western Philosophy*, 87; Popper, *Conjectures and Refutations*, 38; Karl R. Popper, *The Open Universe: An Argument for Indeterminism* (Totowa, N.J.: Rowman and Littlefield, 1982), 2, 90. Bergson also accused Einstein of believing in a four-dimensional block universe, whose original conception was frequently traced back to Parmenides.

42. St. Augustine's position was different from the tradition Aristotelian conception, which was largely interpreted as one that conflated spatial movement with temporality. Poincaré and Bergson were both associated with Augustine's views, whereas Einstein's with Aristotle's: "At the same time as Bergson, Poincaré reopened thus the ancient refutation, that of Augustine, about the possibility of measuring time, that is of the Aristotelian confusion of time and movement." Souriau, *Le Temps*, 9n1. For the comparison of Einstein's conception with Artistotle's, see Abel Rey in Langevin, "Le temps, l'espace et la causalité," 334.

43. David Couzens Hoy, *The Time of Our Lives: A Critical History of Temporality* (Cambridge, Mass.: MIT Press, 2009). Hoy considered "the initial task" of his book "to explain how the 'time of our lives' emerged as a separate problem from the 'time of the universe,'" arguing for the importance of the first category. He mentions the Bergson-Einstein debate on pp. 121, 132. Other key texts exploring these dichotomies are Paul Ricœur, *Time and Narrative*, 3 vols. (Chicago: University of Chicago Press, 1984–1988); Hans Blumenberg, *Lebenszeit und Weltzeit* (Frankfurt am Main: Suhrkamp, 1986).

POSTFACE

1. Dennis Overbye, "Einstein and the Black Hole," *New York Times*, 12 August 2013, D1.

2. Sokal and Bricmont, *Impostures intellectuelles*, 165.

3. Ibid., 165–166.

4. S. Toulmin, "The Emergence of Post-modern Science," in *The Great Ideas Today* (Chicago: Encyclopedia Britannica, 1981); Jean François Lyotard, *The Postmodern Condition: A Report on Knowledge*, Theory and History of Literature (Minneapolis: University of Minnesota Press, 1984).

5. Ilya Prigogine, "Evolution of Physics: Review of Bergson and the Evolution of Physics, edited and translated by P.A.Y. Gunter," *Nature* 234 (1971): 159.

6. Ilya Prigogine and Isabelle Stengers, *Order Out of Chaos: Man's New Dialogue with Nature* (Toronto: Bantam Books, 1984), 301–302.

7. Prigogine, "The Arrow of Time."

8. Isabelle Stengers, *Power and Invention: Situating Science*, Theory Out of Bounds (Minneapolis, Minn.: University of Minnesota Press, 1997), 41. Stengers underlined the

need to include the temporality of different species and agencies in Isabelle Stengers, *Cosmopolitics*, Posthumanities (Minneapolis: University of Minnesota Press, 2010).

9. Stengers, *Power and Invention*, 41–42.

10. Ibid.

11. Ibid., 165.

12. The chapter on Einstein and Bergson did not appear in the English edition of the Sokal and Bricmont book, *Fashionable Nonsense*.

13. Alan D. Sokal and Jean Bricmont. "Un regard sur l'histoire des rapports entre science et philosophie: Bergson et ses successeurs." In *Impostures Intellectuelles* (Paris: Odile Jacob, 1997), 166.

14. Sokal and Bricmont, *Impostures intellectuelles*.

15. Henri Bergson, "Les Temps fictifs et le temps réel," 248.

16. Adolf Grünbaum, "Time and Entropy," *American Scientist* 43 (1955): 565. Grünbaum, who criticized "Bergson's very questionable conception," listed Bridgman, Watanabe, and Costa de Beauregard among those who denied that a purely physical description of time had been successful.

17. Sokal and Bricmont, *Impostures intellectuelles*, 172n214.

18. For Deleuze and Bergson, see P.A.Y. Gunter, "Gilles Deleuze, Deleuze's Bergson and Bergson Himself," in *Deleuze, Whitehead, Bergson: Rhizomatic Connections*, ed. Keith A. Robinson (Basingstoke, England: Palgrave Macmillan, 2009).

19. Gilles Deleuze, "La Conception de la différence chez Bergson," in *Les Études bergsoniennes* (Paris: Presses Universitaires de France, 1956), presented in 1954. This work was followed by Gilles Deleuze, "Bergson. 1859–1941," in *Les Philosophes célèbres*, ed. Maurice Merleau-Ponty (Paris: L. Mazenod, 1956) which includes a comparison of Bergson to phenomenology. For his lectures on *Creative Evolution* in 1960, see Gilles Deleuze, "Lecture Course on Chapter Three of Bergson's 'Creative Evolution.'" *SubStance* 36, no. 3 (2007).

20. Pearson, *Germinal Life*.

21. Gilles Deleuze, *Bergsonism* (New York: Zone Books, 1991), 130n120. In original French: "On a souvent dit que le raisonnement de Bergson impliquait un contresens sur Einstein."

22. Ibid. In original French: "Mais souvent aussi, on a fait un contresens sur le raisonnement de Bergson lui-même."

23. Gilles Deleuze and Félix Guattari, *What Is Philosophy?* (New York: Columbia University Press, 1994), 131.

24. Ibid. They referred to "The Relation of Sense-Data to Physics," in Russell, *Mysticism and Logic*.

25. Deleuze and Guattari, *What Is Philosophy?*, 131.

26. "Here, already, the general orientation of philosophy comes into question, for it is not enough to say that philosophy is at the origin of the sciences and that it was their mother; rather, now that they are grown up and well established, we must ask why there is still philosophy, in what respect science is not sufficient. Philosophy has only ever responded to such a question in two ways, doubtless because there are only two possible responses. One says that science gives us a knowledge of things, that it is therefore in a certain relation with them, and philosophy can renounce its rivalry with science, can leave things to science, and present itself solely in a critical manner, as a reflection on this knowledge of things. On the contrary view, philosophy seeks to establish, or rather

restore, an other relationship to things, and therefore an other knowledge, a knowledge and a relationship that precisely science hides from us, of which it deprives us, because it allows us only to conclude and to infer without ever presenting, giving to us the thing in itself. It is this second path that Bergson takes by repudiating critical philosophies when he shows us in science, in technical activity, intelligence, everyday language, social life, practical need, and, most important, in space—the many forms and relations that separate us from things and from their interiority." Deleuze, "Bergson. 1859–1941." Reprinted in Gilles Deleuze, *Desert Islands and Other Texts, 1953–1974* (Los Angeles: Semiotext(e), 2004), 22.

27. "We know that in the end they lead back to the always rediscovered distinction of matter and duration. Matter and duration are never distinguished as two things but as two movements, two tendencies, like relaxation and contraction." Deleuze, *Desert Islands*, 25.

28. This interpretation is implicit in Merleau-Ponty's Bergson.

29. Deleuze, *The Logic of Sense*, 61.

30. Gilles Deleuze and Félix Guattari, *A Thousand Plateaus: Capitalism and Schizophrenia*, trans. Brian Massumi (University of Minnesota Press, 1987), 484. "Most important": Cesare Casarino and Antonio Negri, "It's a Powerful Life: A Conversation on Contemporary Philosophy," *Cultural Critique* 57 (2004): 159 Michel Foucault once considered "that perhaps one day, this century will be known as Deleuzian." Michel Foucault, *Language, Counter-Memory, Practice: Selected Essays and Interviews* (Ithaca, N.Y.: Cornell University Press, 1977), 65.

31. Deleuze and Guattari, *A Thousand Plateaus*, 484.

32. "But we must go further, if the theme and the idea of purity have a great importance in the philosophy of Bergson, it is because in every case the two tendencies are not pure, or are not equally pure." Deleuze, *Desert Islands*, 25.

33. Latour, "Some Experiments in Art and Politics." For Latour's first solid engagement with Einstein and the theory of relativity see Bruno Latour, "A Relativistic Account of Einstein's Relativity," *Social Studies of Science* 18, no. 1 (1988): 5.

34. Bruno Latour, "Why Has Critique Run Out of Steam? From Matters of Fact to Matters of Concern," *Critical Inquiry* 30, no. 2 (2004).

35. Latour, "Some Experiments in Art and Politics," 5.

36. Bruno Latour, "Trains of Thoughts—Piaget, Formalism and the Fifth Dimension," *Common Knowledge* 6, no. 3 (1997): 183.

37. Callon, "Whose Imposture?," 276.

38. Latour, "Some Experiments in Art and Politics," 5.

39. Ibid.

40. Latour, *We Have Never Been Modern*, 134.

41. Ibid.

42. Derrida considered his project as a new way of understanding the same dichotomies that had characterized much of the Einstein-Bergson debate, in particular, the dichotomy between "the world" and "lived experience." Derrida developed the concept of the "trace" as a way to link these two categories: "The unheard difference . . . (between the 'world' and 'lived experience') is the condition of all other differences, of all other traces, and it is already a trace." Jacques Derrida, *Of Grammatology*, 1st American ed. (Baltimore: Johns Hopkins University Press, 1976), 65.

BIBLIOGRAPHY

Abbott, Edwin Abbott. *Flatland: A Romance of Many Dimensions*. London: Seeley, 1884.

Agamben, Giorgio. *Infancy and History: Essays on the Destruction of Experience*. London: Verso, 1993. First published in 1978.

Alain. *Propos*. Bibliothèque de la Pléiade. Paris: Gallimard, 1956.

Alder, Ken. *The Measure of All Things: The Seven-Year Odyssey and Hidden Error That Transformed the World*. New York: Free Press, 2002.

Andrade, Jules. *Le Mouvement: mesures de l'étendue et mesures du temps*. Paris: Félix Alcan, 1911.

Antliff, Mark. "The Rhythms of Duration: Bergson and the Art of Matisse." In *The New Bergson*, edited by John Mullarkey, 184–208. Manchester: Manchester University Press, 1999.

Arago, François. "Notice scientifique sur la tonnerre." *Annuaire du bureau des longitudes* (1838): 221–618.

Arzeliès, Henri. *La Cinématique relativiste*. Études relativistes. Vol. 1. Paris: Gauthier-Villars, 1955.

Ayer, A.J. "Editor's Introduction." In *Logical Positivism*, edited by A. J. Ayer, 3–28. New York: The Free Press, 1959.

———. *Logical Positivism*. New York: The Free Press, 1959.

Azouvi, François. *La Glorie de Bergson: essai sur le magistère philosophique*. Paris: Gallimard, 2007.

Bachelard, Gaston. *The Dialectic of Duration*. Translated and annotated by Mary McAllester Jones. Manchester: Clinamen Press, 2000.

———. *Épistémologie*. Edited by Dominique Lecourt. Paris: Presses Universitaires de France, 1971.

———. "The Philosophic Dialectic of the Concepts of Relativity." In *The Library of Living Philosophers*, edited by Paul Arthur Schilpp, 565–580. La Salle, Ill.: Open Court, 1949.

———. *La Valeur inductive de la relativité.* Paris: J. Vrin, 1929.

Bain, Jonathan. "Whitehead's Theory of Gravity." *Studies in History and Philosophy of Modern Physics* 29 (1998): 547–574.

Balázs, Béla. "Béla Balázs, Visible Man, or the Culture of Film." *Screen* 48, no. 1 (2007): 91–108. First published in 1924.

Balibar, Françoise, and Jean-Philippe Mathieu. "Einstein-Lorentz: une correspondance scientifique et politique." *Mil neuf cent* 8 (1990): 23–32.

Barker, George F. "Some Modern Aspects of the Life Question: Address of Professor George F. Barker, the Retiring President of the Association." *Science: A Weekly Record of Scientific Progress* 1 (4 September 1880): 112–118.

Barreau, Hervé. "Bergson et Einstein: à propos de Durée et simultanéité." *Les Études bergsoniennes* 10 (1973): 73–134.

Barthes, Roland. "The Brain of Einstein." Translated by Annette Lavers. In *Mythologies*, 68–70. New York: Hill and Wang, 1972.

Bataillon, Marcel. "At the Collège de France." In *The Bergsonian Heritage*, edited by Thomas Hanna, 107–118. New York: Columbia University Press, 1962.

Becquerel, Jean. "Débats sur la relativité: I. critique par M. J. Becquerel de l'ouvrage 'Durée et simultanéité' de M. Bergson." *Bulletin scientifique des étudiants de Paris* (March 1923): 18–29.

———. Préface. In *La Relativité: exposé dans formules des théories d'Einstein et réfutation des erreurs contenues dans les ouvrages les plus notoires*, v–xviii. Paris: Étienne Chiron, 1923.

Benda, Julien. *Belphégor: essai sur l'esthétique de la présente société française*. Paris: Émile-Paul Frères, 1918.

Benjamin, Walter. "Theses on the Philosophy of History." Translated by Harry Zohn. In *Illuminations*, edited by Hannah Arendt, 253–264. New York: Schocken Books, 1969.

———. "Über einige Motive bei Baudelaire." *Zeitschrift für Sozialforschung* 8 (1939): 50–91.

Benoît, J.-René. "De la précision dans la détermination des longueurs en métrologie." In *Rapports présentés au Congrès International de Physique réuni à Paris en 1900*, edited by Ch.-Éd. Guillaume and L. Poincaré, 30–77. Paris: Gauthier-Villars, 1900.

Benrubi, Isaac. *Souvenirs sur Henri Bergson*. Neuchâtel: Delachaux & Niestlé, 1942.

Bensaude-Vincent, Bernadette. "When a Physicist Turns on Philosophy: Paul Langevin (1911–39)." *Journal of the History of Ideas* 49, no. 2 (1988): 319–338.

Bensaude-Vincent, Bernadette, and Eva Telkes-Klein, eds. *Paul Langevin: Propos d'un physicien engagé*. Paris: Vuibert et Société française d'histoire des sciences et des techniques, 2007.

Bergson et nous, Congrès Bergson: Paris, 17–20 mai 1959. Bulletin de la Société française de philosophie: Actes du Xe Congrès des sociétés de philosophie de langue française. Edited by Société française de philosophie. Vol. 10, Paris: A. Colin, 1959.

Bergson, Henri. "Conférence de Madrid: La Personnalité." In *Écrits philosophiques*, 508–535. Paris: Presses Universitaires de France, 2011 (6 May 1916).

———. *Correspondances*. Edited by André Robinet. Paris: Presses Universitaires de France, 2002.

———. *Creative Evolution*. Mineola, N.Y.: Dover, 1998.

———. "Discussion avec Einstein." In *Mélanges*, edited by André Robinet, 1340–1347. Paris: Presses Universitaires de France, 1972.

———. "Durée et simultanéité: a propos de la théorie d'Einstein." In *Mélanges*, edited by André Robinet, 58–244. Paris: Presses Universitaires de France, 1972.

———. *Durée et simultanéité: á propos de la théorie d'Einstein*. Edited by Élie During. 4th ed. Quadrige, edited by Frédéric Worms. Paris: Presses Universitaires de France, 2009.

———. *Écrits et paroles*. Edited by Rose-Marie Mossé-Bastide. Bibliothèque de philosophie contemporaine. Paris: Presses Universitaires de France, 1957.

———. *Écrits philosophiques*. Edited by Arnaud Bouaniche, Arnaud François, Élie During, Frédéric Fruteau de Laclos, Frédéric Keck, Stéphane Madelrieux, Camille Riquier, Ghislain Waterlot and Frédéric Worms. Quadrige, edited by Frédéric Worms. Paris: Presses Universitaires de France, 2011.
———. "Einstein et Bergson, 1922." In *Durée et simultanéité*, edited by Élie During, 391–398. Paris: Presses Universitaires de France, 2009.
———. *Essai sur les données immédiates de la conscience*. Edited by Arnaud Bouaniche. 9th ed. Quadrige, edited by Frédéric Worms. Presses Universitaires de France, 2011.
———. "Identité et réalité. Par Émile Meyerson." *Séances et travaux de l'Académie des sciences morales et politiques* 171 (23 January 1909): 664–666.
———. *La Pensée et le mouvant: essais et conférences*. Edited by Arnaud Bouaniche, Arnaud François, Frédéric Fruteau de Laclos, Stéphane Madelrieux, Claire Marin, Ghislain Waterlot. 16th ed. Quadrige, edited by Frédéric Worms. Paris: Presses Universitaires de France, 2009.
———. "La Perception du changement." In *Mélanges*, edited by André Robinet, 888–914. Paris: Presses Universitaire de France, 1972.
———. "La Philosophie." In *La Science française*, 15–37. Vol. 1. Paris: Ministère de l'Instruction publique, 1915.
———. "La Philosophie française." *Revue de Paris* (15 May 1915): 236–256.
———. *L'Énergie spirituelle*. Edited by Frédéric Worms, Élie During, and Arnaud François. 9th ed. Quadrige, edited by Frédéric Worms. Paris: Presses Universitaires de France, 2009.
———. "Le Rêve." *Revue scientifique* 15, no. 23 (1901): 705–715.
———. "Le Rêve." *Bulletin de l'Institut général psychologique* 1, no. 3 (1901): 97–122.
———. *Les Deux sources de la morale et de la religion*. Edited by Ghislain Waterlot and Frédéric Keck. 10th ed. Quadrige, edited by Frédéric Worms. Paris: Presses Universitaires de France, 2008.
———. "Les Temps fictifs et le temps réel." *Revue de philosophie* 31, no. 3 (1924): 241–260.
———. "Les Temps fictifs et le temps réel." In *Mélanges*, edited by André Robinet, 1432–1449. Paris: Presses Universitaires de France, 1972.
———. "Les Temps fictifs et le temps réel." In *Durée et simultanéité: á propos de la théorie d'Einstein*, edited by Élie During, 417–430. Paris: Presses Universitaires de France, 2010.
———. *Matière et mémoire: Essai sur la relation du corps à l'esprit*. Edited by Camille Riquier, 8th ed. Quadrige, edited by Frédéric Worms. Paris: Presses Universitaires de France, 2009.
———. *Mélanges*. Edited by André Robinet. Paris: Presses Universitaires de France, 1972.
———. "Message au Congrès Descartes." In *Écrits philosophiques*, 696–701. Paris: Presses Universitaires de France, 2011. First published in June 1937.
———. "Presidential Address." *Proceedings of the Society for Psychical Research* 26 (1913): 462–479.
———. "Réponse de M. Henri Bergson." *Revue de philosophie* 31 (1924): 440.
Berlin, Isaiah. "Impressionist Philosophy." *London Mercury* 32, no. 191 (1935): 489–490.
Berteval, W. "Bergson et Einstein." *Revue philosophique de la France et l'étranger* 132 (1943): 17–28.
Biagioli, Mario. "Meyerson: Science and the 'Irrational.'" *Studies in History and Philosophy of Science* 19, no. 1 (1988): 5–42.

Biagioli, Mario. "Stress in the Book of Nature: The Supplemental Logic of Galileo's Realism." *MLN* 118, no. 3 (2003): 557–585.

Biezunski, Michel. *Einstein à Paris: le temps n'est plus . . .* Vincennes: Presses Universitaires de Vincennes, 1991.

———. "Einstein's Reception in Paris in 1922." In *The Comparative Reception of Relativity*, edited by Thomas F. Glick, 169–188. Boston Studies in the Philosophy of Science. Dordrecht: D. Reidel, 1987.

———"La Diffusion de la théorie de la relativité en France." Ph.D. diss., Université Paris Diderot, 1981.

Bigg, Charlotte. "Evident Atoms: Visuality in Jean Perrin's Brownian Motion Research." *Studies In History and Philosophy of Science* 39 (2008): 312–322.

———. "A Visual History of Jean Perrin's Brownian Motion Curves." In *Histories of Scientific Observation*, edited by Lorraine Daston and Elizabeth Lunbeck, 156–179. Chicago: University of Chicago Press, 2011.

Blume, Georg. "Der Philosoph François Fédier über Martin Heidegger." *Die Zeit* (18 January 2014).

Blumenberg, Hans. *Lebenszeit und Weltzeit*. Frankfurt am Main: Suhrkamp, 1986.

B[oas], G[eorge]. "Review." *Journal of Philosophy* 31, no. 25 (1934): 694.

Bohr, Niels. "Discussion with Einstein on Epistemological Problems in Atomic Physics." In *Albert Einstein: Philosopher-Scientist*, edited by Paul Arthur Schilpp, 201–241. La Salle, Ill.: Open Court, 1949.

Boltzmann, Ludwig. *Vorlesungen über Gastheorie*. Leipzig: Johann Ambrosius Barth, 1898.

Bonner, John Tyler. *Life Cycles: Reflections of an Evolutionary Biologist*. Princeton, N.J.: Princeton University Press, 1993.

Borella, Vincent. "L'Introduction de la relativité en France, 1905–1922." Ph.D. diss., Nancy 2, 1998.

Born, Max. *Die Relativitätstheorie Einsteins und ihre physikalische Grundlagen*. Berlin: Springer, 1921.

———. "Einstein's Statistical Theories." In *Albert Einstein: Philosopher-Scientist*, edited by Paul Arthur Schilpp, 161–177. La Salle, Ill.: Open Court, 1949.

Bouasse, Henri. *La Question préable contre la théorie d'Einstein*. Paris: Blanchard, 1923. Originally published in *Scientia: revue internationale de synthèse scientifique* (January 1923).

Bourget, Paul. *Essais de psychologie contemporaine*. Paris: Gallimard, 1883.

Boyle, Robert. "An Examen of Mr. T. Hobbes His Dialogus Physicus de Natura Aëris." In *The Works of the Honourable Robert Boyle*, edited by Thomas Birch, 186–242. London: J. & F. Rivington, 1772. First published in1662.

Bridgman, Percy W. "The Concept of Time." *Scientific Monthly* 35, no. 2 (1932): 97–100.

———. *Einstein's Theories and the Operational Point of View*. The Library of Living Philosophers, edited by Paul Arthur Schilpp. Vol. 7, La Salle, Ill.: Open Court, 1949.

Brillouin, Léon. *Relativity Reexamined*. New York: Academic Press, 1970.

Broglie, Louis de. *Physique et microphysique*. Sciences d'aujourd'hui. Paris: A. Michel, 1947.

Browne, H.C. "Einstein's Paradox." *Nature* 110, no. 2768 (18 November 1922): 668–669.

Brunschvicg, Léon. *L'Expérience humaine et la causalité physique*. Paris: Félix Alcan, 1922.

Buchwald, Diana Kormos, editor. *The Collected Papers of Albert Einstein*. 14 vols. Princeton, N.J: Princeton Press, 1987–.

Bunnin, Nicholas, and Jiyuan Yu. *The Blackwell Dictionary of Western Philosophy*. Malden, Mass.: Blackwell, 2004.

Bush, W. T. "The Paris Philosophical Congress." *Journal of Philosophy* 19, no. 9 (27 April 1922): 241–243.

Calinon, Auguste. "Étude critique sur la mécanique." *Bulletin de la société des sciences de Nancy* 7 (1885): 87–180.

Callandreau, Octave. "Histoire abrégée des déterminations de la parallaxe solaire." *Revue scientifique* 2 (1881): 39–43.

Callon, Michel. "Whose Imposture? Physicists at War with the Third Person." *Social Studies of Science* 29, no. 2 (1999): 261–286.

Campbell, Norman Robert. *Théorie quantique des spectres. La Relativité*. Paris: J. Hermann, 1924.

Campbell, William Wallace, and Joel Stebbins. "Report on the Organization of the International Astronomical Union." *Proceedings of the National Academy of Sciences of the United States of America* 6, no. 6 (1920): 349–396.

Canales, Jimena. "Einstein's Discourse Networks." *Zeitschrift für Medien- und Kulturforschung* 5, no. 1 (2014): 11–39.

———. "Las ficciones de Einstein." *Investigación y ciencia*, no. 453 (June 2014): 50–51.

———. "The Media of Relativity: Einstein and Communications Technologies." *Technology and Culture* (forthcoming).

———. "Movement before Cinematography: The High-Speed Qualities of Sentiment." *Journal of Visual Culture* 5, no. 3 (2006): 275–294.

———. "Photogenic Venus: The 'Cinematographic Turn' and Its Alternatives in Nineteenth-Century France." *Isis* 93 (2002): 585–613.

———. "Recording Devices." In *Companion to the History of Science*, edited by Bernard Lightman. Malden, Mass.: Wiley-Blackwell, forthcoming.

———. *A Tenth of a Second: A History*. Chicago: Chicago University Press, 2009.

Čapek, Milič. "La Pensée de Bergson en Amérique." *Revue international de philosophie* 31 (1977): 329–350.

Cariou, Marie. *Bergson et Bachelard*. Questions. Paris: Presses Universitaires de France, 1995.

Carnap, Rudolf. *Der Raum. Ein Beitrag zur Wissenschaftslehre*. Berlin: Reuther and Reichard, 1922.

———. *Der logische Aufbau der Welt*. Berlin-Schlachtensee: Weltkreis, 1928.

———. "Überwindung der Metaphysik durch Logische Analyse der Sprache." *Erkenntnis* 2 (1932): 219–241.

Carr, Wildon. "The IVth International Congress of Philosophy, Bologna, April 6th–11th, 1911." *Proceedings of the Aristotelian Society* 11 (1910): 223–226.

———. "On Mr. Russell's Reasons for Supposing That Bergson's Philosophy Is Not True." In *The Philosophy of Bergson*, edited by Bertrand Russell 26–32. Cambridge: Bowes and Bowes, 1914.

———. "Symposium: The Problem of Simultaneity: Is There a Paradox in the Principle of Relativity in Regard to the Relation of Time Measured to Time Lived?" *Proceedings of the Aristotelian Society, Supplementary Volumes* 3 (1923): 15–25.

Carr, Wildon. *A Theory of Monads: Outlines of the Philosophy of the Theory of Relativity*. London: Macmillan, 1922.

Carrel, Alexis. "Physiological Time." *Science* 74 (1931): 618–621.

Casarino, Cesare, and Antonio Negri. "It's a Powerful Life: A Conversation on Contemporary Philosophy." *Cultural Critique* 57 (2004): 151–183.

Cassirer, Ernst. *An Essay on Man: An Introduction to a Philosophy of Human Culture*. New Haven: Yale University Press, 1944.

———. *The Philosophy of Symbolic Forms*. Vol. 4. New Haven, Conn.: Yale University Press, 1996.

———. *Substance and Function and Einstein's Theory of Relativity*. Translated by William Curtis Swabey and Marie Collins Swabey. New York: Dover, 1953. First published in 1923.

Cassirer, Toni. *Mein Leben mit Ernst Cassirer*. Hildesheim: Gerstenberg, 1981. Originally published privately in New York,1950.

Challaye, Félicien. *Bergson*. Les Maitres de la pensée moderne. Paris: Librairie Mellottée, 1929.

Champetier de Ribes, Auguste. "Proceedings of the Two Hundred and Fifteenth Day, Friday, 30 August 1946, Morning Session." In *Trial of the Major War Criminals before the International Military Tribunal*, 297–304. Nuremberg, Germany: International Military Tribunal, 1948.

Chang, Hasok. "A Misunderstood Rebellion: The Twin-Paradox Controversy and Herbert Dingle's Vision of Science." *Studies in the History and Philosophy of Science* 24 (1994): 741–790.

Chazy, Jean. *La théorie de la relativité de la mécanique céleste*. Vol. 1. Paris: Gauthier-Villars, 1928.

Chevalier, Jacques. *Entretiens avec Bergson*. Paris: Plon, 1959.

———. *Henri Bergson*. Translated by Lilian A. Clare. London: Rider and Co., 1928.

———. "Le Continu et le Discontinu." *Proceedings of the Aristotelian Society, Supplementary Volumes* 4 (1924): 170–194.

Chimisso, Cristina. "Introduction." In *The Dialectic of Duration*, by Gaston Bachelard, 1–11. Manchester: Clinamen Press, 2000.

Christian, James Lee. *Philosophy: An Introduction to the Art of Wondering*. 7th ed. Fort Worth: Harcourt Brace, 1998.

Clark, Ronald William. *Einstein: The Life and Times*. New York: World, 1971.

Clausberg, Karl. "A Microscope for Time: What Benjamin and Klages, Einstein and the Movies Owe to Distant Stars." In *Given World and Time*, edited by Tyrus Miller, 297–358. Budapest: Central European University Press, 2008.

Cleugh, Mary Frances. *Time and Its Importance in Modern Thought*. London: Methuen & Co., 1937.

Cock, A. G. "Chauvinism and Internationalism in Science: The International Research Council, 1919–1926." *Notes and Records of the Royal Society of London* 37, no. 2 (1983): 249–288.

Cohen, Morris R. "The Insurgence against Reason." *Journal of Philosophy* 22, no. 5 (1925): 113–126.

Cohen, Richard A. "Philo, Spinoza, Bergson: The Ride of an Ecological Age." In *The New Bergson*, edited by John Mullarkey, 18–31. Manchester: Manchester University Press, 1999.

Cook, Gary A. *George Herbert Mead: The Making of a Social Pragmatist*. Urbana: University of Illinois Press, 1993.
Cornu, Alfred. "Détermination nouvelle de la vitesse de la lumière." *Journal de l'École Polytechnique* 27 (1874): 133–180.
Costa de Beauregard, Olivier. "Le Principe de relativité et spatialisation du temps." *Revue de questions scientifiques* 8 (1947): 38–65.
———. *Le Second principe de la science du temps*. Paris: Le Seuil, 1963.
Couturat, Louis. "Compte-rendu critique de B. Russell, 'Essai sur les fondements de la géométrie.'" *Revue de métaphysique et morale* 6 (1898): 354–380.
———. *De l'infini mathématique*. Paris: Félix Alcan,1896.
Crease, Robert P. *World in the Balance: The Historic Quest for an Absolute System of Measurement*. New York: W.W. Norton, 2011.
Crelinsten, Jeffrey. *Einstein's Jury: The Race to Test Relativity*. Princeton, N.J.: Princeton University Press, 2006.
Crookes, William. "Some Possibilities of Electricity." *Fortnightly Review* 51 (1 February 1892): 173–181.
Crystal, Lisa. "Quantum Times: Physics, Philosophy, and Time in the Postwar United States." Ph.D. diss., Harvard University, 2013.
Cunningham, Ebenezer. *The Principle of Relativity*. Cambridge: Cambridge University Press, 1914.
D'Abro, Arthur. *Bergson ou Einstein*. Paris: H. Goulon, 1927.
Darrigol, Olivier. "The Mystery of the Einstein-Poincaré Connection." *Isis* 95, no. 4 (2004): 614–626.
Daston, Lorraine. "Intelligences: Angelic, Animal, Human." In *Thinking with Animals: New Perspectives on Anthropomorphism*, edited by Lorraine Daston and Gregg Mitman, 37–58. Chicago: University of Chicago Press, 2005.
Daston, Lorraine, and Peter Galison. *Objectivity*. New York: Zone Books, 2007.
Daston, Lorraine, and Elizabeth Lunbeck, eds. *Histories of Observation*. Chicago: University of Chicago Press, 2010.
Daudet, Léon. "L'Assaut à l'Académie-Le Juif Bergson et F. Bourgeois." *Action française* 1, no. 43 (12 February 1914).
———. "Le Juif Bergson à l'Académie. Le suprême effort de l'intrigue." *Action française* 1, no. 39 (8 February 1914).
———. *Souvenirs et polémiques*. Bouquins, edited by Robert Laffont. Paris: R. Laffont, 1992.
———. "Un Juif à l'Académie française, l'intrigue Bergson." *Action française* 1, no. 27 (27 January 1914).
———. *Verts d'académie et vers de presse*. Paris: Éditions du Capitole, 1930.
de Broglie, Louis. *Continu et discontinu en physique moderne*. Paris: Albin Michel, 1941.
———. "Les conceptions de la physique contemporaine et les idées de Bergson sur le temps et sur le mouvement." *Revue de métaphyisique et morale* 48 (1941): 241–257.
Delboeuf, Joseph. "Déterminisme et liberté." *Revue philosophique de la France et de l'étranger* 13 (1882): 608–638.
Deleuze, Gilles. "Bergson. 1859–1941." In *Les Philosophes célèbres*, edited by Maurice Merleau-Ponty, 292–299. Paris: L. Mazenod, 1956.
———. *Bergsonism*. New York: Zone Books, 1991. First published in 1966.
———. *Cinema 1: The Movement-Image*. Translated by Hugh Tomlinson and Barbara Habberjam. Minneapolis: University of Minnesota Press, 1986.

———. *Desert Islands and Other Texts, 1953–1974*. Los Angeles: Semiotext(e), 2004.
———. "La Conception de la différence chez Bergson." In *Les Études bergsoniennes*, 77–112. Paris: Presses Universitaires de France, 1956.
———. "Lecture Course on Chapter Three of Bergson's 'Creative Evolution.'" *SubStance* 36, no. 3 (2007): 72–90.
Deleuze, Gilles, and Félix Guattari. *A Thousand Plateaus: Capitalism and Schizophrenia*. Translated by Brian Massumi. Minneapolis: University of Minnesota Press, 1987.
———. *What Is Philosophy?* New York: Columbia University Press, 1994.
Derrida, Jacques. *Of Grammatology*. 1st American ed. Baltimore: Johns Hopkins University Press, 1976.
Desmet, Ronny. "Did Whitehead and Einstein Actually Meet?" In *Researching with Whitehead: System and Adventure*, edited by Franz Riffert and Hans-Joachim Sander, 127–155. Freiburg: Alber Verlag, 2008.
Destouches, Jean-Louis. "La Théorie physique et ses principes fondamentaux." *Bulletin de la Société française de philosophie* 42, no. 1 (1947): 419–449.
Dewey, John. Preface to *A Contribution to a Bibliography of Henri Bergson*, xii. New York: Columbia University Press, 1912.
Di-Capua, Yoav. "Arab Existentialism: An Invisible Chapter in the Intellectual History of Decolonization." *American Historical Review* 117, no. 4 (2012): 1061–1091.
Dingle, Herbert. "History and Philosophy of Science." *Higher Education Quarterly* 6 (1952): 337–345.
———. Introduction to *Duration and Simultaneity*, by Henri Bergson, xv–xlv. Indianapolis: Bobbs-Merrill Company, 1965.
———. *Science at the Crossroads*. London: Martin, Brian and O'Keefe, 1972.
Doumerge, Paul. Préface to *Le Matérialisme actuel*, 1–6. Paris: Ernest Flammarion, 1913.
Dunlap, Knight. "The Shortest Perceptible Time-Interval Between Two Flashes of Light." *Psychology Review* 22 (1915): 226–250.
Du Noüy, Pierre Lecomte. *Biological Time*. London: Methuen and Co., 1936.
———. *L'Homme devant la science*. Bibliothèque de philosophie scientifique. Paris: Flammarion, 1939.
Dunoyer, Louis. "Einstein et la relativité (I)." *La Revue universelle* 9, no. 2 (15 April 1922): 179–188.
Dupont, Paul. "Trois conceptions du temps physique." *Revue philosophique de la France et l'étranger* 96 (1923): 83–108.
During, Élie. *Bergson et Einstein: la querelle du temps*. Paris: Presses Universitaires de France, 2013.
———. "Dossier critique: I. Notes." In *Durée et simultanéité: á propos de la théorie d'Einstein*, edited by Élie During, 219–244. Paris: Presses Universitaires de France, 2009.
———. "Entre Bergson et Saint Thomas: La Correspondance Metz-Meyerson." *Corpus, revue de philosophie* 58 (2010): 237–254.
———. "Introduction au dossier critique." In *Durée et simultanéité: á propos de la théorie d'Einstein*, edited by Élie During, 219–244. Paris: Presses Universitaires de France, 2009.
———. "Notes on the Bergsonian Cinematograph." In *Cine-dispositives. Essays in Epistemology across Media*, edited by F. Albera and M. Tortajada, 115–128.Amsterdam: Amsterdam University Press, 2013.

———. "Philosophical Twins? Bergson and Whitehead on Langevin's Paradox and the Meaning of 'Space-Time.'" In *Alfred North Whitehead's Principles of Natural Knowledge*, edited by G. Durand and M. Weber, 79–104. Frankfurt: Ontos, 2007.

———. "Vie et mort du cinématographe: de L'Évolution créatrice à Durée et Simultanéité." In *Bergson*, edited by Camille Riquier, 139–162. Paris: Cerf, 2012.

Eberty, Felix. *Die Gestirne und die Weltgeschichte; Gedanken über Raum, Zeit und Ewigkeit*. Edited by Gregory Itelson. Berlin: Gregor Rogoff Verlag, 1923.

[———]. *The Stars and the Earth, or Thoughts Upon Space, Time and Eternity*. Boston: Lee and Shepard, 1882. First published in 1846.

Eddington, Arthur Stanley. "Einstein's Theory of Space and Time." *Contemporary Review* 116 (1919): 639–643.

———. *Espace, temps et gravitation*. Translated by J. Rossignol. Paris: J. Hermann, 1921.

———. *The Nature of the Physical World*. Edited by Ernest Rhys. Everyman's Library: Science. London: J.M. Dent & Sons, 1935. First published in 1928.

———. "The Relativity of Time." *Nature* 106, no. 2677 (1921): 802–804.

———. *Space, Time, and Gravitation: An Outline of the General Relativity Theory*. Cambridge Science Classics. Cambridge, UK: Cambridge University Press, 1987. First published in 1920.

Edman, Irwin. Introduction to *Creative Evolution*, xxv. New York: The Modern Library, 1944.

"Een interview met prof. Albert Einstein." *Nieuwe Rotterdamsche Courant* (4 July 1921).

Einstein, Albert. "À propos de la deduction relativiste de M. Émile Meyerson." *Revue philosophique 105* (1928): 161–166.

———. "Autobiographical Notes." In *Albert Einstein: Philosopher-Scientist*, edited by Paul Arthur Schilpp, 3–94. La Salle, Ill.: Open Court, 1949.

———. *Correspondances françaises*. Edited by Michel Biezunski. Vol. 4 of Oeuvres Choisies, edited by Françoise Balibar. Paris: Éditions du Seuil, 1989.

———. "Dialog über Einwände gegen die Relativitätstheorie." *Die Naturwissenschaften* 6, no. 48 (29 November 1918): 697–702.

———. "Die Freie Vereinigung für technische Volksbildung." *Neue Freie Presse*, 24 July 1920, 8.

———. "Die Grundlage der allgemeinen Relativitätstheorie." *Annalen der Physik* 49 (1916): 769–822.

———. "Die Relativitätstheorie." In *Die Kultur der Gegenwart. Ihre Entwicklung und ihre Ziele*, edited by Emil Warburg, 703–713. Leipzig: Teubner, 1915.

———. "Einstein Considers Seeking New Home," *New York Times*, 22 December 1930.

———. "Einsteins amerikanische Eindrüche. Was er wirklich sah." *Vossische Zeitung*, 10 July 1921.

———. "Geometry and Experience." Translated by Sonja Bargmann. In *Ideas and Opinions*, edited by Carl Seelig, 232–246. New York: Three Rivers, 1954. First published in 1921.

———. "Induktion und Deduktion in der Physik." *Berliner Tageblatt*, 25 December 1919.

———. "Kinetische Theorie des Wärmegleichgewichtes und des zweiten Hauptsatzes der Thermodynamik." *Annalen der Physik* 9 (1902): 417–433.

———. "Le principe de relativité et ses conséquences dans la physique moderne." *Archives des sciences physiques et naturelles* (Genève) 29, no. 1–2 (1910): 5–28, 125–144.

———. *Letters to Solovine*. New York: Philosophical Library, 1987.
———. *The Meaning of Relativity: Four Lectures Delivered at Princeton University, May, 1921*. London: Methuen & Co., 1922.
———. "Meine Antwort. Ueber die anti-relativitätstheoretische G.m.b.H." *Berliner Tageblatt* (27 August 1920).
———. "Paul Langevin." *La Pensée* 12 (May–June 1947): 13–14.
———. "The Principal Ideas of the Theory of Relativity." In *The Collected Papers of Albert Einstein*, edited by Diana Kormos Buchwald, vol. 7, pp. 3–6 (Princeton, N.J.: Princeton University Press, 2002.
———. "A Re-examination of Pacifism." *Polity* 3, no. 1 (January 1935): 4–5.
———. "Relativitätstheorie." *Vierteljahrsschrift Naturforschende Gesellschaft Zürich* 56 (1911): 1–14.
———. *Relativity: The Special and the General Theory*. New York: Crown, 1961. First published in 1916.
———. "Remarks on Bertrand Russell's Theory of Knowledge." In *The Philosophy of Bertrand Russell*, edited by Paul Arthur Schilpp, 279–291. La Salle, Ill.: Open Court, 1944.
———. "Reply to Criticisms." In *Albert Einstein: Philosopher-Scientist*, edited by Paul Arthur Schilpp, 665–688. La Salle, Ill.: Open Court, 1970. First published in 1949.
———. "Theoretische Bemerkung über die Brownsche Bewegung." *Zeitschrift fur Elektrochemie und angewandte physikalische Chemie* 13 (1907): 41–42.
———. "Über das Relativitätsprinzip und die aus demselben gezogenen Folgerungen." *Jahrbuch der Radioaktivität und Elektronik* 4 (1907): 411–462.
———. "Über die von der molekularkinetischen Theorie der Wärme geforderte Bewegung von in ruhenden Flüssigkeiten suspendierten Tielchen." *Annalen der Physik* 17 (1905): 549–560.
———. "Über eine Methode zur Bestimmung des Verhältnisses der transversalen und longitudinalen Masse des Elektrons" *Annalen der Physik* 21(1906): 583–586.
———. "Von der Tagung der Völkerbundkommission für intellektuelle Zusammenarbeit aus Genf zuruckgekehrt, habe ich das Bedurfnis." *Frankfurter Zeitung*, 29 August 1924.
———. "Wie ich Zionist wurde." *Jüdische Rundschau* (1921): 351–352.
———. "Zionists Greet Einstein, Here for Palestine," *New York Tribune*, 3 April 1921, 1, 3.
———. "Zum Relativitätsproblem." *Scientia* 15 (1914): 337.
———. "Zur Elektrodynamik bewegter Körper." *Annalen der Physik* 17 (1905): 891–921.
———. "Zur Errichtung der hebräischen Universität in Jerusalem." *Jüdische Pressezentrale Zürich*, 26 August 1921.
Einstein, Albert, and Michele Besso. *Correspondance 1903–1955*. Translated by Pierre Speziali. Paris: Hermann, 1979.
Einstein, Albert, and Max and Hedwig Born. *The Born-Einstein Letters: Friendship, Politics and Physics in Uncertain Times*. New York: Macmillan, 2005.
Esclangon, Ernest. *Les Preuves astronomiques de la relativité*. Paris: Gauthier Villars, 1922.
———. "Sur la relativité du temps." *Comptes rendus de l'Académie des sciences 173* (1921): 1340–1342.
Ewald, Paul Peter, George Uhlenbeck, Thomas S. Kuhn, and Ella Ewald. "Interview with Dr. Léon Brillouin." In *Oral History Transcripts*. American Institute of Physics, 29 March 1962.

Ezrahi, Yaron. "Einstein and the Light of Reason." In *Albert Einstein: Historical and Cultural Perspectives*, edited by Gerald Holton and Yehuda Elkana, 253–278. Princeton, N.J.: Princeton University Press, 1986.
Fabré, Lucien, and Albert Einstein. *Une Nouvelle figure du monde: les théories d'Einstein*. Paris: Payot, 1922.
Fanon, Frantz. *Black Skin, White Masks*. New York: Grove Press, 2008.
Faye, Hervé. "Sur les erreurs d'origine physiologique." *Comptes rendus de l'Académie des sciences* 59 (12 September 1864): 473–480.
Fenves, Peter. *The Messianic Reduction: Walter Benjamin and the Shape of Time*. Stanford, Calif.: Stanford University Press, 2011.
Ferguson, Allan. "Reviews: The Case Against Einstein by Arthur Lynch," *Mathematical Gazette* 17, no. 225 (1933): 272–274.
Ferrari, Giulio Cesare. "IVe Congrès international de philosophie (Bologne, mars-avril 1911)." *Revue de métaphysique et de morale* 18, no. 3 (1910): 35–36.
Feuer, Lewis S. *Einstein and the Generations of Science*. New York: Basic Books, 1974.
Feynman, Richard Phillips. *The Character of Physical Law*. London: British Broadcasting Corp., 1965.
Fine, Arthur. *The Shaky Game: Einstein, Realism, and the Quantum Theory*. Science and its Conceptual Foundations. 2nd ed. Chicago: University of Chicago Press, 1996.
Flusser, Vilém. "Die Auswanderung der Zahlen aus dem alphanumerischen Code." In *Literatur im Informationszeitalter*, edited by Dirk Matejovski and Friedrich Kittler. 9–14. Frankfurt: Campus Verlag, 1996.
Fölsing, Albrecht. *Albert Einstein: A Biography*. New York: Viking, 1997.
Fontana, Paul. "Cours de Bergson au Collège de France: théories de la volonté." In *Mélanges*, edited by André Robinet, 685–722. Paris: Presses Universitaires de France, 1972.
Fontcuberta Famadas, Laura. "La disputa Einstein-Bergson." Ph.D. diss., Universitat Autònoma de Barcelona, 2005.
Fontené, G. *La Relativité restreinte*. Paris: Vuibert, 1922.
Foucault, Michel. *Language, Counter-Memory, Practice: Selected Essays and Interviews*. Ithaca, N.Y.: Cornell University Press, 1977.
François, Arnaud. "Notices et notes: Introduction ii." In *La Pensée et le mouvant: Essais et conférences*, 322–379. Paris: Presses Universitaires de France, 2009.
Frank, Philipp. *Einstein: His Life and Times*. New York: Alfred Knopf, 1947.
Franklin, Benjamin. "Advice to a Young Tradesman." In *Franklin: The Autobiography and Other Writings on Politics, Economics, and Virtue*, 200–202. Cambridge: Cambridge University Press, 2004.
Frewin, Leslie. *La Notion du temps d'apres Einstein*. Paris: F. Alcan, 1921.
Friedman, Michael. *Foundations of Space-Time Theories: Relativistic Physics and Philosophy of Science*. Princeton, N.J.: Princeton University Press, 1983.
———. "Geometry as a Branch of Physics: Background and Context for Einstein's 'Geometry and Experience.'" In *Reading Natural Philosophy*, edited by David Malament, 193–229. Chicago: Open Court, 2002.
Fujita, Hisashi. "Cassirer, lecteur de Bergson." *Annales Bergsoniennes* 3 (2007): 53–70.
Galison, Peter. "Einstein's Clock: The Place of Time." *Critical Inquiry* 26 (Winter 2000): 355–389.

———. *Einstein's Clocks, Poincaré's Maps: Empires of Time*. New York: W.W. Norton and Company, 2003.

———. "Judgment Against Objectivity." In *Picturing Science, Producing Art*, edited by Caroline A. Jones and Peter Galison, 327–359. New York: Routledge, 1998.

Genovesi, Angelo. "Henri Bergson: Lettere a Einstein." *Filosofia* 49, no. 1 (1998): 3–39.

———. *Il Carteggio tra Albert Einstein ed Édouard Guillaume*. Filosofia. Milano: Franco Angeli, 2000.

George, André. "Bergson et Einstein." *Les Documents de la vie intellectuelle* (January 1930): 52–64.

———. "Le Temps, la vie, et la mort." *Les Documents de la vie intellectuelle* 43, no. 1 (1946): 121–146.

Geroulanos, Stefanos. *An Atheism That Is Not Humanist Emerges in French Thought*. Stanford, Calif.: Stanford University Press, 2010.

Gibbons, Gary, and Will M. Clifford. "On the Multiple Deaths of Whitehead's Theory of Gravity." *Studies in History and Philosophy of Modern Physics* 39 (2008): 41–46.

Gilbert, Leo. "A Satire on the Principle of Relativity." *Monist* 24, no. 2 (April 1914): 288–309.

Gingras, Yves. "The Collective Construction of Scientific Memory: The Einstein-Poincaré Connection and Its Discontents, 1905–2005." *History of Science* 46 (2008): 75–114.

Giovanelli, Marco. "Talking at Cross-Purposes: How Einstein and the Logical Empiricists Never Agreed on What They Were Disagreeing About." *Synthese* 190, no. 17 (2013): 3819–3863.

Gitelman, Lisa. *Always Already New: Media, History, and the Data of Culture* Cambridge, Mass.: MIT Press, 2006.

Glick, Thomas F., ed. *The Comparative Reception of Relativity*. Vol. 103. Boston Studies in the Philosophy of Science. Dordrecht: D. Reidel, 1987.

Goenner, Hubert. *Einstein in Berlin, 1914–1933*. Munich: C. H. Beck, 2005.

Goldberg, Stanley. "Poincare's Silence and Einstein's Relativity: The Role of Theory and Experiment in Poincaré's Physics." *British Journal for the History of Science* 5 (1970): 73–84.

Gooday, Graeme J.N. *The Morals of Measurement: Accuracy, Irony and Trust in Late Victorian Electrical Practice*. Cambridge: Cambridge University Press, 2004.

Gordon, Peter E. *Continental Divide: Heidegger, Cassirer, Davos*. Cambridge, Mass.: Harvard University Press, 2010.

———. "Heidegger in Black." *New York Review of Books* 61, no. 15 (2014): 26–28.

Grandjean, Frank. *Une révolution dans la philosophie. La Doctrine de M. Bergson*. Geneva: Atar, 1916.

Granek, Galina. "Poincaré's Light Signaling and Clock Synchronization Thought Experiment and its Possible Inspiration to Einstein." In *Albert Einstein Century International Conference*, edited by J.-M. Alimi and A. Füzfa, 1095–1102. Melville, N.Y.: American Institute of Physics, 2006.

Grattan-Guinness, I. "Russell and Karl Popper: Their Personal Contacts." *Russell: The Journal of the Bertrand Russell Archives*, no. 12 (1992): 3–18.

Gray, Jeremy. *Henri Poincaré: A Scientific Biography*. Princeton, NJ: Princeton University Press, 2013.

Gregory, Frederick. "The Mysteries and Wonders of Natural Science: Aaron Bernstein's Naturwissenschaftliche Volksbücher and the Adolescent Einstein," In *Einstein: The*

Formative Years, 1879–1909, edited by John J. Stacheland and Don Howard, 23–41. Boston: Birkhauser, 2000.

Grogin, R.C. *The Bergsonian Controversy in France: 1900–1914*. Calgary: University of Calgary Press, 1988.

Grossmann, Marcel. "Mise à point mathématique." *Archives des sciences physiques et naturelles* 125, no. 6 (1920): 497–499.

———. "Physikprofessoren." *Neue Schweizer Zeitung*, 15 June 1920.

Grünbaum, Adolf. "Time and Entropy." *American Scientist* 43 (1955): 550–572.

Grundmann, Siegfried. *The Einstein Dossiers: Science and Politics*. Berlin: Springer, 2005.

———. *Einsteins Akte*. Berlin: Springer-Verlag, 1998.

Guillaume, Charles-Édouard. "L'Oeuvre du Bureau International des poids et mesures." In *La Création du Bureau International des poids et mesures et son oeuvre*, edited by Ch.-Éd. Guillaume, 33–258. Paris: Gauthier-Villars, 1927.

Guillaume, Édouard. "La Question du temps, d'après M. Bergson." *Revue générale des sciences pures et appliquées* 33 (1922): 573–582.

———. "La Théorie de la relativité et le temps universel." *Revue de Métaphysique et morale* 25, no. 3 (May–June 1918): 285–323.

———. "La Théorie de la relativité et sa signification." *Revue de Métaphysique et morale* 27, no. 4 (October–December 1920): 423–469.

———. *La Théorie de la relativité: Résumé des conférences faites à l'Université de Lausanne au semestre d'été*. Lausanne, France: F. Rouge, 1921.

———. "Les bases de la théorie de la relativité." *Revue générale des sciences* 31, no. 7 (1920): 200–210.

———. "Représentation et mesure de temps." *Archives des sciences physiques et naturelles* 125, no. 2 (1920): 125–146.

———. "Sur la théorie de la relativité, compte rendu de la séance de la Societé suisse de physique tenue à Berthond le 10 mai 1919." *Archives des sciences physiques et naturelles* 124, no. 3 (1919): 246–250.

Gunter, P.A.Y., ed. *Bergson and the Evolution of Physics*. Knoxville: University of Tennessee Press, 1969.

———. "Bergsonian Method and the Evolution of Science." In *Bergson and the Evolution of Physics*, edited by P.A.Y Gunter, 3–42. Knoxville: University of Tennessee Press, 1969.

———. "Gilles Deleuze, Deleuze's Bergson and Bergson Himself." In *Deleuze, Whitehead, Bergson: Rhizomatic Connections*, edited by Keith A. Robinson, 167–180. Basingstoke, England: Palgrave Macmillan, 2009.

Habermas, Jürgen. *The Structural Transformation of the Public Sphere: An Inquiry Into a Category of Bourgeois Society*. Studies in Contemporary German Social Thought. Cambridge, Mass.: MIT Press, 1989.

Hafele, J. C., and Richard E. Keating. "Around-the-World Atomic Clocks: Predicted Relativistic Time Gains." *Science* 177 (1972): 166–170.

Haldane, Richard B. *The Reign of Relativity*. 3rd ed. London: J. Murray, 1921.

Hanna, Thomas, ed. *The Bergsonian Heritage*. New York: Columbia University Press, 1962.

———. "Introduction." In *The Bergsonian Heritage*, edited by Thomas Hanna, 1–31. New York: Columbia University Press, 1962.

Harding, Sandra G., ed. *Can Theories Be Refuted? Essays on the Duhem-Quine Thesis*. Dordrecht, Netherlands: Reidel, 1976.

Harman, P. M. *The Natural Philosophy of James Clerk Maxwell*. Cambridge: Cambridge University Press, 1998.

Hartshorne, Charles. "Whitehead's Idea of God." In *The Philosophy of Alfred North Whitehead*, edited by Paul Arthur Schilpp, 515–559. Evanston, Ill.: Northwestern University Press, 1941.

Heidegger, Martin. *Being and Time*. New York: Harper-Collins, 1962. First published in 1927.

———. *The Concept of Time*. Translated by William McNeill. Oxford: Blackwell, 1992.

———. *The Concept of Time*. Translated by Ingo Farin and Alex Skinner. London: Continuum, 2011.

———. "The Concept of Time in the Science of History." Translated by Jarry S. Taylor, Hans W. Uffelmann, and John van Buren. In *Supplements: From the Earliest Essays to Being and Time and Beyond*, edited by John van Buren, 49–60. Albany: State University Press of New York, 2002. First published in 1915.

———. *The Fundamental Concepts of Metaphysics: World, Finitude, Solitude*. Translated by William McNiell and Nicholas Walker. Studies in Continental Thought. Bloomington: Indiana University Press, 1995.

———. *History of the Concept of Time: Prolegomena*. Bloomington: Indiana University Press, 1979.

———. *The Metaphysical Foundations of Logic*. Translated by Michael Heim. Studies in Phenomenology and Existential Philosophy. Bloomington: Indiana University Press, 1984.

———. *Parmenides*. Translated by André Schuwer and Richard Rojcewicz. Bloomington: Indiana University Press, 1992.

———. "The Question Concerning Technology." In *The Question Concerning Technology and other Essays*, 3–35. New York: Harper and Row, 1977. First published in 1954.

———. "Wilhelm Dilthey's Research and the Struggle for a Historical Worldview (1925)." In *Supplements: From the Earliest Essays to Being and Time and Beyond*, 112–176. Albany: State University of New York Press, 2002.

Hellman, John. "Jacques Chevalier, Bergsonism, and Modern French Catholic Intellectuals." *Biography* 4, no. 2 (1981): 138–153.

Helmholtz, Hermann von. "On the Interaction of Natural Forces." In *Science and Culture: Popular and Philosophical Essays*, 18–45. Chicago: University of Chicago Press, 1995. First published in 1854.

Henderson, Linda Dalrymple. "Einstein and 20th-Century Art: A Romance of Many Dimensions." In *Einstein for the 21st Century: His Legacy in Science, Art, and Modern Culture*, edited by Peter L. Galison, Gerald Holton, and Silvan S. Schweber, 101–129. Princeton, N.J.: Princeton University Press, 2008.

Henri, Victor. "Études cinématographique du mouvement brownien." *Comptes rendus de l'Académie des sciences* 146 (1908): 1024–1026.

———. "Influence du milieu sur les mouvements browniens." *Comptes rendus de l'Académie des sciences* 147 (1908): 62–65.

Hentschel, Klaus. *Interpretationen und Fehlinterpretationen der speziellen und der allgemeinen Relativitätstheorie durch Zeitgenossen Albert Einsteins*. Science Networks Historical Studies. Basel: Birkhäuser, 1990.

Hermann, Armin. *Einstein: Der Weltweise und sein Jahrhundert; eine Biographie*. Munich: Piper, 1996.

Hickson, J.W.A. "Causality and Recent Physics." *Philosophical Review* 44, no. 6 (1935): 534–543.
———. "Recent Attacks on Causal Knowledge." *Philosophical Review* 47, no. 6 (1938): 595–606.
———. "Review of *Der Kampf um das Kausalgesetz in der jüngsten Physik*, by Hugo Bergmann." Braunschweig, Germany: Vieweg & Sohn. 1929. *Journal of Philosophy* 26, no. 24 (1929): 667–669.
Holton, Gerald. "Einstein and the Cultural Roots of Modern Science." *Daedalus* 127, no. 1 (1998): 1–44.
———. "Einstein, Michelson, and the 'Crucial' Experiment." *Isis* 60 (1969): 133–197.
———. "Ernst Mach and the Fortunes of Positivism in America." *Isis* 83 (1992): 27–60.
———. "On the Origins of the Special Theory of Relativity." In *Thematic Origins of Scientific Thought*, 191–236. Cambridge, Mass.: Harvard University Press, 1973.
———. "On the Thematic Analysis of Science: The Case of Poincaré and Relativity." In *Mélanges Alexandre Koyré, à l'occasion de son soixante-dixième anniversaire*, 257–268. Paris: Hermann, 1964.
———. *Thematic Origins of Scientific Thought; Kepler to Einstein*. Cambridge, Mass.: Harvard University Press, 1973.
Horkheimer, Max. "On Bergson's Metaphysics of Time." *Radical Philosophy* 131 (May/June 2005): 9–19. First published in 1934.
———. "Traditionelle und kritische Theorie." *Zeitschrift für Sozialforschung* 6 (1937): 245–292.
Howard, Don. "Einstein and Duhem." *Synthese* 83 (1990): 363–384.
———. "Realism and Conventionalism in Einstein's Philosophy of Science: The Einstein-Schlick Correspondence." *Philosophia Naturalis* 21, no. 1984 (1993): 618–629.
———. "Was Einstein Really a Realist?" *Perspectives on Science: Historical, Philosophical, Social* 1, no. 2 (1993): 204–251.
Hoy, David Couzens. *The Time of Our Lives: A Critical History of Temporality*. Cambridge, Mass.: MIT Press, 2009.
Husserl, Edmund. *Briefwechsel*. Vol. 3. Dordrecht: Kluwer Academic Publishers, 1994.
———. *The Crisis of European Sciences and Transcendental Phenomenology: An Introduction to Phenomenological Philosophy*. Northwestern University Studies in Phenomenology and Existential Philosophy. Evanston: Northwestern University Press, 1970.
———. *The Phenomenology of Internal Time-Consciousness*. Bloomington: Indiana University Press, 1964.
———. "The Vienna Lecture." Translated by David Carr. In *The Crisis of European Sciences and Transcendental Phenomenology: An Introduction to Phenomenological Philosophy*, edited by John Wild, 269–299. Evanston, Ill.: Northwestern University Press, 1970.
Huxley, Thomas Henry. *Method and Results*. New York: D. Appleton, 1897.
Ingarden, Roman. *On the Motives Which Led Husserl to Transcendental Idealism*. Vol. 64 of Phaenomenologica. Den Haag: Martinus Nijhoff, 1975.
Isaacson, Walter. *Einstein: His Life and Universe*. New York Simon and Schuster, 2007.
James, William. *The Letters of William James*. Edited by Henry James. Vol. 2. Boston: Atlantic Monthly, 1920.
Jammer, Max. *Concepts of Simultaneity: From Antiquity to Einstein and Beyond*. Baltimore: Johns Hopkins University Press, 2006.

Jerome, Fred. *The Einstein File: J. Edgar Hoover's Secret War Against the World's Most Famous Scientist*. New York: St. Martin's Press, 2002.

Johnson, Paul. *Modern Times: The World from the Twenties to the Nineties*. New York: Harper Collins, 1991.

Jones, Caroline A. "Rendering Time." In *Einstein for the 21st Century: His Legacy in Science, Art, and Modern Culture*, edited by Peter L. Galison, Gerald Holton, and Silvan S. Schweber, 130–149. Princeton, N.J.: Princeton University Press, 2008.

Jones, H. Spencer. "The Early History of the ICSU 1919–1946." *ICSU Review* 2 (1960): 169–187.

Joule, James. "On the Calorific Effects of Magneto-Electricity, and On the Mechanical Value of Heat." *Philosophical Magazine* 23, no. 152 (1843): 263–276, 347–352, 435–443.

Judith, Robinson. "Valéry, critique de Bergson." *Cahiers de l'Association internationale des études francaises* 17 (1965): 203–215.

Kafka, Franz. *Gesammelte Werke: Taschenbuchausgabe in sieben Bänden*. Edited by Max Brod. Vol. 7. Frankfurt am Main: Fischer Taschenbuch Verlag, 1976.

———. *Letters to Milena*. Translated by Tania Stern and James Stern. Edited by Willi Haas New York: Schocken, 1953.

Kahan, Théo. "Sur les origines de la théorie de la relativité restreinte." *Revue d'Histoire des Sciences* 12 (1959): 159–165.

Kaiser, David. *How the Hippies Saved Physics: Science, Counterculture, and the Quantum Revival*. New York: Norton, 2011.

Kaufmann, Walter Arnold. *From Shakespeare to Existentialism: Studies in Poetry, Religion, and Philosophy*. Boston: Beacon Press, 1959.

Kessler, Harry Graf. *Tagebücher, 1918–1937*. Frankfurt am Main: Insel-Verlag, 1961.

Kevles, Daniel J. "'Into Hostile Political Camps': The Reorganization of International Science in World War I." *Isis* 62 (1971): 47–60.

Kleinberg, Ethan. *Generation Existential: Heidegger's Philosophy in France, 1927–1961*. Ithaca, N.Y.: Cornell University Press, 2005.

Koyré, Alexandre. "Compte rendu de H. Bergson, Le temps et la durée, 2e éd., 1925." *Versty* 1 (1926): 234–237.

———. "Le 'De motu gravium' de Galilée: de l'expérience imaginaire et de son abus." *Revue d'Histoire des Sciences* 13 (1960): 197–245.

Kuhn, Thomas S. *The Structure of Scientific Revolutions*. 2nd ed. Chicago: University of Chicago Press, 1970.

Lalande, André. "Philosophy in France, 1921." *Philosophical Review* 31, no. 6 (November 1922): 539–563.

———. "Philosophy in France, 1922–1923." *Philosophical Review* 33, no. 6 (1924): 535–559.

———. "Philosophy in France, 1936–37." *Philosophical Review* 47, no. 1 (1938): 1–27.

Landecker, Hannah. "Cellular Features: Microcinematography and Film Theory." *Critical Inquiry* 31 (2005): 903–925.

———. *Culturing Life: How Cells Became Technologies*. Cambridge, Mass.: Harvard University Press, 2007.

Langer, Susanne K. *Philosophy in a New Key*. Cambridge, Mass.: Harvard University Press, 1951. First published in 1942.

Langevin, Jean, and Michel Paty. "Le Séjour d'Einstein en France en 1922." *Fundamenta Scientiae* 93 (1979): 23–41.

Langevin, Paul. "L'Aspect général de la theorie de la relativité." *Bulletin scientifique des étudiants de Paris* 2 (1922): 2–22.
———. "L'Évolution de l'espace et du temps." *Revue de métaphysique et morale* 10 (July 1911): 455–466.
———. "L'Évolution de l'espace et du temps." *Scientia* 10 (1911): 31–54.
———. "Le Temps, l'espace et la causalité dans la physique moderne." *Bulletin de la Société française de philosophie* 12 (1912): 1–46.
Laplace, Pierre-Simon. *Essai philosophique sur les probabilités*. Paris: Courcier, 1814.
Latour, Bruno. "A Relativistic Account of Einstein's Relativity." *Social Studies of Science* 18, no. 1 (1988): 3–44.
———. *Science in Action: How to Follow Scientists and Engineers through Society*. Cambridge, Mass.: Harvard University Press, 1987.
———. "Some Experiments in Art and Politics." *E-Flux* 23 (2011): 1–7.
———. "Trains of Thoughts—Piaget, Formalism and the Fifth Dimension." *Common Knowledge* 6, no. 3 (1997): 170–191.
———. *We Have Never Been Modern*. Cambridge, Mass.: Harvard University Press, 1991.
———. "Why Has Critique Run Out of Steam? From Matters of Fact to Matters of Concern." *Critical Inquiry* 30, no. 2 (2004): 225–248.
Laue, Max von. "Theoretisches über neuere optische Beobachtungen zur Relativitätstheorie." *Physikalische Zeitschrift* 21 (1920): 659–662.
Lecourt, Dominique. *Que sais-je? La Philosophie des sciences*. 4th ed. Paris: Presses Universitaires de France, 2008.
LeFèvre, Frédéric. "Une heure avec Maurice Maeterlinck." *Les Nouvelles littéraires* (7 April 1928).
Le Roy, Édouard. "Science et philosophie." *Revue de métaphyisique et morale* 7 (1899): 375–425.
Levenson, Thomas. *Einstein in Berlin*. New York: Bantam Books, 2003.
Levi, William Albert. *Philosophy and the Modern World*. Bloomington: Indiana University Press, 1970.
Lévy-Leblond, Jean-Marc. "Le Boulet d'Einstein et les boulettes de Bergson." In *Bergson et la science*. Vol. 3 of Annales Bergsoniennes, edited by Frédéric Worms, 237–258. Paris: Presses Universitaires de France, 2007.
Lewis, Wyndham. *Time and Western Man*. Santa Rosa, Calif.: Black Sparrow Press, 1993. First published in 1927.
Linett, Maren Tova. *The Cambridge Companion to Modernist Women Writers*. Cambridge: Cambridge University Press, 2010.
Lippman, W. "The Most Dangerous Man in the World." *Everybody's Magazine* 27 (1912): 100–101.
Lorentz, Hendrik A. "Alte und neue Fragen der Physik." *Physikalische Zeitschrift* 11 (1910): 1234–1257.
———. *Collected Papers*. Edited by Pieter Zeeman and A. D. Fokker. 9 vols. The Hague: M. Nijhoff, 1934.
———. *Das Relativitätsprinzip: Drei Vorlesungen gehalten in Teylers Stiftung zu Haarlem*. Beihefte zur Zeitschrift für mathematischen und naturwissenschaftlichen Unterricht. Leipzig: B.G. Teubner, 1914.
———. *The Einstein Theory of Relativity: A Concise Statement*. New York: Brentano's, 1920. First published in 1919.

———. "Electromagnetic Phenomena in a System Moving With Any Velocity Less Than That of Light." *Proceedings of the Royal Netherlands Academy of Arts and Science, Amsterdam* 6 (1904): 809.

———. "The Principle of Relativity for Uniform Translations." In vol. 3 of *Lectures on Theoretical Physics*, edited by A.D. Fokker, 179–326. London: Macmillan, 1931.

———. *Problems of Modern Physics: A Course of Lectures Delivered in the California Institute of Technology*. Edited by H. Bateman. Boston: Ginn and Company, 1927.

———. "Report." *Astrophysical Journal* 68, no. 5 (1928): 345–352.

———. *The Theory of Electrons and its Applications to the Phenomena of Light and Radiant Heat: A Course of Lectures Delivered in Columbia University, New York, in March and April, 1906*. Sammlung von Lehrbüchern auf dem Gebiete der mathematischen Wissenschaften mit Einschluss ihrer Anwendungen. Vol. 29. Leipzig: B.G. Teubner, 1909.

Lorentz, Hendrik A., Albert Einstein, Hermann Minkowski, Hermann Weyl, and Arnold Sommerfeld. *The Principle of Relativity: A Collection of Original Memoirs on the Special and General Theory of Relativity*. Translated by W. Perrett and G. B. Jeffery. London: Methuen & Co., 1923. First published in 1913.

Lovejoy, Arthur O. "The Dialectical Argument Against Absolute Simultaneity. II." *Journal of Philosophy* 27, no. 24 (20 November 1930): 645–654.

———. "The Problem of Time in Recent French Philosophy." *Philosophical Review* 21 (1912): 10–31, 322–343, 527–545.

———. *The Reason, the Understanding, and Time*. Baltimore: Johns Hopkins Press, 1961.

———. *The Revolt Against Dualism: An Inquiry Concerning the Existence of Ideas*. New York: W. W. Norton and Company, 1930.

———. "The Travels of Peter, Paul, and Zebedee." *Philosophical Review* 41, no. 5 (1932): 498–517.

Lowe, Victor. *Alfred North Whitehead: The Man and His Work*. Vol. 2. Baltimore: Johns Hopkins University Press, 1985.

Lucretius. *On the Nature of Things*. Translated by H.A.J. Munro. London: Bell and Sons, 1919.

Lynch, Arthur. *The Case against Einstein*. London: P. Allan, 1932.

Lyotard, Jean François. *The Postmodern Condition: A Report on Knowledge*. Theory and History of Literature. Minneapolis: University of Minnesota Press, 1984.

MacArthur, Charles. "Einstein Baffled in Chicago: Seeks Pants in Only Three Dimensions, Faces Relativity of Trousers." *Chicago Herald and Examiner* (3 May 1921).

Maire, Gilbert. "Les Années de Bergson à Clermont-Ferrand." *Glanes* 2 (March–April 1949).

Malisoff, William Marias. "The Universe of Operations." *Philosophy of Science* 3, no. 3 (1936): 360–364.

Maritain, Jacques. *Bergsonian Philosophy and Thomism*. New York: Philosophical Library, 1955.

———. *La Philosophie bergsonienne*. Paris: Marcel Rivière, 1913.

———. *Réflexions sur l'intelligence et sur sa vie propre*. Bibliothèque française de philosophie. Paris: Nouvelle Librairie Nationale, 1924.

———. *Réflexions sur l'intelligence et sur sa vie propre*. Paris: Nouvelle Libraire Nationale, 1926.

Maritain, Raïssa. "Henri Bergson." *La Relève* (March 1941): 161–187.

———. *We Have Been Friends Together: Memoirs*. Translated by Julie Kernan. London: Longmans, 1942.

Markosian, Ned. "Time." In *Stanford Encyclopedia of Philosophy*, edited by Edward N. Zalta. Spring 2014 edition. http://plato.stanford.edu/archives/spr2014/entries/time/

Marx, Karl. *Economic and Philosophic Manuscripts of 1844*. Translated by Martin Milligan. New York: International Publishers, 1968.

———. *The Eighteenth Brumaire of Louis Bonaparte*. New York: International Publishers, 1963. First published in 1852.

Masriera Rubio, Miguel. "La verdad sobre Einstein." *La Vanguardia*, 15 January 1925.

Maudlin, Tim. *Philosophy of Physics: Space and Time*. Princeton Foundations of Contemporary Philosophy. Princeton, N.J.: Princeton University Press, 2012.

Maurras, Charles. *Action française* 3 (10 April 1913).

Maxwell, James Clerk. "On a Possible Mode of Detecting a Motion of the Solar System through the Luminiferous Ether." *Nature* 21 (1880): 314–315.

———. *Scientific Papers*. New York: Dover Publications, 1965.

McGilvary, Evander Bradley. "Space-Time, Simple Location, and Prehension." In *The Philosophy of Alfred North Whitehead*, edited by Paul Arthur Schilpp, 211–239. Evanston, Ill.: Northwestern University Press, 1941.

Mead, George Herbert. *The Philosophy of the Act*. Edited by Charles W. Morris, David L. Miller, John Monroe Brewster, and Albert Millard Dunham. Chicago: University of Chicago Press, 1938.

———. *The Philosophy of the Present*. Paul Carus Foundation Lectures, edited by Arthur Edward Murphy. Chicago: Open Court, 1932.

———. "Relative Space-Time and Simultaneity." *Review of Metaphysics* 17 (1963–1964): 514–535.

Merleau-Ponty, Maurice. "At the Sorbonne." In *The Bergsonian Heritage*, edited by Thomas Hanna, 133–149. New York: Columbia University Press, 1962.

———. "Bergson in the Making." In *Signs*, edited by John Wild, 182–191. Evanston, Ill.: Northwestern University Press, 1964.

———. "Einstein and the Crisis of Reason." In *Signs*, edited by John Wild, 192–197. Evanston, Ill.: Northwestern University Press, 1964.

———. *Éloge de la philosophie et autres essais*. Paris: Gallimard, 1960.

———. *La Nature: notes, cours du Collège de France*. Edited by Dominique Séglard. Traces écrites. Paris: Seuil, 1995.

———. *Nature: Course Notes from the Collège de France*. Translated by Robert Vallier. Edited by Dominique Séglard. Northwestern University Studies in Phenomenology and Existential Philosophy. Evanston, Ill.: Northwestern University Press, 2003.

———. *Phenomenology of Perception*. International Library of Philosophy and Scientific Method. New York: Humanities Press, 1962.

Metz, André. "Bergson, Einstein et les relativistes." *Archives de philosophie* 22, no. 3 (1959): 369–384.

———. *La Relativité: exposé dans formules des théories d'Einstein et réfutation des erreurs contenues dans les ouvrages les plus notoires*. Paris: Étienne Chiron, 1923.

———. "Le Temps d'Einstein et la philosophie: à propos de l'ouvrage de M. Bergson, Durée et simultanéité." *Revue de philosophie* 31 (1924): 56–88.

———. "Réplique de M. André Metz," *Revue de philosophie* 31 (1924): 437–439.

Meyerson, Émile. *Du cheminement de la pensée* Paris: Alcan, 1931.

———. *Identité et réalité*. Paris: Vrin, 1951. First published in 1908.

———. *La Déduction relativiste*. Paris: Payot, 1925.

Michelson, Albert A. "Comparison of the International Meter with the Wave-Length of the Light of Cadmium." *Astronomy and Astrophysics* 12 (1893): 556–560.

———. "Experimental Determination of the Velocity of Light." *U.S. Nautical Almanac Office Astronomical Papers* 1, Part 3 (1879): 115–145.

———. "Les Plus récentes applications de la méthode interférentielle." *Bulletin de la Société française de philosophie* 21 (12 May 1921): 68–74.

———. "On the Relative Motion of the Earth and the Ether." *American Journal of Science* 3 (1897): 475–478.

———. "The Relative Motion of the Earth and the Luminiferous Ether." *American Journal of Science* 22 (1881): 120–129.

———. "Report." *Astrophysical Journal* 68, no. 5 (1928): 342–345.

Michelson, Albert A., and Edward Williams Morley. "On the Feasibility of Establishing a Light-Wave as the Ultimate Standard of Length." *American Journal of Science* 38, no. 225 (1889): 181–186.

———. "On the Relative Motion of the Earth and the Luminiferous Ether." *American Journal of Science*, no. 34 (1887): 333–345.

Miller, Arthur I. *Albert Einstein's Special Theory of Relativity: Emergence (1905) and Early Interpretation (1905–1911)*. New York: Springer, 1998.

———. *Einstein, Picasso: Space, Time and the Beauty that Causes Havoc*. New York: Basic Books, 2001.

Minkowski, Hermann. "Raum und Zeit." *Jahresbericht der Deutschen Mathematiker-Vereinigung 18* (1909): 75–88.

———. *Zwei Abhandlungen über die Grundgleichungen der Elektrodynamik, mit einem Einführungswort von Otto Blumenthal*. Vol. 1 of Fortschritte der mathematischen Wissenschaften in Monographien hrsg. von O. Blumenthal. Leipzig: B.G. Teubner, 1910.

Moatti, Alexandre. *Einstein, un siècle contre lui*. Sciences. Paris: Odile Jacob, 2007.

Mondor, Henri. *Propos familiers de Paul Valéry*. Paris: Grasset, 1957.

Montabré, Maurice. "Une heure avec Einstein." *L'Intransigeant*, 10 April 1922, 1.

Montague, William Pepperell. "The Einstein Theory and a Possible Alternative." *Philosophical Review* 33, no. 2 (1924): 143–170.

Moran, Jon S. "Bergsonian Sources of Mead's Philosophy." *Transactions of the Charles S. Pierce Society* 32 (1996): 41–63.

Mossé-Bastide, Rose-Marie. *Bergson éducateur*. Paris: Presses Universitaires de France, 1955.

"Mr. Balfour's Objection to Bergson's Philosophy." *Current Literature* 51, no. 6 (1911): 659–661.

Mullarkey, John. "Introduction: La Philosophie Nouvelle, or Change in Philosophy." In *The New Bergson*, edited by John Mullarkey. Manchester, U.K.: Manchester University Press, 1999.

Murakami, Haruki. *Kafka on the Shore*. Translated by J. Philip Gabriel. New York: Vintage International, 2005.

Nagel, Thomas. *The View from Nowhere*. New York: Oxford University Press, 1986.

Nathan, Otto, and Heinz Norden, eds. *Einstein on Peace*. New York: Avenel, 1960.

Newcomb, Simon. *Astronomy for Everybody: A Popular Exposition of the Wonders of the Heavens*. London: Isbiter and Company, 1903.

———. *The Elements of the Four Inner Planets and the Fundamental Constants of Astronomy*. Supplement to the American Ephemeris and Nautical Almanac for 1897. Washington, D.C.: U. S. Government Printing Office, 1895.

———. *Popular Astronomy*. New York: Harper & Bros., 1887.
Nobel Lectures in Physics (1901–1921). Singapore: World Scientific, 1998.
Nordmann, Charles. "Avec Einstein dans les régions dévastées." *L'Illustration* 80, no. 4128 (15 April 1922): 328–331.
———. "Einstein à Paris." *Revue des deux mondes* 8 (1922): 926–937.
———. *Einstein et l'univers*. Le Roman de la science. 1922 ed. Paris: Hachette, 1921.
———. "Einstein expose et discute sa théorie." *Revue des deux mondes* 9 (1 May 1922): 129–166.
———. *Notre maître le temps*. Le Roman de la science. Paris: Hachette, 1924.
Northrop, Filmer S.C. "Whitehead's Philosophy of Science." In *The Philosophy of Alfred North Whitehead*, edited by Paul Arthur Schilpp, 167–207. Evanston, Ill.: Northwestern University Press, 1941.
Norton, John D. "Einstein's Investigations of Galilean Covariant Electrodynamics prior to 1905." *Archive for History of Exact Sciences* 59 (2004): 45–105.
"Notes." *Philosophical Review* 31, no. 3 (1922): 317–323.
Nye, Mary Jo. "The Boutroux Circle and Poincaré's Conventionalism." *Journal of the History of Ideas* 40, no. 1 (1979): 107–120.
———. "Science and Socialism: The Case of Jean Perrin in the Third Republic." *French Historical Studies* 9, no. 1 (1975): 141–169.
Nyman, Alf. "Einstein-Bergson-Vaihinger." *Annalen der Philosophie und philosophischen Kritik* 6 (1927): 178–204.
Oersted, Hans Christian. *Die Naturwissenschaft und die Geistesbildung*. Leipzig: Carl B. Lorch, 1851.
Ostwald, Wilhelm. *Die Überwindung des wissenschaftlichen Materialismus. Vortrag gehalten in der 3. Allgemeinen Sitzung der Versammlung der Gesellschaft deutscher Naturforscher und Ärzte zu Lübeck am 20. September 1895*. Vol. 1 of Versammlung Deutscher Naturforscher und Ärzte zu Lübeck. Leipzig: Von Veit, 1895.
Overbye, Dennis. "Einstein and the Black Hole." *New York Times*, 12 August 2013, D1.
———. *Einstein in Love*. New York: Penguin, 2000.
Pais, Abraham. *Einstein Lived Here*. Oxford: Clarendon Press, 1994.
———. *"Subtle is the Lord . . .": The Science and the Life of Albert Einstein*. Oxford: Clarendon Press, 1982.
Papanicolaou, Andrew C., and P.A.Y. Gunter. *Bergson and Modern Thought: Towards a Unified Science*. Models of Scientific Thought. Chur, Switzerland: Harwood Academic, 1987.
Paty, Michel. "Einstein et la philosophie en France: à propos du sejour de 1922." *La Pensée* 210 (1980): 3–11.
———. *Einstein philosophe: La physique comme practique philosophique*. Philosophie d'aujourd'hui, edited by Paul-Laurent Assoun. Paris: Presses Universitaires de France, 1993.
———. "The Scientific Reception of Relativity in France." In *The Comparative Reception of Relativity*, edited by Thomas F. Glick, 113–167. Vol. 103 of Boston Studies in the Philosophy of Science. Dordrecht: D. Reidel, 1987.
Pearson, Keith Ansell. *Germinal Life: The Difference and Repetition of Deleuze*. London: Routledge, 1999.
Péguy, Charles. "Lettres et entretiens." *Cahiers de la quinzaine* 1 (1927).
———. "Note sur M. Bergson et la philosophie bergsonienne." *Cahiers de la quinzaine* 8 (1914).

Perrin, Jean. *Atoms*. Woodbridge, Conn.: Ox Bow Press, 1990. First published in 1913.

———. "La Réalité des molécules." *Revue scientifique* 49 (1911): 774–784.

———. "Mouvement brownien et réalité moléculaire." *Annales de chimie et de physique* 8, no. 18 (1909): 1–114.

Pesic, Peter. "Einstein and the Twin Paradox." *European Journal of Physics* 24 (2003): 585–590.

Pestre, Dominique. *Physique et physiciens en France, 1918–1940*. Histoire des sciences et des techniques. Paris: Éditions des archives contemporaines, 1984.

Piaggio, H.T.H. "Geometry and Relativity." *Mathematical Gazette* 11, no. 159 (July 1922): 97–100.

Piéron, Henri. "2. Rhythm: sens du temps." *L'Année psychologique* 21, no. 1 (1914): 383–384.

Planck, Max. *Die Stellung der neueren Physik zur mechanischen Naturanschauung. Vortrag gehalten am 23 September 1910*. Versammlung Deutscher Naturforscher und Ärzte in Königsberg. Leipzig: Verlag von S. Hirzel, 1910.

———. *A Survey of Physical Theory*. New York: Dover Publications, 1993.

Poincaré, Henri. "Étude de la propagation du courant en période variable sur une ligne munie de récepteur." *Éclairage électrique* 40 (1904): 121–128, 161–167, 201–212, 241–250.

———. "La Mesure du temps." *Revue de metaphisique et de morale* 6 (1898): 1–13.

———. "La Théorie de Lorentz et le principe de réaction." *Archives néerlandaises des sciences exactes et naturelles* 5 (1900): 252–278.

———. *La Valeur de la science*. Bibliothèque de philosophie scientifique. Paris: Ernest Flammarion, 1908. First published in 1905.

———. "L'Espace et le temps." In *Dernières pensées*, 35–54. Paris: Flammarion, 1917. First published in 1912.

———. "L'Hypothèse des quanta." In *Dernières pensées*, 163–192. Paris: Flammarion, 1917. First published in 1912.

———. "The Quantum Theory." Translated by John W. Bolduc. In *Last Essays*, 75–88. New York: Dover, 1963.

———. "Rapport sur les travaux de H.A. Lorentz, ca. 31 January 1910." In *La Correspondance entre Henri Poincaré et les physiciens, chimistes et ingénieurs*. Basel: Birkhäuser, 2007.

———. *Science and Hypothesis*. New York: Dover, 1952. First published in 1902.

———. *Science and Method*. New York: Dover, 1952. First published in 1908.

———. "Sur la dynamique de l'électron." *Rendiconti del circolo matematico di Palermo* 21 (1906): 1–55.

———. *The Value of Science*. New York: Dover Publications, 1958.

Politzer, Georges. "Après la mort de Bergson." *La Pensée libre*, no. 1 (February 1941).

———. *La Fin d'une parade philosophique: le bergsonisme*. Vol. 3 of Libertés nouvelles. Paris: J. J. Pauvert, 1967.

Popper, Karl R. *Conjectures and Refutations: The Growth of Scientific Knowledge*. New York: Basic Books, 1962.

———. *The Open Universe: An Argument for Indeterminism*. Totowa, N.J.: Rowman and Littlefield, 1982.

Prigogine, Ilya. "The Arrow of Time: Inaugural Lecture." Pescara, Italy: International Center for Relativistic Astrophysics, 1999.

———. "Evolution of Physics." Review of *Bergson and the Evolution of Physics*, edited and translated by P.A.Y. Gunter." *Nature* 234 (1971): 159–160.

Prigogine, Ilya, and Isabelle Stengers. *Order Out of Chaos: Man's New Dialogue with Nature*. Toronto: Bantam Books, 1984.

Proust, Marcel. "Letter to George de Lauris." In *Letters of Marcel Proust*, edited and translated by Mina Curtis, 197–198. New York: Random House, 1949.

Putnam, Hilary. "Time and Physical Geometry." *Journal of Philosophy* 64, no. 8 (1967): 240–247.

Reichenbach, Hans. "Der gegenwärtige Stand der Relativitätsdiskussion. Eine kritische Untersuchung." *Logos* 10 (1921): 316–378.

———. *The Direction of Time*. Berkeley: University of California Press, 1956.

———. "The Philosophical Significance of the Theory of Relativity." In *Albert Einstein: Philosopher-Scientist*, edited by Paul Arthur Schilpp, 289–311. La Salle, Ill.: Open Court, 1949.

———. *Philosophie der Raum-Zeit-Lehre*. Berlin: W. de Gruyter & Co., 1928.

———. *The Rise of Scientific Philosophy*. Berkeley: University of California Press, 1951.

———. "La Signification philosophique de la théorie de la relativité." *Revue philosophique de la France et de l'étranger* 93 (1922): 5–61.

———. *Was ist Radio? Mit 27 Abbildungen und einer Tafel der Sendestationen*. Das Radio-Reihe, edited by Hans Reichenbach. Vol. 1. Berlin: Richard Carl Schmidt & Co., 1924.

Reichenbach, Hans, and Maria Reichenbach. *The Direction of Time*. Mineola, N.Y.: Dover Publications, 1999.

Reiser, Oliver L. "The Problem of Time in Science and Philosophy." *Philosophical Review* 35, no. 3 (1926): 236–252.

Renn, Jürgen, M. Janssen, and Matthias Schemmel. *The Genesis of General Relativity*. 4 vols. Boston Studies in the Philosophy of Science. Dordrecht: Springer, 2007.

Renoliet, Jean-Jacques. *L'UNESCO oubliée: La Société des Nations et la coopération intellectuelle (1919–1946)*. Paris: Publications de la Sorbonne, 1999.

Rey, Abel. *La Théorie de la physique chez les physiciens contemporains: exposé des théories*. 2nd ed. Paris: Félix Alcan, 1923.

Reyes, Alfonso. *Einstein: Notas de Lectura*. Mexico: Fondo de Cultura Económica, 2009.

Richard-Foy, Émile. "Le Temps et l'espace du sens commun et les théories d'Einstein." *Revue philosophique de la France et l'étranger* 94 (1922): 153–200.

Richardson, O.W. "Hendrik Antoon Lorentz." *Journal of the London Mathematical Society* 4, no. 1 (1929): 183–192.

Richet, Charles. "L'Unité psychologique du temps." *Comptes rendus de l'Académie des sciences* 173, no. 25 (1921): 1313–1317.

———. *Our Sixth Sense*. Translated by Fred Rothwell. London: Rider, 1928.

Ricœur, Paul. *Time and Narrative*. 3 vols. Chicago: University of Chicago Press, 1984–1988.

Riley, W. I. "La Philosophie française en Amérique III. Le bergsonisme." *Revue philosophique de la France et de l'étranger* 91 (1921): 75–107, 234–269.

Roberts, Lissa. "The Death of the Sensuous Chemist: The 'New' Chemistry and the Transformation of Sensuous Technology." In *The Empire of the Senses*, edited by David Howes, 106–127. Oxford: Berg, 2005.

Romanell, Patrick. "Bergson in México: A Tribute to José Vasconcelos." *Philosophy and Phenomenological Research* 21 (1961): 501–513.

Roosevelt, Theodore. "The Search for Truth in a Reverent Spirit." *Outlook* 99 (1911): 819–826.

Rousseaux, André. "De Bergson à Louis de Broglie." In *Henri Bergson: Essais et témoignages inédits*, edited by Albert Béguin and Pierre Thévenaz, 279–286. Neuchatel: La Baconnière, 1941.

Rowe, David E., and Robert Schulmann, eds. *Einstein on Politics: His Private Thoughts and Public Stands on Nationalism, Zionism, War, Peace and the Bomb* Princeton, N.J.: Princeton University Press, 2007.

Russell, Bertrand. *The ABC of Relativity*. Harpers Modern Science Series. New York: Harper, 1925.

———. *The Analysis of Matter*. New York: Harcourt, 1927.

———. *An Essay on the Foundations of Geometry*. Cambridge: Cambridge University Press, 1897.

———. *A History of Western Philosophy*. New York: Simon and Schuster, 1945.

———. *Mysticism and Logic* Garden City, N.Y.: Doubleday and Co., 1957.

———. *Mysticism and Logic*. New York: Longmans, Green and Co., 1918.

———. "On the Relations of Universals and Particulars." *Aristotelian Society Proceedings, New Series* 12 (1911–1912): 1–24.

———. "The Philosophy of Bergson." *Monist 22* (1912): 321–347.

———. *The Philosophy of Bergson*. Cambridge: Bowes and Bowes, 1914.

Ryckman, Thomas. "Einstein Agonists: Weyl and Reichenbach on Geometry and the General Theory of Relativity." In *Origins of Logical Empiricism*, edited by Ronald N. Giere and Alan W. Richardson, 165–209. Minneapolis: University of Minnesota Press, 1996.

———. *The Reign of Relativity: Philosophy in Physics 1915–1925*. Oxford Studies in Philosophy of Science. Oxford: Oxford University Press, 2005.

Santayana, George. *Winds of Doctrine: Studies in Contemporary Opinion*. New York: C. Scribner's Sons, 1913.

Schaffer, Simon. "Metrology, Metrification and Victorian Values." In *Victorian Science in Context*, edited by Bernard Lightman, 438–474. Chicago: Chicago University Press, 1997.

Schilpp, Paul Arthur. Preface to *The Philosophy of Alfred North Whitehead*, edited by Paul Arthur Schilpp, xiii–xviii. Evanston, Ill.: Northwestern University Press, 1941.

Schlick, Moritz. "On the Relation Between Psychological and Physical Concepts." Translated by Peter Heath. In *Philosophical Papers*, edited by Henk L. Mulder and Barbara F.B. van de Velde-Schlick, 420–436. Dordrecht: Reidel, 1979. First published in 1935.

Schmidt, James. "The Eclipse of Reason and the End of the Frankfurt School in America." *New German Critique* 100 (2007): 47–76.

Schroeder-Gudehus, Brigitte. "Challenge to Transnational Loyalties: International Scientific Organizations after the First World War." *Science Studies* 3, no. 2 (1973): 93–118.

Seelig, Carl, ed. *Albert Einstein: eine dokumentarische Biographie*. Zürich: Europa Verlag, 1954.

Sertillanges, Antonin-Dalmace. *Henri Bergson et le catholicisme*. Paris: Ernest Flammarion, 1941.

Shapin, Steven. *A Social History of Truth: Civility and Science in Seventeenth-Century England*. Science and Its Conceptual Foundations, edited by David L. Hull Chicago: University of Chicago Press, 1994.

Slatoff, Walter J. *Quest for Failure: A Study of William Faulkner*. Ithaca, N.Y.: Cornell University Press, 1960.

Sokal, Alan D., and Jean Bricmont. *Fashionable Nonsense: Postmodern Intellectuals' Abuse of Science*. New York: Picador, 1998.

———. *Impostures intellectuelles*. Paris: O. Jacob, 1997.

Sommerfeld, Arnold. "To Albert Einstein's Seventieth Birthday." In *Albert Einstein: Philosopher-Scientist*, edited by Paul Arthur Schilpp, 99–105. La Salle, Ill.: Open Court, 1949.

Soulez, Philippe. *Bergson politique*. Paris: Presses Universitaires de France, 1989.

Souriau, Michel. *Le Temps*. Nouvelle Encyclopédie philosophique, edited by H. Delacroix Paris: Félix Alcan, 1937.

Stachel, John J. Introduction to the Centenary Edition of *Einstein's Miraculous Year: Five Papers that Changed the Face of Physics*, edited by John J. Stachel, xv–lxxii. Princeton, N.J.: Princeton University Press, 2005.

Staley, Richard. *Einstein's Generation: The Origins of the Relativity Revolution*. Chicago: University of Chicago Press, 2008.

———. "Michelson and the Observatory." In *The Heavens on Earth: Observatory Techniques in Ninteenth-Century Science*, edited by David Aubin, Charlotte Bigg and H. Otto Sibum. Durham, N.C.: Duke University Press (2010).

Stanley, Matthew. "An Expedition to Heal the Wounds of War: The 1919 Eclipse and Eddington as Quaker Adventurer." *Isis* 94 (2003): 57–89.

———. *Practical Mystic: Religion, Science, and A. S. Eddington*. Chicago: University of Chicago Press, 2007.

Starkie, Enid. "Bergson and Literature." In *The Bergsonian Heritage*, edited by Thomas Hanna, 74–99. New York: Columbia University Press, 1962.

Stebbins, Joel. "The International Astronomical Union." *Popular Astronomy* 27 (1919): 601–612.

Stengers, Isabelle. *Cosmopolitics*. Posthumanities. Minneapolis: University of Minnesota Press, 2010.

———. *Power and Invention: Situating Science*. Theory Out of Bounds. Minneapolis: University of Minnesota Press, 1997.

Stewart, Balfour. *The Conservation of Energy: Being an Elementary Treatise on Energy and Its Laws*. London: H. S. King & Co., 1874.

Taillens, Ch. "À propos des conférences du Dr. Guillaume." *Gazette de Lausanne*, 10 June 1920.

Tanpinar, Ahmet Hamdi. *The Time Regulation Institute*. Translated by Alexander Dawe. New York: Penguin, 2013.

Telkes-Klein, Eva. "Meyerson dans les milieux intellectuels français dans les années 1920." *Archives de Philosophie* 70, no. 3 (2007): 359–373.

"La Théorie de la physique chez les physiciens contemporains." *Bulletin de la Société française de philosophie* 9 (1909): 161–191.

"La Théorie de la relativité: séance du 6 avril 1922." *Bulletin de la Société française de philosophie* 22 (1922): 349–370.

Thiebault, J. "Les Fantômes du professeur Charles Richet devant Einstein." *Revue contemporaine* (August–September 1922): 715–718.

Thomson, E. P. "Time, Work-Discipline, and Industrial Capitalism." *Past and Present* 38 (1967): 56–97.

Thomson, G. P. *The Foreseeable Future*. Cambridge: Cambridge University Press, 1955.
Thomson, William. "The Sorting Demon of Maxwell." *Proceeding of the Royal Institution* 9 (1879): 113–114.
Toulmin, S. "The Emergence of Post-Modern Science." In *The Great Ideas Today*, 68–114. Chicago: Encyclopedia Britannica, 1981.
Ushenko, Paul Andrew. "Einstein's Influence on Contemporary Philosophy." In *Albert Einstein: Philosopher-Scientist*, edited by Paul Arthur Schilpp, 606–645. La Salle, Ill.: Open Court, 1949.
Vagt, Christina. "Im Äther. Einstein, Bergson und die Uhren der Mikrobe." In *Übertragsungsräume*, edited by Eva Johach and Diethard Sawicki, 133–144. Weisbaden: Ludwig Reichert, 2013.
Valéry, Paul. "Allocution de Paul Valéry à l'occasion du décès de M. Henri Bergson prononcée à la séance tenue par l'Académie française le 9 janvier 1941." *Revue philosophique de la France et de l'étranger* 131 (1941): 121–124.
———. *L'Idée fixe: Socrate et son médecin*. Oeuvres de Paul Valéry. Paris: Éditions de la N.R.F., 1936.
Vallentin, Antonina. *The Drama of Albert Einstein*. New York: Doubleday, 1954.
Vuillermoz, Émile. "Devant l'écran." *Le Temps*, 10 October 1917, 3–4.
Wagner, Pierre. "Introduction." In *Les Philosophes et la science*, edited by Pierre Wagner, 9–65. Paris: Gallimard, 2002.
Wahl, Jean. "At the Sorbonne." In *The Bergsonian Heritage*, edited by Thomas Hanna, 150–154. New York: Columbia University Press, 1962.
Walter, Scott. "Minkowski, Mathematicians, and the Mathematical Theory of Relativity." In *The Expanding Worlds of General Relativity*, edited by Hubert Goenner, Jürgen Renn, Jim Ritter, and Tilman Sauer, 45–86. Boston: Birkäuser, 1999.
Walter, Scott and Étienne Bolmont, André Coret, eds. *La Correspondance entre Henri Poincaré et les physiciens, chimistes et ingénieurs*. Basel: Birkhäuser, 2007.
Warwick, Andrew. *Masters of Theory: Cambridge and the Rise of Mathematical Physics*. Chicago: University of Chicago Press, 2003.
Watanabe, Satosi. "Le Concept de temps en physique moderne et la durée pure de Bergson." *Revue de métaphysique et de morale* 56 (1951): 128–142.
———. "Contribution du point de vue de la mécanique ondulatoire, à l'étude du deuxième théorème de la thermodynamique." Ph.D. diss., Université de Paris, 1935.
Wazeck, Milena. *Einsteins Gegner: die öffentliche Kontroverse um die Relativitätstheorie in den 1920er Jahren*. Frankfurt: Campus Verlag, 2009.
———. "The 1922 Einstein Film: Cinematic Innovation and Public Controversy." *Physics in Perspective* 12 (2010): 163–179.
Weyl, Hermann. "Erwiederung auf Einsteins Nachtrag zu H. Weyl, Gravitation und Elektrizität." *Sitzungsberichte der Königlich Preussischen Akademie der Wissenschaften* (1918): 478–480.
———. *Philosophie der Mathematik und Naturwissenschaft*. München: R. Oldenbourg, 1966.
———. *Philosophy of Mathematics and Natural Science*. Translated by Olaf Helmer. Revised and augmented ed. Princeton, N.J.: Princeton University Press, 1949.
———. *Space-Time-Matter*. London: Methuen and Co., 1922.
Whitehead, Alfred North. *Adventures of Ideas*. Cambridge: Cambridge University Press, 1933.

———. "Autobiographical Notes." In *The Philosophy of Alfred North Whitehead*, edited by Paul Arthur Schilpp, 1–14. Evanston, Ill.: Northwestern University Press, 1941.

———. *The Concept of Nature*. 1920.

———. "Discussion: The Idealistic Interpretation of Einstein's Theory." *Proceedings of the Aristotelian Society* 22 (1922): 130–134.

———. *Modes of Thought*. New York: Macmillan Company, 1956.

———. *The Principle of Relativity with Applications to Physical Science*. Cambridge: Cambridge University Press, 1922.

———. *Process and Reality*. New York: Macmillan, 1929.

———. *Science and the Modern World*. 1925.

———. "Symposium: The Problem of Simultaneity: Is There a Paradox in the Principle of Relativity in Regard to the Relation of Time Measured to Time Lived?" In *Relativity, Logic, and Mysticism. Proceedings of the Aristotelian Society, Supplementary Volumes* 3 (1923): 34–41.

Whitrow, G. J., ed. *Einstein: The Man and His Achievement*. New York: Dover Publications, 1963.

Wiener, Norbert. *Cybernetics: Or, Control and Communication in the Animal and the Machine*. 2nd ed. New York: MIT Press, 1961.

———. *The Human Use of Human Beings: Cybernetics and Society*. New York: Avon Books, 1967.

Worms, Frédéric. "Entre critique et métaphysique: la science chez Bergson et Brunschvicg." In *Les Philosophes et la science*, edited by Pierre Wagner, 403–446. Folio essais. Paris: Gallimard, 2002.

Worms, Frédéric, and Jean-Jacques Wunenburger. *Bachelard et Bergson, continuité et discontinuité?: actes du Colloque international de Lyon, 28–30 septembre 2006*. Paris: Presses Universitaires de France, 2008.

Wünsch, Danielle. "Einstein et la Comission internationale de coopération intellectuelle." *Revue d'histoire des sciences* 57, no. 2 (2004): 509–520.

Yourgrau, Palle. *A World Without Time: The Forgotten Legacy of Gödel and Einstein*. New York: Basic Books, 2005.

Zhao, Jiabi. "Xieshizhuyizhe de Shitaiyin [Gertrude Stein the realist]." *Wenyi fengjing* [Literary landscape] 1, no. 1 (1934): 89.

INDEX